Ecotechnologies for Wastewater Treatment

In Focus – Special Book Series

Ecotechnologies for Wastewater Treatment: Impacting the Environment with Innovation in Wastewater Treatment

Editors

Manuela Antonelli, Andrea Turolla, Elena Ficara, Juan Lema, Roberto Canziani and Francesca Malpei

IWA PUBLISHING

Published by **IWA Publishing**
Unit 104-105, Export Building
1 Clove Crescent
London E14 2BA, UK
Telephone: +44 (0)20 7654 5500
Fax: +44 (0)20 7654 5555
Email: publications@iwap.co.uk
Web: www.iwapublishing.com

First published 2022

British Library Cataloguing in Publication Data
A CIP catalogue record for this book is available from the British Library

ISBN: 9781789063684 (paperback)

Contents

doi: 10.2166/wst.2022.231

Editorial: Ecotechnologies for Wastewater Treatment – Impacting the environment with innovation in wastewater treatment

Acknowledging that resource efficiency as well as ecosystem protection are crucial goals to tackle has led to a deep rethinking in the way wastewater has been dealt with up to the 20th century. Indeed, pollution reduction to harmless levels has been the main focus of conventional wastewater treatment plants which have deployed effective and reliable solutions to get rid of suspended and dissolved pollutants. Nowadays, those same wastewaters are looked at as a mine of valuable resources that are to be harvested from the sewage through efficient technologies operated within water and resource recovery facilities (formerly known as wastewater treatment plants).

Indeed, it has been recognized that urban wastewater contains significative amounts of key resources. Nitrogen, phosphorus and potassium, essential nutrients in agriculture, though diluted, are abundant in wastewaters. To name the most relevant ones, the amount of the P load in human excreta accounts for more than 20% of its worldwide demand. Similarly, more chemical energy is available in the wastewater than the energy demand for its processing, while reclaimed water can play a significant role in facing water scarcity from climate change. In addition to this, new classes of pollutants have emerged that require the development of prevention strategies as well as adequate treatment technologies. For these reasons, rethinking the way we deal with wastewater has become the new challenge and a key tool to implement the Circular Economy Action Plan, to address numerous UN Sustainable Development Goals, as well as to provide support tools to improve resilience to environmental and global threats, including climate change, water and food security, epidemic outbreaks and political instability.

To tackle these challenging goals, new technologies as well as multidisciplinary tools are needed to pave the way toward a new water-economy, including: innovative technologies, integrated and risk-based approaches, and adequate regulatory and economic guidelines to promote technological uptake and societal acceptance.

Within this framework, this special issue collects scientists' experiences shared during the 5th International Conference on Ecotechnologies for Wastewater Treatment that was held in Milan, Italy in June 2021. Enabling technologies for resource recovery and energy efficiency are discussed as well as tools for the integrated techno-economic and environmental assessment for an holistic approach to wastewater treatment innovation.

Guest Editors

Manuela Antonelli
Politecnico di Milano, Department of Civil and Environmental Engineering, Piazza L. da Vinci 32, 20133, Milano, Italy
Andrea Turolla
Politecnico di Milano, Department of Civil and Environmental Engineering, Piazza L. da Vinci 32, 20133, Milano, Italy
Elena Ficara
Politecnico di Milano, Department of Civil and Environmental Engineering, Piazza L. da Vinci 32, 20133, Milano, Italy
Juan Lema
CRETUS, Department of Chemical Engineering, Universidade de Santiago de Compostela, 15782 Santiago de Compostela, Spain
Roberto Canziani
Politecnico di Milano, Department of Civil and Environmental Engineering, Piazza L. da Vinci 32, 20133, Milano, Italy
Francesca Malpei
Politecnico di Milano, Department of Civil and Environmental Engineering, Piazza L. da Vinci 32, 20133, Milano, Italy

doi: 10.2166/wst.2022.132

Phosphorus recovery from a pilot-scale grate furnace: influencing factors beyond wet chemical leaching conditions

G. Boniardi[a], A. Turolla[a,*], L. Fiameni[b], E. Gelmi[a], E. Bontempi[b] and R. Canziani[a]

[a] Department of Civil and Environmental Engineering (DICA), Politecnico di Milano, Piazza Leonardo da Vinci 21, 20133, Milano, Italy
[b] INSTM and University of Brescia, Via Branze 38, 25123, Brescia, Italy
*Corresponding author. E-mail: andrea.turolla@polimi.it

AT, 0000-0002-2902-5004

ABSTRACT

Phosphorus is a non-renewable resource going to be exhausted in the future. Sewage sludge ash is a promising secondary raw material due to its high phosphorus content. In this work, the distribution of 19 elements in bottom and cyclone ashes from pilot-scale grate furnace have been monitored to determine the suitability for the phosphorus acid extraction. Moreover, the influence of some parameters beyond wet chemical leaching conditions were investigated. Experimental results showed that bottom ash presented lower contamination in comparison to cyclone ash and low co-dissolution of heavy metals (especially Cr, Pb and Ni), while high phosphorus extraction efficiencies (76–86%) were achieved. High Al content in the bottom ash (9.4%) negatively affected the phosphorus extraction efficiency as well as loss on ignition, while the particle size reduction was necessary for ensuring a suitable contact surface. The typology of precipitating agents did not strongly affect the phosphorus precipitation, while pH was the key parameter. At pH 3.5–5, phosphorus precipitation efficiencies higher than 90% were achieved, with a mean phosphorus content in the recovered material equal to 16–17%, comparable to commercial fertilizers. Instead, the co-precipitation of Fe and Al had a detrimental effect on the recovered material, indicating the need for additional treatments.

Key words: grate furnace, heavy metals contamination, phosphorus recovery, resource recovery, sewage sludge ash, wet chemical extraction and precipitation

HIGHLIGHTS

- Bottom ashes from pilot-scale grate furnace were suitable for phosphorus extraction.
- Influence of parameters beyond chemical extraction conditions were investigated.
- Al content and particle size significantly influenced phosphorus extraction efficiency.
- pH was the most relevant precipitation parameter unlike the type of precipitating agent.
- pH between 3.5 and 5 led to high phosphorus content in the recovered product.

GRAPHICAL ABSTRACT

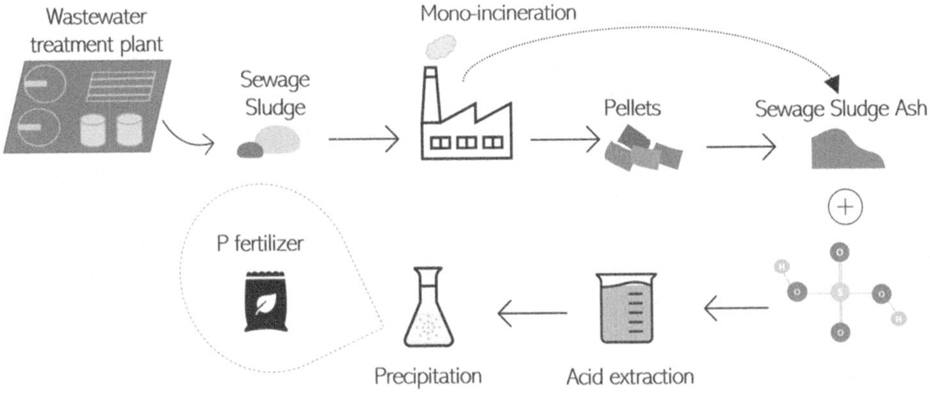

1. INTRODUCTION

Phosphorus is a vital nutrient for all living organisms, including human beings, animals, and crops. It is an irreplaceable and non-renewable resource commonly extracted from natural phosphate rocks. Due to its increasing consumption, mostly attributed to fertilizer manufacturing, phosphate rocks have been introduced in the list of critical raw materials by the European Commission (EC 2017). The global population growth, intensive farming methods and changes in diet in developing countries will accelerate the depletion of ore sources, which is expected in 100–400 years (Desmidt *et al.* 2015). At the same time, the low utilization efficiency of phosphorus in agriculture (below 20%) causes eutrophication, impacting the quality of surface water bodies (Li *et al.* 2019a). Nowadays, the exploration of alternative sources to decrease the pressure on phosphate rocks and to minimise the environmental impact is of paramount importance.

Phosphorus is available in the sewage sludge, in which 80–90% of the discharged phosphorus concentrates during wastewater treatment (Lee & Kim 2017). Moreover, the concerns related to sewage sludge application in agriculture, mostly due to the high heavy metals content, favoured other options for sludge management and disposal such as the mono-incineration. The produced sewage sludge ash comes as enriched phosphorus source, with a mean phosphorus content of 5–11% w/w (Ottosen *et al.* 2013), which is comparable to the content in medium ore grade. According to Donatello & Cheeseman (2013), about 1.7 million tonnes of ashes are produced worldwide every year, and the majority is subsequently landfilled.

During the last few years, in the view of circular economy and a zero-waste approach, more than 30 technological solutions for phosphorus recovery have been developed, among which the wet chemical extraction from sewage sludge ashes has proven to be one of the most effective (Fahimi *et al.* 2021). Its simple setup, low energy consumption and applicability at large scale are the main advantages (Fang *et al.* 2020). Few technological solutions have been scaled-up so far: the available ones at full-scale in Europe are Remondis Tetraphos, Phos4life and EcoPhos processes (Chrispim *et al.* 2019; Fang *et al.* 2020). Beyond high operating costs, the main reasons making technology development and scale-up challenging are related to variations in phosphorus concentration in the sewage sludge ash and recovered product contamination (e.g., by heavy metals) (Li *et al.* 2019b).

Looking at the related literature, due to the maturity of fluidized bed technology, the majority of the studies published refer to wet chemical extraction from fly ash, while very few studies focused on the ashes obtained from grate furnaces (Cheng *et al.* 2020; Fang *et al.* 2020). Data about the mean composition of sewage sludge ashes specifically used for phosphorus extraction is provided in Table S1, in which the type of incineration technology is also considered. As pointed out by Gorazda *et al.* (2007), the incineration technology plays an important role since operating parameters (such as the temperature) and the conditions of each thermal process significantly affect the physical and chemical properties of the ashes produced. From 72 samples collected, only 6% referred to grate furnace, while 68% were derived from fluidized bed incinerators. For 17% of samples, the type of combustion technology was not provided nor easily available. Cheng *et al.* (2020) highlighted that most studies focused on the fluidized bed incinerators with limited insights into the behaviours of heavy metals/metalloids during step grate incineration. For these reasons, we decided to investigate the phosphorus recovery from bottom ashes, which are less enriched with heavy metals compared to fly ashes. The application of grate furnaces can become widespread in the following years (at least in some areas) since critical raw material recovery from ashes has recently gained much more attention than in the past. In addition, the simplification and reduction of the flue gas line and treatment provided by this kind of technology can promote its diffusion.

Furthermore, most of the previous research efforts addressed the influence of extraction conditions (e.g, acid type, acid concentration, liquid-to-solid ratio, contact time, temperature) on the phosphorus extraction efficiency (PEE), while less attention was posed on the influence of upstream treatments on the downstream phosphorus recovery (Liu *et al.* 2021).

Overall, the type of extractants tested for the leaching process were inorganic acids (H_2SO_4, HCl, HNO_3), organic acids ($H_2C_2O_4$, $C_6H_8O_7$), chelating agents (EDTMP, EDTA) and bases (e.g., NaOH). The highest phosphorus extraction efficiencies (70–100%) were obtained by acids, while bases and complexing agents led to lower extraction efficiencies (9–40%). In general, strong acids can extract phosphorus by breaking the metal-phosphorus bonds, inevitably causing the co-dissolution of unwanted elements such as heavy metals. In literature, to lower the heavy metal content in the recovered material and thus to increase its quality, three main strategies have been considered so far. The first one deals with prevention from leachate contamination, the second one with leachate purification, while the third one is focused on the optimisation of selective precipitation conditions.

Fang *et al.* (2018) investigated the first strategy by introducing a pre-treatment of sewage sludge ash with EDTA (0.02 mol/L) followed by extraction with sulfuric acid (0.2 mol/L). However, even if the purity of the extract was increased, the phosphorus content in the leachate decreased by 29%. Then, the first attempts at leachate purification were made by ion exchange resins (Franz 2008; Xu *et al.* 2012; Wang *et al.* 2018), adsorbent materials (Biswas *et al.* 2009; Yu *et al.* 2021), nanofiltration (Niewersch *et al.* 2014) and liquid-liquid extraction (Hong *et al.* 2005; Baldi *et al.* 2021). Considering the recent literature, Yu *et al.* (2021) proposed the use of a phosphorus selective adsorbent (based on zirconium) to recover 70% of phosphorus in ash as hydroxyapatite, while Baldi *et al.* (2021) found to be effective keeping in contact the acid leachate with an organic solvent solution (isoamyl alcohol) with a ratio of 1:1. They indeed found a phosphorus extraction rate of 81% with low amount of iron and aluminium (extraction rate of 25% and 10%, respectively). Looking at industrial scale, ion-exchange and selective nanofiltration are the actual techniques adopted (e.g., Remondi Tetraphos and EcoPhos), but alternatives to lower the costs are required (Baldi *et al.* 2021).

Considering instead the third strategy, Baldi *et al.* (2021) investigated the purification of leachate by pH variation. They found that this way was impractical due to the simultaneously precipitation of phosphorus with aluminium and iron at low values of pH (2–3.7). Similar results were found by Fang *et al.* (2018). At pH=3 almost the totality of iron precipitated, while the same happened for aluminium at pH=4. In addition, they studied the influence of precipitating agents on metals extraction and observed a slightly delay in precipitation of iron and aluminium in comparison to phosphorus by using $Ca(OH)_2$ instead of NaOH. However, a low significative reduction in terms of iron and aluminium content in the recovered material was detected and a high amount of gypsum was found. As a result, further efforts should be addressed to the optimisation of the phosphorus precipitation to increase the quality of the recovered material (Baldi *et al.* 2021; Liu *et al.* 2021).

Taking into account all the discussed current challenges, the aims of the present research work are: (i) the characterization of sewage sludge ash obtained from pilot-scale grate furnace in the view of phosphorus recovery; (ii) the comparison of the physical and chemical composition of bottom and cyclone ashes to determine the suitability of each type of ash for the phosphorus extraction; (iii) the investigation of factors beyond wet chemical extraction conditions that affect the phosphorus dissolution (upstream treatment characteristics, particle size dimension, loss on ignition – LOI, pH); (iv) the optimization of precipitation process conditions (pH, aging time, type of precipitating agents).

2. METHODS

2.1. Samples collection

Pelletized and dried sewage sludge from municipal wastewater was incinerated at 850 °C in a pilot-scale grate furnace located near the city of Milan (San Giuliano Milanese), in Italy, and engineered by VOMM Impianti e Processi SpA (Rozzano, Lombardy, Italy). The pelleting was adopted to simplify and reduce the fly ash capture systems and thus to protect the tube bundles of the boiler and the heat exchanger from erosion and fouling. The flat grid technology allows the treatment of dried sludge (85–90% dry matter content), which is deposited on the flat surface of the moving grid, while the combustion air is blown under the grate with a centrifugal fan. The combustion chamber is constituted by an adiabatic oven capable of withstanding the high temperatures (1,000 °C). As a first step, a dehydration phase takes place, leading to the removal of residual moisture content. Then, an intermediate pyrolysis phase occurs, followed by a complete oxidation (combustion). To prevent an excessive NOx production, a flue gas recirculation system is applied and a selective non-catalytic reduction system (SNCR) is adopted. Ninety percent of ash accumulates at the end of the grid (bottom ash), thus favouring collection, while fly ash is subsequently captured and treated. Indeed, flue gas cleaning system consists of cyclones, bag filters, scrubber-type washing towers and activated carbons to fully meet the limits of emissions reported in the EU 2010/75 legislation. For this work, three experimental campaigns were conducted from August 2020 to April 2021. Each campaign lasted about 10 days, in which the furnace was operated continuously treating about 150 kg per hour. For each campaign, ashes were sampled three times a day during the sampling days (which were fixed at the beginning, at the middle and at the end of the operation). All samples were collected in PE plastic bottles and kept at room temperature. The total amount collected was about 2 kg of bottom ash (BA) and 2 kg of cyclone ash (CYC). Mean samples of BA collected during the operation period were named BA,1°, BA,2° and BA,3°, where the number denotes the experimental campaign (1°, 2° and 3°). In a similar way, CYC samples were named CYC,1°, CYC,2° and CYC,3° for the three campaigns. Three sub-samples of BA were collected during the third experimental campaign at the beginning, in the middle and at the end (named BA1,3°, BA2,3° and BA3,3°, respectively), to assess the influence of changes in the operating conditions of the mono-incineration plant. Sampling points are schematically shown in Figure S1.

2.2. Characterization of sewage sludge ash

BA samples were pounded by a manual mortar before characterization, while CYC samples were analysed as received. The ash humidity was measured by drying the samples in an oven at 105 °C overnight, while the LOI was evaluated by heating the dried samples in a laboratory standard muffle (LT 9/12 SKM Nabertherm, Germany) at 900 °C for 2 h. Particle size distribution was determined using a W. S. Tyler RO-TAP® Sieve shaker (Model RX-29) on a total ash mass of 200 g. Larger sieves used (850, 600, 425, 212 and 180 μm) met the Standard Test Sieve method (ASTM International 2020), while smaller sieves (0.125, 0.063 and 0.025 mm) were compliant with the ISO 3310–1:2016 method (ISO 2016). Ash pH was determined after 2 h leaching test using deionized water at a L/S (liquid-to-solid) ratio of 10:1, following the guideline of EN 12457-2 method (European Committee for Standardization 2002) and the optimized procedure described by Bontempi *et al.* (2010).

Ash total phosphorus content was evaluated using Hach Lange LCK 348 analytical kits, after acid mineralization with concentrated HNO_3 (65% w/w) at a sample-to-HNO_3 volume ratio of 5:2, according to EN 15956 (European Committee for Standardization 2011). The presence of most abundant elements (Al, Ca, Fe, K, Mg, Na) and metals/metalloids (As, Cd, Co, Cr, Cu, Hg, Ni, Pb, Sb, Tl, V, Zn) was determined by aqua regia mineralization followed by ICP-MS (Agilent, 7700 Series) analysis, according to UNI EN ISO 17294-1 (2007).

The crystalline composition of ash samples (BA and CYC) was investigated by X-ray diffraction (XRD) using a PANalytical X'Pert PRO diffractometer (Malvern Panalytical), at a voltage of 40 kV and a current of 40 mA, equipped with a Cu Kα anode. Phase identification was processed with Philips X'Pert software, associated with the crystallography open database (COD).

Morphological characterization of ash samples was performed by means of scanning electron microscopy combined with energy-dispersive X-ray spectrometry (SEM-EDXS) using a microscope LEO EVO 40 (Carl Zeiss), coupled with a probe for elemental and semiquantitative chemical analysis (Oxford Instruments). Secondary electron mode was used to investigate the sample morphology.

2.3. Phosphorus wet chemical extraction

Acid leaching was performed with 0.2 M H_2SO_4 prepared by dilution of high purity reagent (96% w/w) with ultra-pure water. BA sample collected in the first experimental campaign was used as received (raw), milled (with ball mill), pounded (with manual mortar) or split into four classes (class I, <0.063 mm; class II, 0.063–0.125 mm; class III, 0.180–0.212 mm; class IV, 0.425–0.600 mm). Mean BA samples collected in the second and third campaigns were directly pounded by manual mortar, while particle size lower than 0.212 mm was considered for the three sub-samples of BA from the third experimental campaign. CYC samples were always used as received.

For each experiment, 20 ± 0.1 g of dried ash were treated with 400 mL of 0.2 M H_2SO_4 solution. Agitation was provided by a magnetic stirrer (Variomag Poly 15) at a rotation speed of 650 rpm. Beakers were covered with watch glasses to prevent water evaporation. The suspension was then centrifuged at 3,500 rpm for 10 min and immediately filtered at 0.45-μm Whatman filter paper. Tests were performed at least in duplicate. PEE and extraction rates of metals/metalloids were calculated as reported by Boniardi *et al.* (2021).

While PEE was determined on all samples collected, only BA samples were used for the investigation of the influence of ash characteristics on phosphorus extraction. This choice was taken because about the 90% of total ash produced is constituted by bottom ash. In detail, BA,1° and BA,3° samples were selected to investigate the influence of particle size dimension and LOI.

2.4. Phosphorus precipitation

Precipitation tests were carried out using 30 mL of leachate obtained by the acid leaching of pounded BA (second experimental campaign). Only BA,2° sample was used to assess the influence of factors affecting the precipitation process. Preliminary precipitation tests were performed in triplicate to evaluate the amount of precipitated material and target pH at phosphorus thermodynamic equilibrium, as well as required aging time. Precipitation tests were conducted afterwards in duplicate under the following operating conditions: room temperature, 1 h as reaction time, 2 h as aging time. Different values of pH were studied: 2.5, 3.5, 5 and 8. To assess the influence of the precipitating agent on precipitated material characteristics, various procedures were compared. In the first trial, 0.1 M NaOH was gradually dosed until pH 3.5 followed by the addition of 0.5% $Ca(OH)_2$ until the target pH was reached. Such procedure was defined to prevent excessive co-precipitation of gypsum ($CaSO_4$) in the early stage of pH increase. In the second trial, only 0.1 M KOH was dosed. Obtained suspensions were filtered through a 0.45-μm Whatman filter paper and the phosphate content in filtered precipitate solution was determined. Overall, phosphorus precipitation efficiency (PPE) was calculated according to the procedure by Boniardi *et al.* (2021). In addition, the content of Al, Ca, Fe, Cu and Zn in the filtered solution was quantitatively determined by ICP-MS. Crystallinity of the precipitates was investigated using XRD. The same equipment previously described was used.

3. RESULTS AND DISCUSSION

3.1. Characterization of sewage sludge ash: grading curves, LOI, humidity, pH, crystalline forms and morphology

Grading curves of the investigated samples are depicted in Figure S2. As expected, CYC had a median particle size of 0.05 mm, while pounded BA samples were coarser with a median particle size of 0.1–0.2 mm. Raw pelletized BA instead had 90% of particles above 2 mm due to the type of configuration adopted for the sludge incineration. Similarly, Lin *et al.* (2018) investigated the particle size distribution of fly ash obtained from a hearth type incinerator and found a mean particle size of 0.07 mm, while bottom ash had a mean diameter of 0.8 mm. Li *et al.* (2017) instead dealt with sewage sludge ashes from fluidized bed incinerator and found a mean particle size of 0.056 mm, while Lee & Kim (2017) reported a mean particle size of 0.153 mm for fly ash collected from the bag filters. Donatello (2009) investigated the mean particle diameters of ashes collected from different UK incinerators and found a range of 0.008–0.263 mm, highlighting the influence of the type of incineration technology on particle size distribution. The large variability of size dimension was also reported by Ma & Rosen (2021) with a range of 0.05–1.5 mm. Thus, experimental results obtained in this study meet the mean size range reported in literature.

Considering the measure of unburned carbon content, CYC had LOI values below 1.6% for all campaigns, thus complying with the Directive 2010/75/EU (maximum value set at 5%). Instead, LOI values for BA,1°, BA,2°, and BA2,3° were above 5%, reaching a maximum value of 20.3% for BA,2°. This can be attributed to sub-optimal incineration conditions, which were optimized during the third experimental campaign by controlling flow rates of fed sludge, air, and residence time. As a result, LOI values for BA1,3° and BA3,3° were equal to 2.4% and 1.1%, respectively, while BA2,3° had still a higher LOI equal to 5.7%. Similarly, Zabielska-Adamska (2019) reported a LOI value of 4.72% for bottom ash obtained from a grate furnace. Krüger *et al.* (2014) provided different LOI values for ashes depending on combustion technique: 30 samples obtained from grate furnace showed a mean LOI of 4.8% with a maximum value of 13.1%. In addition, Cyr *et al.* (2007) reported an LOI value of 5.5% for ash obtained by fluidized bed incineration and indicated a mean literature value of 6.1% by analysing more than 20 sources. LOI values are affected by the combustion conditions and the type of incineration technology, so that low values can be achieved for good combustion.

As for pH, values in the range of 6.8–8.1 were measured for BA and 6.7–8.8 for CYC. Similarly, Li *et al.* (2017) reported a mean pH of 8.45 for ashes obtained from a fluidized bed incinerator. Humidity content was in the range of 0.1–0.8%, with a peak of 3.6% for BA,2°, probably due to storage conditions. Experimental data for LOI, pH and humidity are summarized in Table S2.

Experimental results of XRD analysis in Figure S3 revealed a similar crystalline composition for all BA samples: quartz [SiO_2] was detected as main phase, followed by a calcium iron phosphate phase, namely calcium hydrogen iron phosphate

[$Ca_9FeH(PO_4)_7$] in BA,1° and BA,3°, and whitlockite [$Ca_8FeOH(PO_4)_6 \cdot 10H_2O$] in BA,2°. Iron oxide [$Fe_2O_3$] was identified in BA,1° and BA,3°, while potassium hydrogen phosphate [KH_2PO_4] was present only in BA,2°. Moreover, anorthite [$CaAl_2Si_2O_8$] was detected in BA,1°. The same analysis performed on CYC samples (Figure S4) showed again quartz as the main crystalline phase in all samples, with a little part of tridymite (another polymorphic form of [SiO_2]) in CYC,3°. $Ca_9FeH(PO_4)_7$ was identified in CYC,1° while $Ca_8FeOH(PO_4)_6 \cdot 10H_2O$ in CYC,2° and CYC,3°. Fe_2O_3 was detected in CYC,1° and CYC,3°, and another phosphate phase was found in CYC,1° and CYC,2°: sodium hydrogen phosphate [$Na_5H_{2.5}PO_3$]. In addition, small peaks revealed the presence of anhydrite [$CaSO_4$] in CYC,3°. Similar results were reported by Gorazda *et al.* (2017): phosphorus in sewage sludge ash was present as calcium phosphate, iron as hematite, and aluminium as anorthite. Quartz and tridymite were also identified. In line with literature and presented experimental results, Abis *et al.* (2018) indicated quartz, hematite and whitlockite as the main crystallised minerals.

Morphologically, the CYC samples are very interesting. SEM images for CYC,1° and CYC,2° at different magnifications are shown in Figures 1 and 2, with the corresponding EDXS results reported in Table S3. The presence of spheres, mainly composed of Al, Si, P, Ca, Fe (areas 2-3-5 in CYC,1°-A and point 7 in CYC,1°-B), can be observed, as already reported by Fiameni *et al.* (2021). Area 1 in CYC,1°-A reveals an agglomerate mainly consisting of P and Fe, while area 4 in CYC,1°-A is a particular agglomerate composed of Mg and Si. Point 6 in CYC,1°-B is a Si-Al agglomerate, such as point 8, with a major presence of K. In CYC,2°-A, area 1 shows the absence of P, and a high percentage of Al and Si, while the chemical composition of the area 2 is similar to that of the spheres cited above, even if the surface is not smooth. Indeed, the image of CYC,2°-B shows that spheres have a rough surface and are internally hollow.

3.2. Characterization of sewage sludge ash: elemental composition

Tables 1 and 2 show the elemental composition of BA and CYC samples. Major elements, excluding heavy metals, were distributed as follows: *P*>Ca>Al>Fe>Mg>K>Na for BA samples, and Ca>*P*>Al>Fe>Mg>K>Na for CYC samples.

Mean concentrations of elements in pelletized and dried sewage sludge were compared to ash to calculate the enrichment factor, being reported in Figure S5. Most elements had an enrichment factor between 1 and 5 in BA and CYC samples, due to mass reduction during combustion. As, Cd, Pb, Sb and Zn showed an enrichment factor higher than 10 in CYC samples. Due to the higher volatility of those metals, they tend to concentrate in fine size ash, which is collected after flue gas treatments. Similar results were obtained by Cheng *et al.* (2020), that reported experimental data from multi-cyclone dust ash. On the contrary, observed enrichment factor for Fe (0.6–1.1) was much lower than expected. Indeed Cheng *et al.* (2020) observed an enrichment factor of 5–6 for all non-volatile elements including Fe. However, since the data refer to a fluidized bed incinerator, the enrichment factor for grate furnace can be different. To give a better insight of the enrichment factor for Fe, further investigation is needed.

Phosphorus content in sewage sludge ash samples ranged between 7.2% and 10.4%, thus perfectly aligned with literature data (4.9–11.9%) reported by Krüger & Adam (2015) for municipal sludge derived ash samples. In their study, they investigated ash chemical composition obtained by 24 incineration facilities equipped with different combustion technology (e.g.,

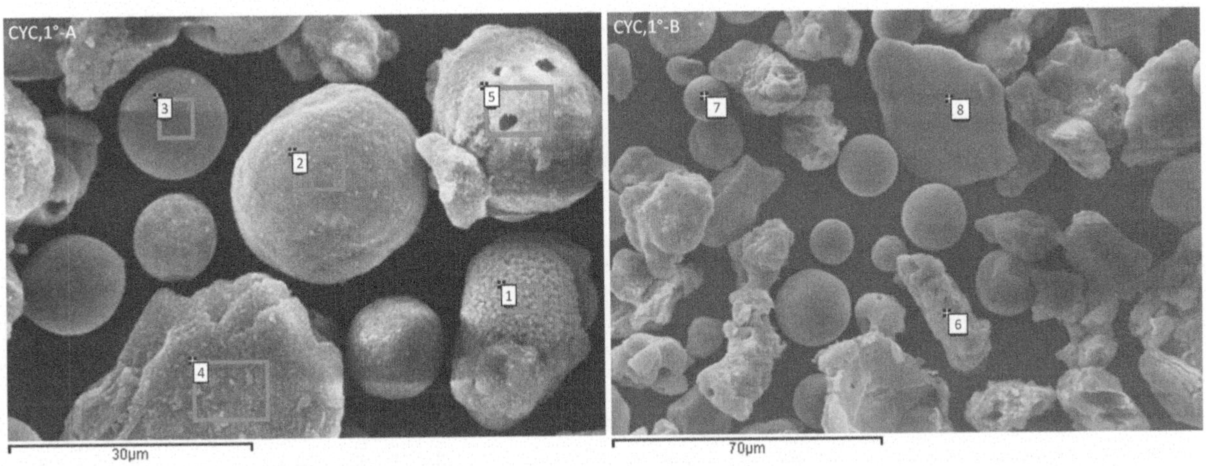

Figure 1 | SEM images (two magnification values: -A and -B) of cyclone ash obtained from the first experimental campaign (CYC,1°).

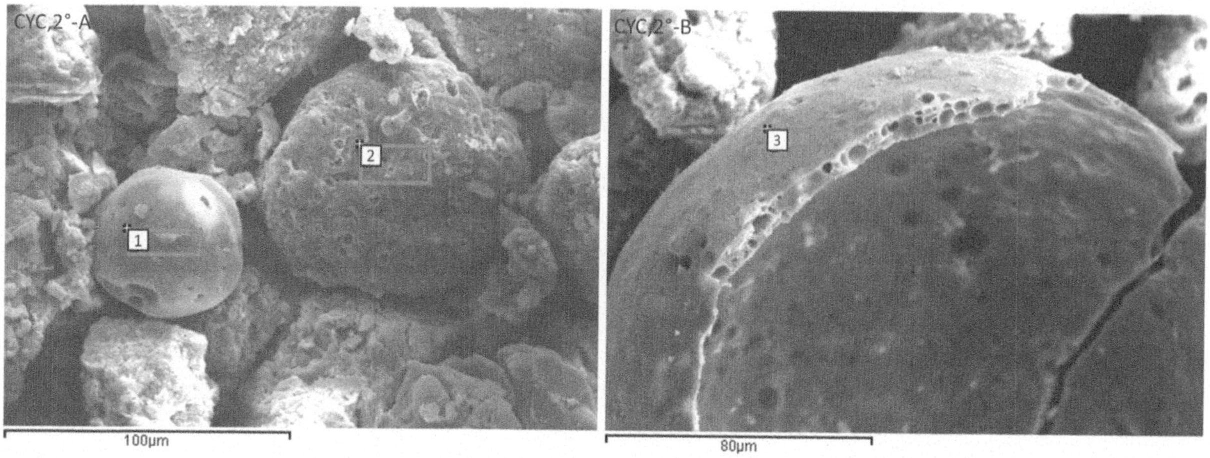

Figure 2 | SEM images (two magnification values: -A and -B) of cyclone ash obtained from the second experimental campaign (CYC,2°).

Table 1 | Element concentration (% or mg/kg) in bottom ash (BA) and cyclone ash (CYC) samples from the first and second experimental campaigns

	U.M.	BA,1°	CYC,1°	BA,2°	CYC,2°	Limit
P	%	7.2±1.0	7.5±1.1	7.2±1.0	8.5±1.2	–
Fe	%	3.4±0.5	3.5±0.5	2.2±0.3	2.0±0.3	–
Al	%	6.1±0.9	6.7±1.0	5.8±0.8	6.1±0.9	–
Ca	%	8.5±1.2	8.9±1.3	8.9±1.3	9.1±1.3	–
K	%	0.7±0.1	0.8±0.1	0.5±0.1	0.6±0.1	–
Mg	%	1.4±0.2	1.2±0.2	0.9±0.1	1.0±0.1	–
Na	%	0.32±0.05	0.3±0.04	0.2±0.02	0.2±0.03	–
As	mg kg^{-1}	32.9±4.7	**92.2±13.0**	25.8±3.7	**49.7±7.0**	40
Cd	mg kg^{-1}	0.31±0.04	**76.8±10.9**	0.51±0.07	**37.1±5.2**	1.5
Co	mg kg^{-1}	31.8±4.5	26.3±3.7	16.6±2.4	20.9±3	–
Cr	mg kg^{-1}	**441.6±62.5**	**468.8±66.3**	**426.4±60.3**	**640.2±90.5**	2
Cu	mg kg^{-1}	**1,014.6±143.5**	**1,274.0±180.2**	**1,179.4±166.8**	**1,202.6±170.1**	600
Hg	mg kg^{-1}	0.49±0.07	0.64±0.09	0.54±0.08	0.54±0.07	1
Ni	mg kg^{-1}	**552.0±78.1**	**560.4±72.3**	**605.5±85.6**	**514.1±72.7**	100
Pb	mg kg^{-1}	**124.5±17.6**	**2,522.8±356.8**	**213.9±30.3**	**802.5±113.5**	120
Sb	mg kg^{-1}	6.6±0.9	6.1±0.8	7.6±1.1	6.7±0.9	–
Tl	mg kg^{-1}	0.19±0.03	0.36±0.05	0.20±0.03	0.35±0.05	–
V	mg kg^{-1}	66.3±9.4	72.7±10.3	49.9±7.1	52±7.4	–
Zn	mg kg^{-1}	**2,381.6±336.8**	**6,244.8±883.2**	**3,229.7±456.8**	**3,197.2±452.2**	1,500

Reported limits refer to European regulation for fertilizer application (EU 2019/1009). Limit for chromium (Cr) refers to Cr (VI), while the content in the investigated ashes refer to total chromium. Values in bold indicate the exceedance of the limits

fluidized bed, grate furnace, multiple hearth firing). The highest phosphorus concentrations were observed in the third experimental campaign (April), reaching a maximum concentration of about 9% in BA and 10.4% in CYC. The lowest phosphorus concentrations (7.2% for BA and 7.5% for CYC) were measured in the second experimental campaign (August). The same trend was reported by Krüger *et al.* (2014), who analysed the monthly variations of the phosphorus content in ashes for different facilities. They reported peaks of phosphorus concentration in February, March and April, and a decline of phosphorus content from June to August. Variation in phosphorus concentration can be explained by the common fluctuations of inlet

Table 2 | Element concentration (% or mg/kg) in bottom ash (BA) and cyclone ash (CYC) samples from the third experimental campaigns

	U.M.	BA1,3°	BA2,3°	BA3,3°	CYC,3°	Limit
P	%	9.9±1.4	9.0±1.3	8.0±1.1	10.4±1.5	–
Fe	%	1.9±0.3	2.0±0.3	2.1±0.3	1.7±0.2	–
Al	%	9.0±1.3	7.4±1.0	8.6±1.2	9.4±1.3	–
Ca	%	8.5±1.2	5.8±0.8	8.2±1.2	8.8±1.2	–
K	%	0.48±0.06	0.57±0.08	0.49±0.07	0.54±0.08	–
Mg	%	1.3±0.2	1.2±0.2	1.3±0.2	1.3±0.2	–
Na	%	0.34±0.05	0.33±0.05	0.39±0.06	0.29±0.04	–
As	mg kg^{-1}	26.6±3.8	**60.2±8.5**	27.8±3.9	**227.6±32.2**	40
Cd	mg kg^{-1}	0.19±0.03	0.19±0.03	0.20±0.03	**139.1±19.7**	1.5
Co	mg kg^{-1}	13.3±1.9	12.3±1.7	12.7±1.8	12.1±1.7	–
Cr	mg kg^{-1}	**226.4±32.0**	**218.5±31.0**	**209.7±29.7**	**423.3±59.9**	2
Cu	mg kg^{-1}	**1,174.8±166.1**	**1,244.6±176.0**	**1,418.7±200.6**	**1,184.8 ±167.6**	600
Hg	mg kg^{-1}	0.58±0.08	0.56±0.08	0.52±0.07	0.57±0.08	1
Ni	mg kg^{-1}	**606±85.7**	**197.5±27.9**	**604.4±85.5**	**586.2±82.9**	100
Pb	mg kg^{-1}	104.3±14.8	83.4±11.8	66.1±9.4	**6,317.7±893.5**	120
Sb	mg kg^{-1}	7.4±1.0	7.2±1.0	6.9±1.0	9.3±1.3	–
Tl	mg kg^{-1}	0.20±0.03	0.20±0.03	0.20±0.03	0.56±0.08	–
V	mg kg^{-1}	49.5±7.0	57.6±8.1	52.2±7.4	45.5±6.4	–
Zn	mg kg^{-1}	**1,810±256.0**	**1,604±226.8**	1,479.3±209.2	**14,792.4±2,092.0**	1,500

Limits refer to European regulation for fertilizer application fertilizer application (EU 2019/1009). Values in bold indicate the exceedance of the limits.

phosphorus concentration during the year and different phosphorus removal performances of the wastewater treatment plant (WWTP). Moreover, Krüger *et al.* (2014) suggested that the variation of food habits, leisure time and higher temperature in summer may be responsible for the drop of phosphorus concentration. It is reasonable to imagine that during summer people have more leisure time and those who are living in metropolitan cities such as Milan can travel to touristic locations. This can reduce the influent phosphorus loading rate at the WWTP, affecting the phosphorus content in the sludge and in the derived ash (Krüger *et al.* 2014). In addition, seasonal changes in P removal efficiency at WWTP can be related to phosphorus accumulating organisms (PAOs) performances fluctuations. Thus, temperature can change the dynamics of the microbial communities, raising possible competition of phosphorus accumulating organisms (PAOs) with other bacteria (Erdal *et al.* 2003; Ni *et al.* 2021).

Considering the two types of sewage sludge ashes collected within the same experimental campaign, CYC samples had a slightly higher phosphorus content compared to BA samples. Despite the non-volatility of phosphorus, it tended to concentrate in the finer fraction. This could be explained by the presence of reducing combustion conditions in the upper layer of the fuel bed, promoting the vaporisation of phosphorus (Han *et al.* 2009). As highlighted by Razmjoo *et al.* (2016), the complete consumption of oxygen takes place in the combustion layer of the fuel bed, resulting in reducing conditions in the fuel layers above. Instead, Gorazda *et al.* (2017) found a phosphorus concentration in BA (9.9%) slightly higher than CYC (9.7%).

As for the content of Fe and Al in ashes, higher concentrations of Al were observed in all experimental campaigns due to the use of Al salts in the WWTP. Within the experimental campaigns, higher concentration of Fe, Al and Ca were observed in CYC samples in comparison to BA samples. An exception was observed for Fe content in the second and third experimental campaigns. Similar results were reported by Gorazda *et al.* (2017): Fe content at 6.0% and 6.7%, Al content at 4.6% and 7.3%, and Ca content at 11.2% and 12% in BA and CYC were measured, respectively. High concentrations of Fe and Al in ashes are critical as they limit the quality of the recovered material, due to their precipitation with phosphorus.

Concerning the content of metals and metalloids in BA and CYC samples, Zn was the most abundant metal (1,479–14,792 mg/kg) followed by Cu (1,014–1,418 mg/kg). It clearly emerged from Table 2 that the Zn content of 14,792±2,092 mg/kg (CYC,3°) is out of the range (644–2,535 mg/kg) reported in literature (Ottosen *et al.* 2013; Krüger *et al.* 2014; Herzel *et al.* 2016; Wang *et al.* 2018; Benassi *et al.* 2019; Liang *et al.* 2019). Such outlier can be explained by an exceptional industrial load at the WWTP, as confirmed by the WWTP manager. Table S4 summarizes the heavy metal concentrations in sewage sludge ashes reported in literature.

In general, regarding the distribution of metals and metalloids in samples, it was highlighted that semi-volatile elements such as Zn, Pb, Cd and As tended to concentrate in the finer particles (CYC), while the non-volatile elements (Cr, Cu, Ni and V) equally distributed among the two types of ashes (Toledo *et al.* 2005; Van de Velden *et al.* 2008; Nzihou & Stanmore 2013). Shao *et al.* (2008) found similar results, indicating that Pb and Zn were mostly present in fly ash, while other heavy metals such as Cr, Ni, V resulted equally distributed between bottom and fly ash. Same observations were reported by Cheng *et al.* (2020), pointing out that higher enrichments of Pb and Cd were found for cyclone ash because of temperature and particle size.

As a result, BA samples, corresponding to the 90% of total ash produced during combustion, showed a lower level of metals contamination if compared to CYC samples, whose hazardousness is mainly due to high concentrations of As, Cd, Cr, Pb and Zn. Overall, both ash typologies do not comply with the limits for fertilizer applications (EU 2019/1009), resulting in a ban for agricultural use.

3.3. Factors affecting the phosphorus extraction efficiency

Experimental results demonstrated that not only the operating conditions of the chemical wet extraction can affect the PEE, but also the chemical-physical characteristics of sewage sludge ash, as particle size dimension, LOI and pH.

The influence of particles size dimension was studied considering the sample BA,1°. As expected, BA in the form of pellets (90% of particles above 2 mm) had the lowest PEE (56%), while BA milled (with ball mill) and pounded (with manual mortar) showed higher PEE (80% and 78%, respectively). Small particle size ensures a larger contact area between solid particles and extractant. Thus, the extractant can probably travel into pores and reach the reaction zone in a shorter time. Additional tests were performed considering four different particle size classes of pounded BA,1° (class I, <0.063 mm; class II, 0.063–0.125 mm; class III, 0.180–0.212 mm; class IV, 0.425–0.600 mm). Experimental results are reported in Figure S6. Particle size dimension lower than 0.212 mm ensured a good PEE (>70%), while higher particle size dimension (0.425–0.600 mm) resulted in lower PEE. In conclusion, very fine particle size (lower than 0.063 mm) is not strictly required. Instead, too small particles can be tightly packed, decreasing the mechanisms of diffusion and mass transport.

Furthermore, the influence of LOI on PEE was investigated considering the three BA sub-samples collected in the third experimental campaign, in which combustion conditions (mass flow rate and residence time) were slightly changed for process optimization. To avoid the influence of particle size dimension, sewage sludge ashes with particle size lower than 0.212 mm were considered, while acid extraction conditions were kept constant. As a result, the highest PEE (85.6%) was reached by BA3,3° sample, which had the lowest LOI (1.1%). Thus, unburned fractions may subtract extracting capacity to the acid solution, causing a decrease of the process selectivity towards phosphorus. In addition, formation of different species containing the unburned fraction can affect the solubility of phosphate-bearing compounds. These experimental results evidence the importance of optimizing the combustion conditions in the design of a process aimed at phosphorus recovery.

Experimental results from leaching tests in deionized water revealed that the pH of BA samples was between 6.8 and 8.1, while it was between 6.7 and 8.8 for CYC samples. In general, sewage sludge ash samples with pH around 7 (BA,3°, BA2,3° and CYC,1°) showed the highest PEE. Instead, basic pH slightly lowers the extraction efficiency possibility due to ash basicity that tends to neutralize the acidic pH.

The chemical composition of sewage sludge ash, which is related to the incineration technology, can also affect the PEE. Liu *et al.* (2021) highlighted indeed the importance of upstream treatment conditions in the phosphorus speciation in sewage sludge ash. Figure 3 shows the experimental results of leaching tests conducted on the investigated samples.

Overall, mean phosphorus concentration in leachate varied between 2,823 and 3,775 mg/L, while PEE ranged between 71% and 86%. The first and second experimental campaigns had comparable phosphorus concentrations in leachate for both ashes (2,823 and 3,325 mg/L), while the third campaign showed higher phosphorus concentration in leachate (from 3,425 to 3,775 mg/L). The highest concentrations were measured in the samples that had the highest phosphorus content in the ash (BA1,3° and CYC,3°), while the highest PEE was reached by BA3,3° (85.6%) and CYC,1° (83.0%) samples. Instead, lowest PEE was reached by BA,2° (75.0%) and CYC,3° (69.1%) samples. Thus, a high phosphorus concentration in the leachate (e.g., observed in CYC,3°) does not necessarily correspond to a high PEE, since the extraction efficiency is calculated as the ratio between the mass of phosphorus in the leachate and the mass of phosphorus in the ash.

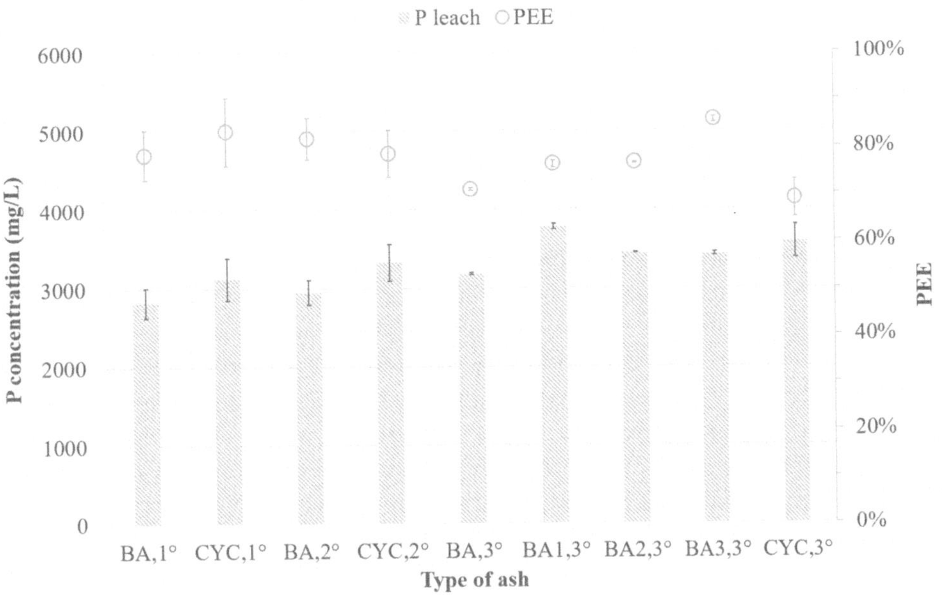

Figure 3 | Phosphorus concentration in leachate (as mg/L) and PEE for the investigated samples.

The decrease of PEE for CYC samples can be attributed to the higher content of Al in CYC,3° sample (9.4%) if compared to CYC,1°–2° sample (6.0–6.7%). Kalmykova & Karlfeldt Fedje (2013) pointed out that a high content of Al in sewage sludge ash can significantly decrease the PEE, while Gorazda *et al.* (2017) reported that high concentration of Al may cause precipitation of phosphate compounds during extraction. Therefore, phosphate compounds can be present in larger extent in the acidic solid residue, decreasing the phosphorus concentration in leachate. On the contrary, when Fe is immobilised in the hematite phase, which is slightly insoluble in strong acid, the extraction becomes more selective towards phosphorus compounds (Gorazda *et al.* 2017). Indeed, Ma & Rosen (2021) recommended the acid extraction only for Fe-rich sewage sludge ashes. However, even if the iron co-dissolution could be limited, the level of Fe content in the ash is the main parameter affecting the Fe concentration in the leachate and in the precipitated product. Thus, the use alone of biological P removal at the WWTP would be the best solution for decreasing the Al and Fe content in the sludge and thus in the ash.

3.4. Co-dissolution of undesired elements

Because of strong acidic conditions achieved during leaching tests (pH between 1.3 and 2), undesired elements such as metals and metalloids can dissolve. Figure S7 shows the percentage of elements passing in solution, while Table S5 summarises their concentrations in liquid phase. In general, Cr, Ni and Pb tended to concentrate in the acidic solid residues, while 58–91% of As and 54–100% of Cd easily passed in solution. Kim & Kazonich (2004) highlighted that the non-silicate phase present in sewage sludge ashes can be responsible for this higher solubility. In addition, Fang *et al.* (2018) reported a nearly completely dissolution of As, and Li *et al.* (2017) obtained an extraction rate of 75%. Considering the most abundant metals (Fe and Al), the extraction rates observed were between 26–74% for Fe and 30–100% for Al. The lowest extraction rate of Al was observed for BA,3° and CYC,3° samples, which had also a low PEE. Regarding the most present metals (Zn and Cu), extraction rates between 25% and 81% were observed. Overall, leachates derived from CYC samples showed higher contamination in comparison to leachates from BA samples. Thus, BA samples can be preferred for phosphorus extraction and subsequent recovery due to the lower contamination.

3.5. Phosphorus precipitation

Some flocculation phenomena started at nearly pH 2, increasing its intensity with the addition of basic compounds. At pH 3, about 80–90% of the total mass of precipitated material was formed, while the remaining part (20–10%) precipitated within pH 7. Experimental results of preliminary tests, shown in Figure S8 and Figure S9, revealed that aging time of 2 h and pH of 3.5 were sufficient to reach the phosphorus thermodynamic equilibrium, to have a high PPE and phosphorus content.

Precipitation tests conducted afterwards revealed that the type of precipitating agent (NaOH, $Ca(OH)_2$, KOH) did not strongly affect the amount of precipitated phosphorus. Thus, more than 90% of phosphorus precipitates from leachate within pH 3.5. This is in accordance with Fang *et al.* (2018), who reported the occurrence of phosphorus precipitation within pH 3.2–3.9.

The precipitated material showed high phosphorus content (16–21%) in the first stage of pH increase (Figure 4), followed by a low decline due the co-precipitation of undesirable elements. Thus, the increase of the mass of the recovered material implied a higher phosphorus dilution. As a result, pH 3.5 and 5 can be considered to be the most suitable values for ensuring both a high PPE (>90%) and phosphorus content (17%).

For all samples, XRD analysis revealed the absence of crystalline forms. As a result, most of the phosphorus in the recovered material is present in amorphous phases, thus possibly easily available for the plant root. It could be interesting to further study the crystallization kinetics of the precipitates, to investigate the formation of crystalline Ca, Al, or Fe phosphates.

Considering the co-precipitation of other elements, as highlighted by Ito *et al.* (2013), aluminium and iron phosphates combined with their hydroxides are the major compounds formed between pH 3 and 4, while calcium phosphates prevail from pH 4. Similarly, Baldi *et al.* (2021) assumed the forms of phosphorus precipitated as $Ca_3(PO4)_2$, $FePO_4$ and $AlPO_4$ at pH 3.7. In addition, Fang *et al.* (2018) reported that the Fe precipitated at pH lower than 2.6 may correspond to goethite or hematite. The analysis of the filtrate solution obtained after precipitation tests confirmed the complete precipitation of Al and Fe at pH 5. Instead, the complete precipitation of Zn and Cu occurred at pH 8. Figure S10 shows the precipitation efficiency of investigated elements (P, Al, Fe, Ca, Zn and Cu) for the different precipitation tests. It can be noted that the type of precipitating agent (NaOH followed by $Ca(OH)_2$ and KOH) did not strongly affect the precipitation efficiencies. An unexpected behaviour was observed for Zn, since a small delay in its precipitation at pH 2.5 occurred, while dosing NaOH instead of KOH. Overall, at pH 3.5, 40% of Zn and 20% of Cu precipitated, while at pH 5 about 65% of Zn and 80% of Cu precipitated. At the same time, Ito *et al.* (2013) reported 40% of Zn precipitation and 50% Cu at pH 4, confirming the co-precipitation of those elements with phosphorus.

In conclusion, the quality of the recovered material decreased while increasing the precipitation pH. The presence of Al and Fe can reduce the plant availability of the recovered product, as reported by Liang *et al.* (2019). In addition, Liu *et al.* (2021) highlighted the need for process optimization for Ca-P product recovery, as innovative strategies for leachate decontamination are required. They suggested the use of new bio-based adsorbent to remove the interfering ions before precipitation.

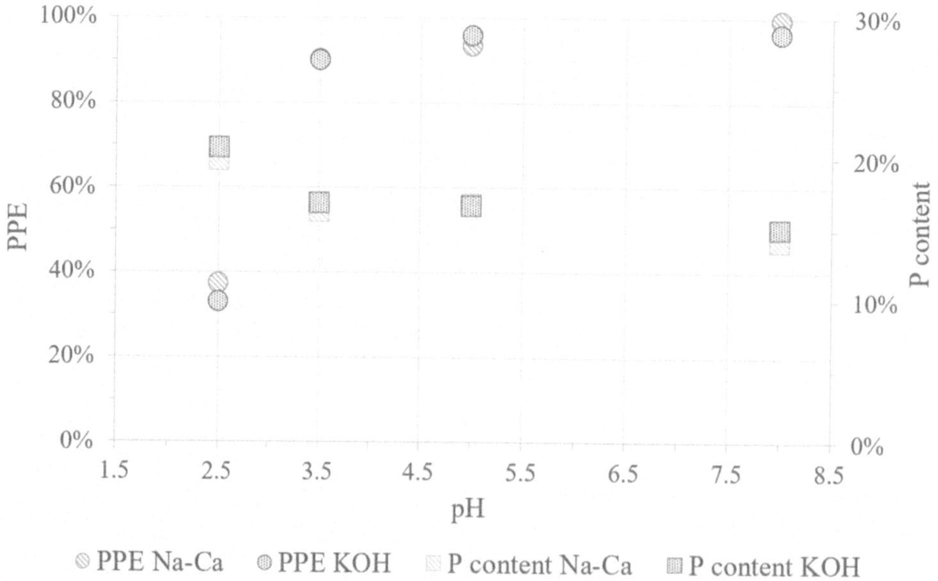

Figure 4 | Influence of the type of precipitating agent on PEE, phosphorus content and pH (sample BA,2° sample). Na-Ca refers to dosage of NaOH until pH 3.5 followed by dosage of $Ca(OH)_2$ for the higher pH ranges (from pH 3.5 to 5 and from pH 3.5 to 8).

4. CONCLUSIONS

- Grate furnace is a viable option for sewage sludge ash mono-incineration in the view of phosphorus recovery. Acid leaching of bottom ash (BA) and cyclone ash (CYC) showed a mean phosphorus extraction efficiency (PEE) between 69% and 86% at 0.2 M H_2SO_4, L/S of 20:1, 2 h as contact time and room temperature.
- BA samples showed a slightly lower phosphorus content (7.2–9.9%) and a much lower level of heavy metal contamination with respect to CYC samples. As a result, BA, representing the 90% of the ashes produced, is more suitable for the phosphorus extraction process.
- Upstream treatment characteristics such as the type of phosphorus-removal technology used at WWTP can affect the PEE and the quality of the recovered material. High concentration of Al in ash can be responsible for a decrease of the extraction capacity. The lowest PEE was indeed achieved by the sample with the highest Al content in the ash (9.4%). At the same time, Fe content in the ash of 3.5% can affect the quality of the recovered material, due to formation of iron phosphates compounds in the final product.
- Ash physical characteristics influence the PEE. Particle size dimension lower than 0.212 mm turned out to be sufficiently fine to provide a good contact surface between solid particles and the extractant. At the same time, a good combustion efficiency showed better extraction performances: the lower the LOI, the higher the PEE.
- Positioning the recovery of P as first design parameter is of paramount importance. Thus, the type of combustion technology can strongly influence the chemical composition and morphology of ashes.
- Key factor for ensuring a high phosphorus precipitation efficiency (PPE) and phosphorus content is pH, while precipitating agents do not have a strong influence. pH in the range of 3.5–5 ensures a PPE higher than 90%, a P content between 16–17% and a low co-dissolution of Cu and Zn.
- Further investigations and strategies on leachate decontamination (e.g., using bio-based adsorbents or performing ashes inertization prior to acid leaching) are needed for ensuring a high-quality fertilizer material. Thus, the minimization of heavy metals, Fe and Al content is required.

ACKNOWLEDGEMENTS

This work was partially funded by Region Lombardy (Italy) within the research project 'PerformWater_2030' (Decree DGR N. 5245/2016 – AZIONE I.1.B.1.3 – ASSE I POR FESR 2014–2020 – CUP E46D17000120009). The authors would like to thank VOMM Impianti e Processi for the assistance in ash sampling and for providing information about the incineration plant. The authors would also like to gratefully thank Rea Dalmine (Greenthesis Group, Dalmine, Italy) for funding the PhD grant of Gaia Boniardi.

DATA AVAILABILITY STATEMENT

All relevant data are included in the paper or its Supplementary Information.

REFERENCES

Abis, M., Calmano, W. & Kuchta, K. 2018 Innovative technologies for phosphorus recovery from sewage sludge ash. *Detritus* **1**, 23–29. https://doi.org/10.26403/detritus/2018.23.

ASTM International 2020 ASTM E-11: Standard Specification for Woven Wire Test Sieve Cloth and Test Sieves.

Baldi, M., Martinotti, A., Sorlini, S., Katsoyiannis, I., Abbà, A., Carnevale Miino, M. & Collivignarelli, M. 2021 Extraction and purification of phosphorus from the ashes of incinerated biological sewage sludge. *Water* **13**, 1102. https://doi.org/10.3390/w13081102.

Benassi, L., Zanoletti, A., Depero, L. E. & Bontempi, E. 2019 Sewage sludge ash recovery as valuable raw material for chemical stabilization of leachable heavy metals. *Journal of Environmental Management* **245**, 464–470. https://doi.org/10.1016/j.jenvman.2019.05.104.

Biswas, B. K., Inoue, K., Harada, H., Ohto, K. & Kawakita, H. 2009 Leaching of phosphorus from incinerated sewage sludge ash by means of acid extraction followed by adsorption on orange waste gel. *Journal of Environmental Sciences* **21**, 1753–1760. https://doi.org/10.1016/S1001-0742(08)62484-5.

Boniardi, G., Turolla, A., Fiameni, L., Gelmi, E., Malpei, F., Bontempi, E. & Canziani, R. 2021 Assessment of a simple and replicable procedure for selective phosphorus recovery from sewage sludge ashes by wet chemical extraction and precipitation. *Chemosphere* **285**, 131476. https://doi.org/10.1016/j.chemosphere.2021.131476.

Bontempi, E., Zacco, A., Borgese, L., Gianoncelli, A., Ardesi, R. & Depero, L. E. 2010 A new method for municipal solid waste incinerator (MSWI) fly ash inertization, based on colloidal silica. *Journal of Environmental Monitoring* **12**, 2093. https://doi.org/10.1039/c0em00168f.

Cheng, Y., Oleszek, S., Shiota, K., Oshita, K. & Takaoka, M. 2020 Comparison of sewage sludge mono-incinerators: mass balance and distribution of heavy metals in step grate and fluidized bed incinerators. *Waste Management* **105**, 575–585. https://doi.org/10.1016/j.wasman.2020.02.044.

Chrispim, M. C., Scholz, M. & Nolasco, M. A. 2019 Phosphorus recovery from municipal wastewater treatment: critical review of challenges and opportunities for developing countries. *Journal of Environmental Management* **248**, 109268. https://doi.org/10.1016/j.jenvman.2019.109268.

Cyr, M., Coutand, M. & Clastres, P. 2007 Technological and environmental behavior of sewage sludge ash (SSA) in cement-based materials. *Cement and Concrete Research* **37**, 1278–1289. https://doi.org/10.1016/j.cemconres.2007.04.003.

Desmidt, E., Ghyselbrecht, K., Zhang, Y., Pinoy, L., Van der Bruggen, B., Verstraete, W., Rabaey, K. & Meesschaert, B. 2015 Global phosphorus scarcity and full-scale P-recovery techniques: a review. *Critical Reviews in Environmental Science and Technology* **45**, 336–384. https://doi.org/10.1080/10643389.2013.866531.

Directive 2010/75/EU. Available from: http://data.europa.eu/eli/dir/2010/75/2011-01-06

Donatello, S. 2009 Characteristics of incinerated sewage sludge ashes: potential for phosphate extraction and re-use as a pozzolanic material in construction products. https://doi.org/10.13140/RG.2.2.33926.63040

Donatello, S. & Cheeseman, C. R. 2013 Recycling and recovery routes for incinerated sewage sludge ash (ISSA): a review. *Waste Management* **33**, 2328–2340. https://doi.org/10.1016/j.wasman.2013.05.024.

Erdal, U. G., Erdal, Z. K. & Randall, C. W. 2003 The competition between PAOs (phosphorus accumulating organisms) and GAOs (glycogen accumulating organisms) in EBPR (enhanced biological phosphorus removal) systems at different temperatures and the effects on system performance. *Water Science and Technology* **47**, 1–8. https://doi.org/10.2166/wst.2003.0579.

European Commission (EC) 2017. Available from: https://ec.europa.eu/growth/sectors/raw-materials/specific-interest/critical_en

European Committee for Standardization 2002 EN 12457-2 Characterisation of waste – leaching – compliance test for leaching of granular waste materials and sludges – Part 2: One stage batch test at a liquid to solid ratio of 10 L/kg for materials with particle size below 4 Mm (without or with size reduction).

European Committee for Standardization 2011 EN 15956 Fertilizers – extraction of phosphorus soluble in mineral acids.

Fahimi, A., Federici, S., Depero, L. E., Valentim, B., Vassura, I., Ceruti, F., Cutaia, L. & Bontempi, E. 2021 Evaluation of the sustainability of technologies to recover phosphorus from sewage sludge ash based on embodied energy and CO2 footprint. *Journal of Cleaner Production* **289**, 125762. https://doi.org/10.1016/j.jclepro.2020.125762.

Fang, L., Li, J., Donatello, S., Cheeseman, C. R., Wang, Q., Poon, C. S. & Tsang, D. C. W. 2018 Recovery of phosphorus from incinerated sewage sludge ash by combined two-step extraction and selective precipitation. *Chemical Engineering Journal* **348**, 74–83. https://doi.org/10.1016/j.cej.2018.04.201.

Fang, L., Wang, Q., Li, J., Poon, C. S., Cheeseman, C. R., Donatello, S. & Tsang, D. C. W. 2020 Feasibility of wet-extraction of phosphorus from incinerated sewage sludge ash (ISSA) for phosphate fertilizer production: a critical review. *Critical Reviews in Environmental Science and Technology* 1–33. https://doi.org/10.1080/10643389.2020.1740545.

Fiameni, L., Assi, A., Fahimi, A., Valentim, B., Moreira, K., Predeanu, G., Slăvescu, V., Vasile, B. Ş., Nicoară, A. I., Borgese, L., Boniardi, G., Turolla, A., Canziani, R. & Bontempi, E. 2021 Simultaneous amorphous silica and phosphorus recovery from rice husk poultry litter ash. *RSC Advances* **11**, 8927–8939. https://doi.org/10.1039/D0RA10120F.

Franz, M. 2008 Phosphate fertilizer from sewage sludge ash (SSA). *Waste Management* **28**, 1809–1818. https://doi.org/10.1016/j.wasman.2007.08.011.

Gorazda, K., Wzorek, Z. & Jodko, M. 2007 The influence of thermal processing of sewage sludge on the usage properties of the formed ash. *Polish Journal of Chemical Technology* **9**, 21–27. https://doi.org/10.2478/v10026-007-0083-y.

Gorazda, K., Tarko, B., Wzorek, Z., Kominko, H., Nowak, A. K., Kulczycka, J., Henclik, A. & Smol, M. 2017 Fertilisers production from ashes after sewage sludge combustion – a strategy towards sustainable development. *Environmental Research* **154**, 171–180. https://doi.org/10.1016/j.envres.2017.01.002.

Han, J., Kanchanapiya, P., Sakano, T., Mikuni, T., Furuuchi, M. & Wang, G. 2009 The behaviour of phosphorus and heavy metals in sewage sludge ashes. *IJEP* **37**, 357. https://doi.org/10.1504/IJEP.2009.026054.

Herzel, H., Krüger, O., Hermann, L. & Adam, C. 2016 Sewage sludge ash – a promising secondary phosphorus source for fertilizer production. *Science of The Total Environment* **542**, 1136–1143. https://doi.org/10.1016/j.scitotenv.2015.08.059.

Hong, K.-J., Tarutani, N., Shinya, Y. & Kajiuchi, T. 2005 Study on the recovery of phosphorus from waste-activated sludge incinerator ash. *Journal of Environmental Science and Health, Part A* **40**, 617–631. https://doi.org/10.1081/ESE-200046614.

International Organization for Standardization (ISO) 2016 Test Sieves – Technical Requirements and Testing Test Sieves of Metal Wire Cloth (ISO 3310-1:2016).

Ito, A., Yamada, K., Ishikawa, N. & Umita, T. 2013 Separation of metals and phosphorus from incinerated sewage sludge ash. *Water Science and Technology* **67**, 2488–2493. https://doi.org/10.2166/wst.2013.142.

Kalmykova, Y. & Karlfeldt Fedje, K. 2013 Phosphorus recovery from municipal solid waste incineration fly ash. *Waste Management* **33**, 1403–1410. https://doi.org/10.1016/j.wasman.2013.01.040.

Kim, A. G. & Kazonich, G. 2004 The silicate/non-silicate distribution of metals in fly ash and its effect on solubility. *Fuel* **83**, 2285–2292. https://doi.org/10.1016/j.fuel.2004.06.005.

Krüger, O. & Adam, C. 2015 Recovery potential of German sewage sludge ash. *Waste Management* **45**, 400–406. https://doi.org/10.1016/j.wasman.2015.01.025.

Krüger, O., Grabner, A. & Adam, C. 2014 Complete survey of German sewage sludge ash. *Environmental Science & Technology* **48**, 11811–11818. https://doi.org/10.1021/es502766x.

Lee, M. & Kim, D.-J. 2017 Identification of phosphorus forms in sewage sludge ash during acid pre-treatment for phosphorus recovery by chemical fractionation and spectroscopy. *Journal of Industrial and Engineering Chemistry* **51**, 64–70. https://doi.org/10.1016/j.jiec.2017.02.013.

Li, J., Xue, Q., Fang, L. & Poon, C. S. 2017 Characteristics and metal leachability of incinerated sewage sludge ash and air pollution control residues from Hong Kong evaluated by different methods. *Waste Management* **64**, 161–170. https://doi.org/10.1016/j.wasman.2017.03.033.

Li, B., Boiarkina, I., Yu, W., Huang, H. M., Munir, T., Wang, G. Q. & Young, B. R. 2019a Phosphorous recovery through struvite crystallization: challenges for future design. *Science of The Total Environment* **648**, 1244–1256. https://doi.org/10.1016/j.scitotenv.2018.07.166.

Li, B., Udugama, I. A., Mansouri, S. S., Yu, W., Baroutian, S., Gernaey, K. V. & Young, B. R. 2019b An exploration of barriers for commercializing phosphorus recovery technologies. *Journal of Cleaner Production* **229**, 1342–1354. https://doi.org/10.1016/j.jclepro.2019.05.042.

Liang, S., Chen, H., Zeng, X., Li, Z., Yu, W., Xiao, K., Hu, J., Hou, H., Liu, B., Tao, S. & Yang, J. 2019 A comparison between sulfuric acid and oxalic acid leaching with subsequent purification and precipitation for phosphorus recovery from sewage sludge incineration ash. *Water Research* **159**, 242–251. https://doi.org/10.1016/j.watres.2019.05.022.

Lin, W. Y., Ng, W. C., Wong, B. S. E., Teo, S. L.-M., Sivananthan, G. d., Baeg, G. H., Ok, Y. S. & Wang, C.-H. 2018 Evaluation of sewage sludge incineration ash as a potential land reclamation material. *Journal of Hazardous Materials* **357**, 63–72. https://doi.org/10.1016/j.jhazmat.2018.05.047.

Liu, H., Hu, G., Basar, I. A., Li, J., Lyczko, N., Nzihou, A. & Eskicioglu, C. 2021 Phosphorus recovery from municipal sludge-derived ash and hydrochar through wet-chemical technology: a review towards sustainable waste management. *Chemical Engineering Journal* **417**, 129300. https://doi.org/10.1016/j.cej.2021.129300.

Ma, P. & Rosen, C. 2021 Land application of sewage sludge incinerator ash for phosphorus recovery: a review. *Chemosphere* **274**, 129609. https://doi.org/10.1016/j.chemosphere.2021.129609.

Ni, M., Pan, Y., Chen, Y., Zhang, X., Huang, Y. & Song, Z. 2021 Effects of seasonal temperature variations on phosphorus removal, recovery, and key metabolic pathways in the suspended biofilm. *Biochemical Engineering Journal* **176**, 108187. https://doi.org/10.1016/j.bej.2021.108187.

Niewersch, C., Battaglia Bloch, A. L., Yüce, S., Melin, T. & Wessling, M. 2014 Nanofiltration for the recovery of phosphorus – development of a mass transport model. *Desalination* **346**, 70–78. https://doi.org/10.1016/j.desal.2014.05.011.

Nzihou, A. & Stanmore, B. 2013 The fate of heavy metals during combustion and gasification of contaminated biomass – a brief review. *Journal of Hazardous Materials* **256–257**, 56–66. https://doi.org/10.1016/j.jhazmat.2013.02.050.

Ottosen, L. M., Kirkelund, G. M. & Jensen, P. E. 2013 Extracting phosphorous from incinerated sewage sludge ash rich in iron or aluminum. *Chemosphere* **91**, 963–969. https://doi.org/10.1016/j.chemosphere.2013.01.101.

Razmjoo, N., Sefidari, H. & Strand, M. 2016 Measurements of temperature and gas composition within the burning bed of wet woody residues in a 4 MW moving grate boiler. *Fuel Processing Technology* **152**, 438–445. https://doi.org/10.1016/j.fuproc.2016.07.011.

Regulation (EU) 2019/1009. Available from: https://eur-lex.europa.eu/legal-content/EN/TXT/?uri=CELEX%3A32019R1009

Shao, J., Yan, R., Chen, H., Yang, H., Lee, D. H. & Liang, D. T. 2008 Emission characteristics of heavy metals and organic pollutants from the combustion of sewage sludge in a fluidized bed combustor. *Energy Fuels* **22**, 2278–2283. https://doi.org/10.1021/ef800002y.

Toledo, J. M., Corella, J. & Corella, L. M. 2005 The partitioning of heavy metals in incineration of sludges and waste in a bubbling fluidized bed. *Journal of Hazardous Materials* **126**, 158–168. https://doi.org/10.1016/j.jhazmat.2005.06.021.

Van de Velden, M., Dewil, R., Baeyens, J., Josson, L. & Lanssens, P. 2008 The distribution of heavy metals during fluidized bed combustion of sludge (FBSC). *Journal of Hazardous Materials* **151**, 96–102. https://doi.org/10.1016/j.jhazmat.2007.05.056.

Wang, Q., Li, J., Tang, P., Fang, L. & Poon, C. S. 2018 Sustainable reclamation of phosphorus from incinerated sewage sludge ash as value-added struvite by chemical extraction, purification and crystallization. *Journal of Cleaner Production* **181**, 717–725. https://doi.org/10.1016/j.jclepro.2018.01.254.

Xu, H., He, P., Gu, W., Wang, G. & Shao, L. 2012 Recovery of phosphorus as struvite from sewage sludge ash. *Journal of Environmental Sciences* **24**, 1533–1538. https://doi.org/10.1016/S1001-0742(11)60969-8.

Yu, X., Nakamura, Y., Otsuka, M., Omori, D. & Haruta, S. 2021 Development of a novel phosphorus recovery system using incinerated sewage sludge ash (ISSA) and phosphorus-selective adsorbent. *Waste Management* **120**, 41–49. https://doi.org/10.1016/j.wasman.2020.11.017.

Zabielska-Adamska, K. 2019 Sewage sludge bottom ash characteristics and potential application in road embankment. *Sustainability* **12**, 39. https://doi.org/10.3390/su12010039.

First received 11 January 2022; accepted in revised form 11 April 2022. Available online 20 April 2022

doi: 10.2166/wst.2022.129

Granulation strategies applied to industrial wastewater treatment: from lab to full-scale

M. Caluwé, K. Goossens, K. Seguel Suazo, E. Tsertou and J. Dries*

Research group BioWAVE, Biochemical Wastewater Valorization and Engineering, Faculty of Applied Engineering, University of Antwerp, Groenenborgerlaan 171, Antwerp 2020, Belgium

*Corresponding author. E-mail: jan.dries2@uantwerpen.be

JD, 0000-0002-0204-5697

ABSTRACT

About one third of the industrial activated sludge (AS) plants worldwide suffer from bad settling sludge, often caused by filamentous bulking phenomena. The present study investigated the effectiveness of a sludge granulation/densification strategy, based only on a metabolic selection mechanism, to eliminate sludge bulking in a sequencing batch reactor (SBR) treating real industrial wastewater. The wastewater originated from a tank truck cleaning company transporting chocolate and beer. The proposed strategy involved the introduction of a slow unaerated/anaerobic feeding step in the SBR operation. On lab-scale, the new feeding strategy resulted in (1) excellent settling with a sludge volume index (SVI) decreasing from more than 300 mL·g^{-1} to 100 mL·g^{-1} and lower, (2) the elimination of sludge bulking genera and (3) the significant enrichment of glycogen-accumulating organisms (GAO), mainly *Defluviicoccus* and *Candidatus* Competibacter, and this in less than 80 days. The feeding strategy was then applied to the full-scale installation, yielding similar results: a stable average SVI of 37 mL·g^{-1} was reached after approximately 150 days. Full granulation was however not reached, which warrants further optimization. The present study shows that the proposed strategy can easily be applied in existing SBR systems to solve the problem of sludge bulking.

Key words: anaerobic selector, feeding strategy, filamentous bulking, glycogen accumulating organisms (GAO), sequencing batch reactor, storage

HIGHLIGHTS

- Filamentous bulking in industrial activated sludge was eliminated by introducing an anaerobic feeding step.
- Anaerobic feeding selected for glycogen-accumulating organisms (GAO).
- Lab-scale results were successfully translated to full-scale implementation.
- Excellent settling sludge was obtained, but full granulation was not achieved.

GRAPHICAL ABSTRACT

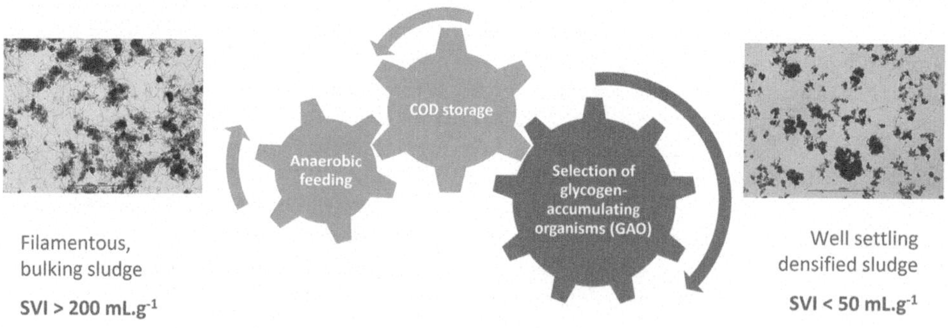

INTRODUCTION

Activated sludge (AS) systems represent the main biological treatment in many industrial wastewater treatment plants. In its simplest form, AS consists of an aeration tank where the biological conversions take place, followed by a secondary settler to separate sludge from the treated water. The settled sludge is then returned to the aeration tank. In a recent survey in Flanders, Cornelissen *et al.* (2018) reported that about one third of the industrial AS plants are facing sludge separation problems due to bad settling sludge. These findings agree with more global reports on sludge bulking occurring in more than 30% of the biological treatment plants (Rossetti *et al.* 2017). Besides being linked to nutrient deficiency and oxygen limitation, sludge bulking can often be related to the operation of the AS plant, more specifically the feeding pattern (Jenkins *et al.* 2003). Slow feeding of readily degradable chemical oxygen demand (rbCOD) under aerobic conditions, without build-up of a significant organic substrate concentration, promotes filamentous growth and negatively impacts sludge settling (Martins *et al.* 2004b).

To counter sludge bulking, both curative (e.g. chlorination) as preventive solutions (e.g. introduction of a selector) have been proposed (Jenkins *et al.* 2003), the latter being preferred as they provide a longer-term solution to the bulking problem. A selector is a separate zone in the reactor that receives both the return sludge and the influent wastewater. Depending on the retention time within the selector, the sludge is exposed to high substrate concentrations and feast–famine conditions which selects for floc-forming bacteria according to the kinetic selection theory. Aerobic selectors need to be operated at high sludge loading rate, 3–4 kg COD·(kg MLSS·d)$^{-1}$, and with high dissolved oxygen (DO) concentrations to achieve the kinetic selection (Jenkins *et al.* 2003). Under anaerobic conditions, however, the main mechanism in a selector is not a kinetic but a metabolic selection, and lower sludge loading rates can be applied. Even without build-up of substrate in the selector, substrate storing organisms, such as polyphosphate and/or glycogen accumulating organisms (PAO and/or GAO) are stimulated, leading to well settling sludge without filamentous bulking (Yao *et al.* 2019). Martins *et al.* (2004a) showed the formation of well settling sludge when synthetic acetate-based wastewater was slowly fed under anaerobic conditions in the presence of phosphate. Under aerobic conditions, slow feeding of acetate led to severe filamentous bulking (Martins *et al.* 2003).

The metabolic selection of PAO and GAO also forms the main mechanism for aerobic granulation in wastewater treatment plants. PAO and GAO are referred to as 'slow-growing' organisms as they convert rbCOD to internal storage polymers such as polyhydroxyalkanoates (PHAs) under anaerobic conditions. Under aerobic (or anoxic) conditions, these PHAs are subsequently slowly degraded without electron acceptor diffusion limitations. Since the whole volume of the flocs/granules is then metabolically active, compact sludge aggregates are formed and maintained (Pronk *et al.* 2015b). In full-scale aerobic granular sludge (AGS) reactors of the Nereda® type treating domestic wastewater, full granulation is achieved through a combination of (1) PAO selection by a plug-flow anaerobic feeding regime, and (2) hydraulic selection of well settling sludge aggregates through selective withdrawal of slow settling floccular sludge (Pronk *et al.* 2015a). Regarding industrial wastewater treatment by AGS, most lab studies report the application of a pulse feed of wastewater to build up substrate during the first minutes of a cycle, followed by prolonged anaerobic mixing phase to promote PAO/GAO metabolism (e.g. Caluwé *et al.* 2017; Stes *et al.* 2019; De Vleeschauwer *et al.* 2021).

Pulse feeding of influent wastewater, i.e. feeding the influent in a few minutes time, like is often done in lab-scale studies (e.g. De Vleeschauwer *et al.* 2021), is, to the best of our knowledge, not technically feasible at full-scale. Conversion of existing full-scale installations to Nereda® type reactors with an upflow batch feeding regime, on the other hand, can be done, but at a significant investment cost. There are but a few studies reporting granulation using a slow feeding strategy applicable in practice. Meunier *et al.* (2016) reported on the successful implementation of a granulation/densification strategy to improve the sludge quality in a lab-scale sequencing batch reactor (SBR) treating phosphate-rich dairy wastewater. By introducing a slow anaerobic feeding step (90 min) in the process sequence, PAO were stimulated, filamentous bulking sludge was suppressed, and granular sludge was formed. No hydraulic selective pressure was applied. Stes *et al.* (2020) applied an anaerobic feast/aerobic famine strategy only, using slow anaerobic feeding (127 min), to form granules in a SBR treating chocolate processing effluent low in phosphorus. Stable AGS was formed when a partial static feeding step was introduced in the SBR operation.

Building upon the lab research of Meunier *et al.* (2016), Henriet *et al.* (2017) investigated the impact of an anaerobic feast/aerobic famine sequence on the sludge morphology in an existing full-scale SBR treating dairy wastewater. The full-scale plant experienced severe bulking caused by *Thiothrix* species. The fully aerobic feed step was replaced by an extended unaerated mixed feeding step (95 min), leading to the promotion of GAO and PAO, and the formation of densified granular-like sludge. Full granulation was, however, not achieved.

The present study investigates the application of this novel anaerobic feeding strategy to improve sludge settling properties in an SBR treating wastewater at a tank truck cleaning site (TTC) and experiencing severe sludge bulking. In this specific case, the wastewater originates from trucks transporting liquid food, i.e. chocolate and beer. The nutrient-deficient influent is high in readily biodegradable chemical oxygen demand (rbCOD). An additional aim is to assess whether mere implementation of an unaerated feeding step and the associated metabolic selection would also lead to formation of AGS, as defined by de Kreuk *et al.* (2007).

METHODS

Description of the full-scale installation

The case study involves a transport and tank truck cleaning company (TTC) transporting liquid feedstuff, mostly chocolate, beer and glucose. Wastewater, resulting from the cleaning activities of tank truck interiors, is first stored in a buffer tank, then pre-treated in a chemically enhanced primary dissolved air flotation (DAF) unit, and finally biologically treated in two parallel sequencing batch reactors (SBR). At the start of each SBR cycle, the sludge volume is approximately 382 m^3 in both reactors. The average daily flow to each SBR is 40 m^3. The 24 h SBR cycle consists of 4 h aerobic feeding, 16 h aerobic reaction, 2 h settling and 2 h discharge (Table 1). During the study reported here, the average sludge loading rate, the food to microbes ratio (F/M), was 0.10 ± 0.03 kg COD·(kg MLSS·d)$^{-1}$. During aeration, the dissolved oxygen (DO) level was controlled between 1 and 4 mg·L^{-1} using an on/off control strategy. The sludge retention time (SRT) was set at 30 d. Due to excessive loss of suspended solids from the SBRs, the bio-treated effluent is post-treated by the DAF.

Lab-scale reactor set-up and operation

A fully automated double-walled lab-scale SBR with a working volume of 5.5 L and an H/D ratio of approximately 1, was operated at a controlled temperature (22 °C) for 125 days. The SBR was equipped with a mechanical mixer (RZR2020, Heidolph Instruments, Germany) to keep the sludge in suspension, a feeding pump (ES-B16VC-3, IWAKI, Japan), a discharge valve (Eriks RX10.X33.S00, ERIKS, Belgium) and an aeration disc (Aquadistri) connected to an aeration pump (Super Fish Koi Flow 60, AQUADISTRI, China). Process operation was controlled and monitored by a Siemens PLC and a custom-built LabVIEW program (National Instruments). The reactor was equipped with a luminescent dissolved oxygen (LDO) sensor (Hach Lange, USA). During aerobic phases, the DO level was controlled between 1 and 4 mg·L^{-1} using an on/off control strategy.

The reactor was seeded with activated sludge from the full-scale plant, and fed daily with 750 mL of wastewater collected from the plant after DAF treatment. The SBR cycle consisted of a feeding phase, a reaction phase and ended with settling and discharge. During the lab-scale test, the cycle duration (12 h or 24 h), the feeding regime (anaerobic or aerobic, mixed or static) and the application of an anaerobic mixed reaction phase varied (Table 1). The sludge retention time (SRT) was kept at 30 d by daily wasting of sludge from the mixed liquor.

Industrial wastewater

During the lab-scale test, 16 different influent batches were fed to the SBR. The main composition of the wastewater is summarized in Table 2. On average, the soluble COD (sCOD) represented 87% of the total COD, and the average ratio of volatile fatty acids (VFA) to total COD was 35% (on COD basis). The average ratio of influent sCOD to dissolved organic carbon (DOC) was 3.3. The nitrogen (N) and phosphorus (P) content of the nutrient-deficient influent was adapted to reach a desired COD/N/P ratio of 100/3.0/0.5 (adapted to 100/2.5/0.4 during period 5) by adding NH_4Cl and K_2HPO_4. The average pH of

Table 1 | Modes of operation of the lab-scale SBR and the full-scale SBRs (FS-SBR1 and FS-SBR2; Ana: anaerobic, Aer; aerobic)

SBR	Day (Period P)	Cycles (#·d^{-1})	Feeding regime	Feed (h)	Ana mixing (h)	Aer react	Settling + Discharge
LAB-SBR	P1: 0–7	1	Ana, Mixed	4	–	17 h 55 min	2 h + 5 min
	P2: 8–27	1	Ana, Static	3	1	17 h 55 min	2 h + 5 min
	P3: 28–64	2	Ana, Static	1.5	1	8 h 25 min	1 h + 5 min
	P4: 65–86	1	Ana, Static	3	2	16 h 55 min	2 h + 5 min
	P5: 87–125	1	Aer, Mixed	3	–	18 h 55 min	2 h + 5 min
FS-SBR1	P1: 0–158	1	Aer, Mixed	4	–	16 h	2 h + 2 h
FS-SBR1	P2: 160–250	1	Ana, Mixed	4	1	15 h	2 h + 2 h
FS-SBR2	0–250	1	Ana, Mixed	4	1	15 h	2 h + 2 h

Table 2 | Composition of the influent wastewater during the lab-scale SBR test ($n = 16$), and the applied sludge loading rate (F/M ratio) in the lab-scale SBR (LAB) and in both full-scale SBRs (FS-SBR1 and FS-SBR2)

Parameter		Min	Max	Avg. ± SD	CV (%)
COD (mg L^{-1})		2,120	6,590	4,988 ± 1,142	23
sCOD (mg L^{-1})		1,840	6,000	4,350 ± 1,066	25
F/M ratio (kg COD· (kg MLSS·d)$^{-1}$)	LAB	0.04	0.19	0.14 ± 0.04	29
	FS-SBR1	0.02	0.31	0.12 ± 0.06	47
	FS-SBR2	0.02	0.21	0.10 ± 0.03	35

the influent was 4.3 ± 0.3, so the influent pH was adjusted to 7.0 by adding sodium hydroxide (1M NaOH). The influent wastewater was kept in a fridge (at 4 °C) to prevent COD degradation.

Implementation in the full-scale plant

After the conclusion of the lab-scale experiment, the granulation strategy, i.e. the anaerobic feeding strategy, was implemented at full-scale. The implementation was done in one of the two full-scale SBRs, FS-SBR2, while FS-SBR1 acted as a reference with an unchanged fully aerobic process operation (Table 1). To this end, FS-SBR2 was first completely emptied by transferring sludge from FS-SBR2 to FS-SBR1. Then, a submerged mixer (1.5 kW) was installed, at a height of 3 m above the SBR floor, to keep the sludge in suspension during the unaerated feeding step. Finally, half of the mixed liquor from FS-SBR1 was transferred to FS-SBR2, resulting in a similar initial situation in both reactors. The new operational strategy in FS-SBR2 consisted of 4 h unaerated mixed feeding, 1 h unaerated mixing, 15 h aerobic reaction, 2 h settling and 2 h discharge.

After about 150 days of operation, the new strategy was also implemented in FS-SBR1. During a two-week period, sludge wasting from FS-SBR2 was stopped to allow sludge growth, while the loading rate to FS-SBR1 and the sludge height in FS-SBR1 were gradually lowered. An identical submerged mixer (1.5 kW) as in FS-SBR2 was installed, and FS-SBR1 was seeded with sludge from FS-SBR2. A similar unaerated feeding strategy was then applied in both FS-SBRs (Table 1).

Analytical methods

Total and soluble COD were measured with micro-COD test tubes (HI 93754A-25 and HI 93754B-25) from Hannah Instruments (Temse, Belgium). NH_4-N (HI 93715-01), NO_3-N (HI 93766-50) and PO_4-P (HI 93717-01) were measured with test kits from Hannah Instruments, after filtration over a 1.2 µm glass microfibre filter (VWR International, Belgium). The DOC was analysed with a Sievers InnovOx laboratory total organic carbon analyser. The sludge mixed liquor suspended (volatile) solids (ML(V)SS) and the sludge volume index (SVI) were measured according to the standard methods (APHA 1998). Microscopic images were taken with a MOTICBA310 microscope (Opti-service, Belgium).

Anaerobic carbon substrate removal

The removal of dissolved carbon substrate, expressed as DOC or sCOD, was evaluated via in-situ and ex-situ cycle measurements. In-situ refers to the actual substrate removal during a lab or full-scale SBR cycle, while ex-situ refers to the substrate removal capacity of sludge sampled from the full-scale SBR. The in-situ SBR cycle measurements were regularly performed to evaluate the DOC removal during the anaerobic mixing phases in the lab-scale SBR. Grab samples were taken at the end of the cycle (DOC_base), and at the start and the end of the anaerobic mixing phase (respectively DOC_start and DOC_end).

$$\text{Anaerobic DOC removal (\%)} = 100 \times \frac{DOC_{start} - DOC_{end}}{DOC_{start} - DOC_{base}}$$

The in-situ SBR cycle measurements were also performed to evaluate sCOD removal during the anaerobic mixing phases in the full-scale installation. To this end, grab samples were taken every 30 min during the anaerobic feeding phase, which happened while the sludge was in suspension. The sCOD removal at each sampling point was estimated relative to the amount of sCOD added during feeding.

$$\text{Anaerobic sCOD removal at time t (\%)} = 100 \times \frac{(sCOD_{added\_t} + sCOD_{t0}) - sCOD_t}{sCOD_{added\_t} - sCOD_{t0}}$$

With $sCOD_{added-t}$ the amount of sCOD added at time t during feeding (estimated based on influent sCOD concentration, reactor volume and influent flow rate), $sCOD_{t0}$ the measured initial sCOD concentration in the reactor, and $sCOD_t$ the measured sCOD concentration at time t.

Ex-situ SBR cycle measurements were regularly performed to evaluate the anaerobic DOC removal capacity of the sludge from the full-scale installation. To this end, endogenous sludge samples from the full-scale SBR were transported to the lab for an ex-situ batch experiment. 400 mL of the sampled sludge was washed (3 times) with a wash buffer (0.1M $NaHCO_3$ + 0.05M KCl) and transferred to a beaker on a magnetic stirring plate. In two parallel experiments, respectively 40 and 100 mL of wastewater was then added to the suspended sludge. Samples for DOC analysis were then taken at the start and at the end of the test (after 180 min) to calculate the DOC removal.

Microbial community composition analysis

Biomass samples were collected, centrifuged, and pellets were then stored at −80 °C until DNA extraction. DNA was isolated from triplicate samples using FastDNA® SPIN kit. This kit was used because of its high cell-breaking capability (Guo & Zhang 2013) needed for compact flocs or granular sludge following the adapted extraction protocol for AS samples (van Loosdrecht *et al.* 2016). The DNA concentration was then measured using a Qubit dsDNA Assay kit (Invitrogen, USA) following manufacturers protocol.

The bacterial 16S rRNA gen hypervariable region, targeting V1–3, was amplified by polymerase chain reaction (PCR) in a CFX96 Thermocycler (Bio-Rad). The V1–3 region was chosen since it is been described as the most suitable region for AS samples (Albertsen *et al.* 2015; Dueholm *et al.* 2021). Barcoded primers were used 27F (5′-AGAGTTTGATCCTGGCT-CAG-3′), 534R (5′-ATTACCGCGGCTGCTGG-3′) and KAPA HiFi HotStart PCR kit (Boston, MA, USA). The sequencing of the amplified libraries was carried out on an Illumina MiSeq platform at the Center for Medical Genetics (Antwerp, Belgium), using a MiSeq Reagent kit v (Illuminia) following standard guidelines.

The reads were then processed using USEARCH v.11.0.667. Processed reads were clustered into Operational Taxonomy Units (OTUs) based on 97% sequence similarity threshold (Edgar 2013). Taxonomy was assigned using the MiDAS4 data base (Dueholm *et al.* 2021), and the data analysis was performed in R v. 3.2.3 using the ampvis2 package v. 1.24.0 (Andersen *et al.* 2018).

RESULTS AND DISCUSSION

Present study investigates the implementation of an anaerobic feeding strategy to improve the sludge settling properties and sludge morphology in a SBR treating industrial wastewater, first on lab-scale, then on full-scale. The results of the lab-scale test are reported first, followed by the results of the full-scale implementation. The impact of the anaerobic feeding strategy is discussed in the third section. The last section discusses the degree of granulation achieved in the lab-scale and full-scale SBRs.

Lab-scale testing

The seeding sludge for the lab-scale test, collected form the full-scale installation, had a MLSS and MLVSS concentration of 3.71 $g{\cdot}L^{-1}$ and 3.26 $g{\cdot}L^{-1}$ respectively. The SVI after 30 min was about 330 $mL{\cdot}g^{-1}$, indicative of very poor settling sludge, which was confirmed by the filamentous nature of the sludge morphology (Figures 1 and 2).

After seeding the lab-scale SBR, an anaerobic/aerobic process operation was implemented. During the first week of operation (P1 in Table 1), a mixed unaerated feeding pattern was applied, which was subsequently changed to a static feeding regime (P2 to P4 in Table 1). The latter was done to create a build-up of substrate in the SBR, resulting in a feast–famine regime (Stes *et al.* 2020).

The first 25 days of operation were characterized by a significant loss of total suspended solids (TSS) from the reactor, bad settling behaviour (SVI values above 250 $mL{\cdot}g^{-1}$) and a predominant filamentous floc morphology (Figure 1). After 50 days of operation, a clear effluent was obtained, and sludge settling improved, reaching 100 $mL{\cdot}g^{1}$ and lower after approximately 75 days. Filaments disappeared almost completely, and more compact sludge aggregates were formed. In the same period, between 25 and 50 days after start-up, a rapid increase in the DOC removal, up to approximately 80%, during the anaerobic mixing phase was observed (Figure 3). Although intracellular storage polymers, such as PHA, were not measured in this study, it is very likely that a large part of the DOC, especially the VFA fraction, was taken up and stored anaerobically. This was confirmed by the enrichment of carbon-storing organisms, mostly GAO, in the sludge (see below).

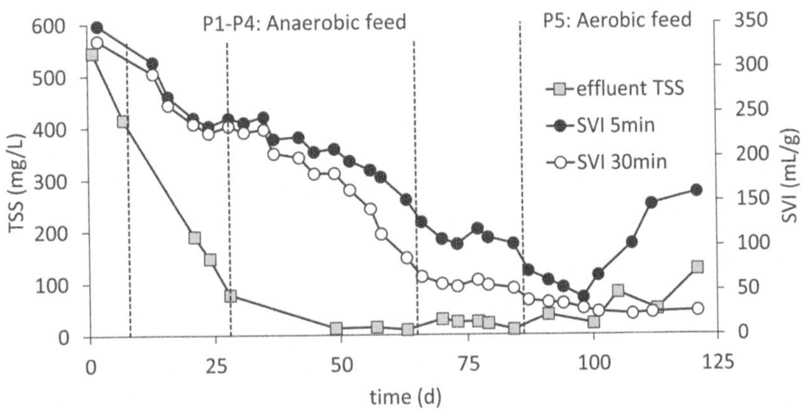

Figure 1 | Evolution of effluent TSS, and SVI after 5 and 30 min during the lab-scale experiment. The vertical dotted lines delineate the periods (P1 to P5) with the different operational conditions described in Table 1. After 86 days, the feeding regime was switched from an anaerobic (P1 to P4) to an aerobic feed (P5).

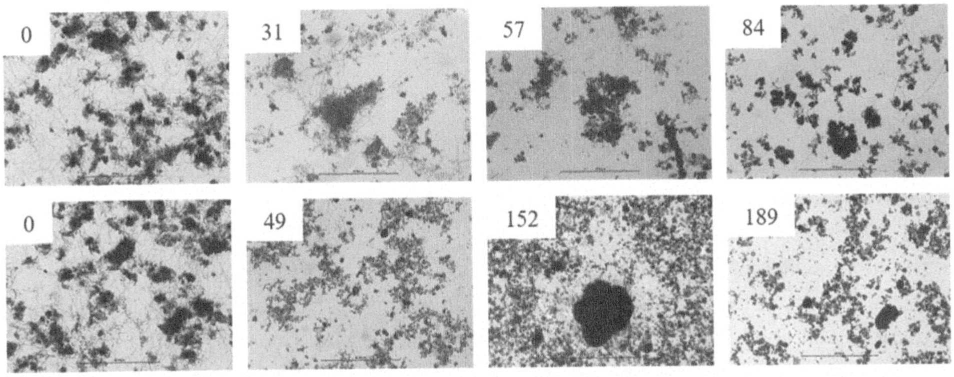

Figure 2 | Evolution of the sludge morphology at different time points during the lab-scale test (top) and full-scale FS-SBR2 test (bottom) (scale bar = 1,000 μm; boxed numbers refer to the number of days after start-up).

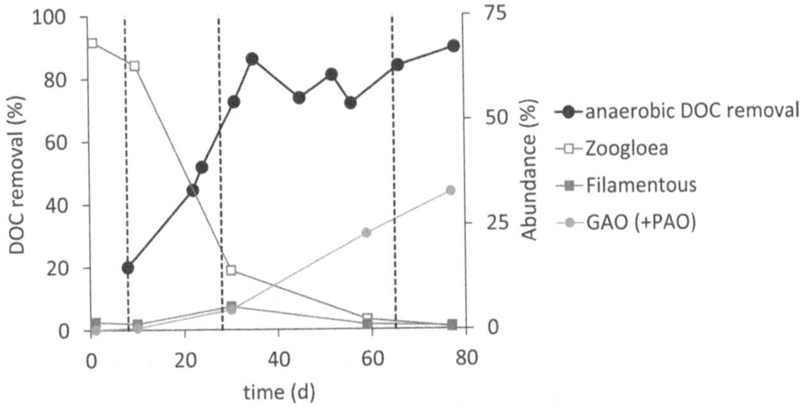

Figure 3 | Evolution of the DOC removal during the anaerobic phase of the lab-scale SBR cycle, and changes in the relative abundance, based on 16S rRNA gene sequencing, of *Zoogloea*, filamentous organisms and GAO, throughout periods P1 to P4 when the SBR was operated with an unaerated feeding step (the vertical dotted lines delineate the periods (P1 to P4) with the different operational conditions described in Table 1).

Although the sludge morphology of the seed sludge was dominated by filamentous organisms (Figure 2), filamentous bacteria were only present in low numbers: read abundances of known filamentous genera were below 5% (Figure 3). The main filamentous genera found were *Thiothrix* and *Kouleothrix*. Both organisms have been associated with serious bulking problems in (industrial) activated sludge plants (Nielsen *et al.* 2000; Nittami *et al.* 2020).

Non-filamentous bulking organisms, i.e. *Zoogloea*, dominated the seed sludge composition. *Zoogloea* bulking is caused by nutrient deficiency, most probably due to a suboptimal nutrient dosing in the full-scale installation (Jenkins *et al.* 2003). During the lab test, nutrients were added to fulfil the requirements for growth. A COD/N/P ratio of 100/3.0/0.5 was adopted.

The implementation of an unaerated feeding step resulted in a gradual increase in the degree of anaerobic DOC removal, coinciding with the enrichment of carbon-storing GAO at the expense of the bulking organisms (Figure 3). The abundance of PAO remained low throughout the entire study, although there was a slight increase in the abundance of *Candidatus* Accumulibacter, up to 1%. Two known genera of GAO were mainly enriched, i.e. *Defluviicoccus* and *Candidatus* Competibacter, from less than 1% abundance at the start to respectively 7% and 25% after 77 days. Both GAO can store short-chain VFA, such as acetate and propionate, and convert them to PHA under anaerobic conditions (Burow *et al.* 2007). VFAs represent an important COD fraction (about 35% on COD basis) in the influent. Storage by GAO prevents 'leakage' of this rbCOD fraction into the aerobic phase which could otherwise lead to stimulation of filamentous organisms, specially under oxygen-limited conditions (Pronk *et al.* 2015b; Haaksman *et al.* 2020).

After 86 days of operation, the feeding regime of the lab-scale SBR was switched to a fully aerobic feeding. Although the SVI30 remained low until the end of the experiment, a gradual increase in SVI5 and effluent TSS was recorded (Figure 1), pointing towards a deterioration of the reactor performance. The fraction of GAO dropped to less than 4% abundance after 100 days (data not shown), although neither *Zoogloea* nor other known filamentous organisms significantly increased in abundance. The lab-scale experiment was terminated after 125 days.

Full-scale implementation

The promising results obtained in the short lab-scale experiment prompted the TTC company to implement the new feeding strategy in one of the two parallel full-scale SBRs. FS-SBR2 was operated using an anaerobic/aerobic sequence, while FS-SBR1 acted as a fully aerobic reference.

The start sludge in both reactors had an MLSS of about 4.5 $g \cdot L^{-1}$ and a SVI of 200 $mL \cdot g^{-1}$, indicating the poor settling properties of the sludge. The morphology of the sludge was dominated by filamentous bacteria (Figure 2). In the first 50 days of operation, the SVI remained high in both reactors with average values between 200 and 270 $mL \cdot g^{-1}$. Seventy-five days after the implementation of the new feeding strategy, a clear difference between the SVI values in both reactors was observed: SVI values dropped consistently below 100 $mL \cdot g^{-1}$ in FS-SBR2, with an average of 75 $mL \cdot g^{-1}$ between days 75 and 150, while the average SVI in FS-SBR1 was 220 $mL \cdot g^{-1}$. The improved SVI values in FS-SBR2 were associated with a significant increase in the anaerobic DOC removal capacity of the sludge, measured in ex-situ batch test (Figure 4). In this batch test, the sludge from FS-SBR2 was exposed to a pulse feed of wastewater at a volume exchange rate (VER) of 20%, which is higher than the VER in the full-scale plant, i.e. 10%. Hence, the results of the test indicate the maximal removal

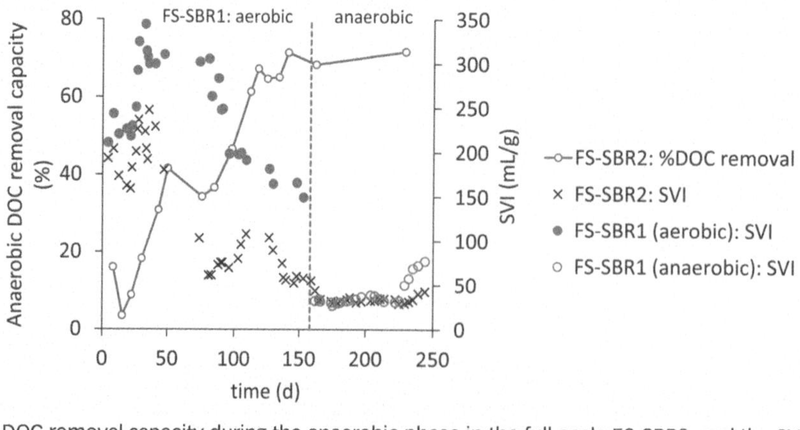

Figure 4 | Evolution of the DOC removal capacity during the anaerobic phase in the full-scale FS-SBR2, and the SVI (after 30 min) in both full-scale SBRs (FS-SBR2 was operated using an anaerobic/aerobic sequence, while FS-SBR1 was operated under fully aerobic conditions, until day 158, when the cycle was switched to an anaerobic/aerobic sequence as well).

capacity of the sludge. The anaerobic DOC removal capacity increased from less than 20% initially to almost 70% after 100 days of operation, suggesting sludge adaptation to the new feeding regime. The in-situ COD removal was always above 80% (data not shown). The evolution of the sludge morphology in FS-SBR2 was in line with changes in the settling properties. Filaments disappeared after about 50 days, and the sludge evolved into floccular sludge with an open structure (Figure 2). In the last 100 days, the sludge morphology consisted of both small flocs and larger, denser, granular-like structures.

In the last 100 days of the full-scale test, the anaerobic feeding strategy was applied to both SBRs. FS-SBR1 was emptied and restarted with sludge from FS-SBR2. This resulted in stable and low SVI values in both reactors, with an average of 37 $mL{\cdot}g^{-1}$.

At the start of the full-scale test, the sludge from both SBRs was high in abundance in *Zoogloea*, and the filamentous genera *Thiothrix* and *Kouleothrix*, much like the sludge in the lab-scale test (see above). In FS-SBR2, operated with an extended anaerobic feeding step, a decline in *Zoogloea* abundance, and a rise followed by a decline in filamentous genera, was observed during the first 100 days of operation (Figure 5(a)). The abundance of known bulking organisms was consistently well below 1% during the last 100 days of the test. The abundance of carbon-storing organisms, GAO and PAO, rapidly increased to average values of approximately 7%. As in the lab-scale test, the enriched GAO genera were *Defluviicoccus* and *Candidatus* Competibacter. The average abundance of PAO remained below 1% (data not shown).

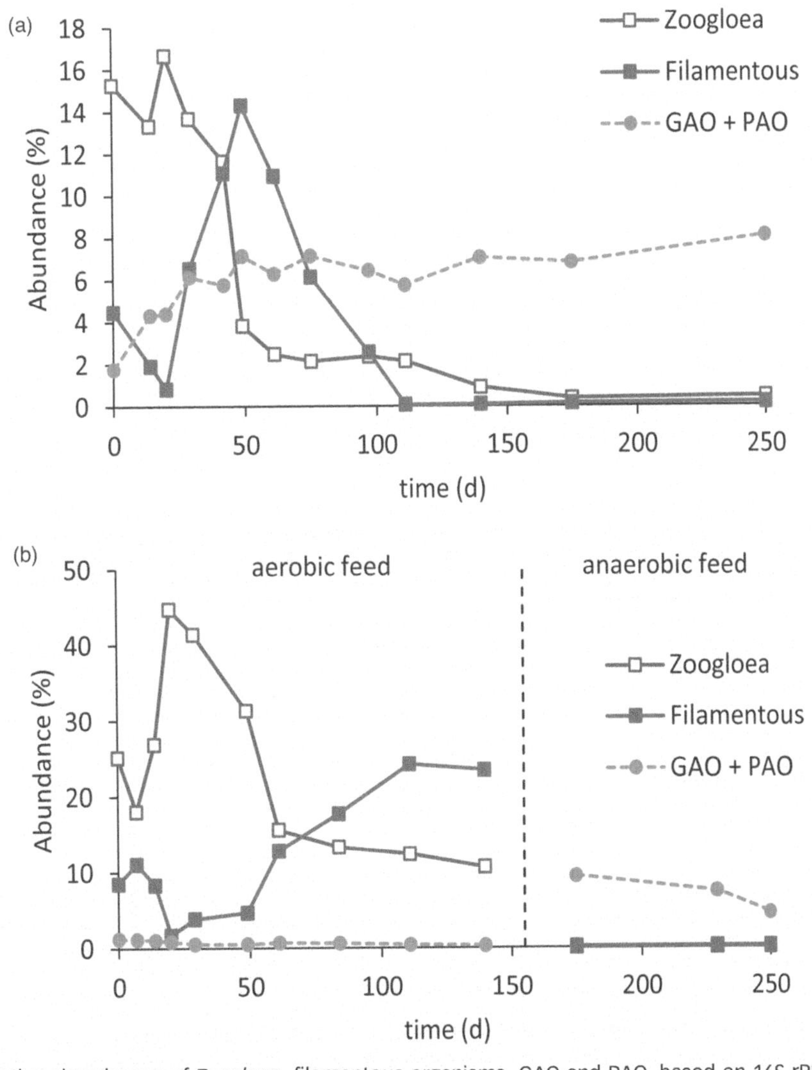

Figure 5 | Changes in relative abundances of *Zoogloea*, filamentous organisms, GAO and PAO, based on 16S rRNA gene sequencing, in (a) FS-SBR2, operated with an anaerobic feed, and (b) FS-SBR1, operated with an aerobic feed until day 158. After day 158, FS-SBR1 was seeded with sludge from FS-SBR2, and operated with an anaerobic feed.

In FS-SBR1 the abundance of bulking organisms remained high as long as the reactor was operated using an aerobic feeding strategy, although the abundance of *Zoogloea* and filamentous organisms did vary over time (Figure 5(b)). After 158 days, when the reactor was seeded with sludge from FS-SBR2 and operated using an anaerobic feeding regime for the remainder of the study, the population of GAO and PAO was maintained at similar levels (approximately 7% abundance).

Impact of the anaerobic feeding strategy

The implementation of an unaerated feeding regime in the SBR process is equivalent to the introduction of an anaerobic selector zone. The average sludge loading rate during the anaerobic phase was 0.6 ± 0.1 kg COD·(kg MLSS·d)$^{-1}$ in the full-scale SBR and 0.8 ± 0.2 kg COD·(kg MLSS·d)$^{-1}$ in the lab-scale SBR, values which are at the low end of typical design criteria for anaerobic selectors (Jenkins *et al.* 2003). The metabolic selection of carbon-storing GAO, both in lab and full-scale, is effective in combating filamentous bulking in the presented industrial case (Figure 4). The main advantage of the anaerobic selector strategy is that influent can be fed at a slow rate into the AS system. The anaerobic fill time ratio (FTR), describing the ratio of the anaerobic feeding time to (active) cycle time was 13.6% in the lab-scale test, and 20% in the full-scale installation. These FTRs are comparable to values reported earlier: Martins *et al.* (2004a) applied an anaerobic FTR of >35% and achieved good settling sludge fed with acetate-based synthetic wastewater, while Stes *et al.* (2020) and Meunier *et al.* (2016) applied FTRs of respectively 25.6% and 17% yielding AGS fed with real industrial wastewaters. On a full-scale dairy wastewater treatment plant, Henriet *et al.* (2017) applied an anaerobic FTR of 31% and obtained well-settling densified sludge. In contrast, under fully aerobic conditions, an aerobic FTR > 6.2% led to low substrate concentrations and bad sludge settleability (Martins *et al.* 2003).

In the lab-scale test, a static feeding was employed in order to create a substrate build up, and ensure full penetration of the substrate into the whole sludge volume during the subsequent mixing phase (Iorhemen & Liu 2021). The low influent pH (4.3 ± 0.3), however, precluded the application of such a static feeding step in the full-scale installation as it would require a significant amount of base to neutralize the wastewater before feeding, depending on the variable influent alkalinity (data not shown). Therefore, a slow mixed feeding was implemented in the full-scale installation. The results indicate that there was no build-up of substrate in the full-scale (not shown), but the fed COD was removed, and probably stored by GAO, almost completely while being fed in the SBR. As Martins *et al.* (2004a) showed for acetate-based synthetic wastewater, this metabolic selection was sufficient to yield well settling sludge (Figures 4 and 5).

Incomplete granulation

An additional aim of the present study was the achievement of AGS by implementation of the metabolic selection strategy as the main driver for granulation (de Kreuk & van Loosdrecht 2004). According to de Kreuk *et al.* (2007), the quantitative definition of AGS includes the SVI5/SVI30 or SVI10/SVI30 ratios, which should be close to 1, to characterize the sludge settleability, and the minimum granule size of 0.2 mm. The average SVI5/SVI30 ratio of the full-scale sludge was 1.85. The lowest average SVI5/SVI30 ratio at the end of the lab-scale test was 1.41. Since particle size determination was not performed, the average sludge size could not be determined. Microscopy analysis shows that the average size was below 0.2 mm (Figure 2). Based on both criteria, it is clear that full granulation was not achieved in the timeframe of the study.

Microscopy analysis showed a difference in morphology between the lab and full-scale sludge (Figure 2). The lab-scale sludge showed the clear formation of dense sludge aggregates consisting of both small and larger particles. The authors hypothesize that the duration of the lab test was too short to result in full granulation. Other works, only applying a metabolic selection for granulation in systems fed with real industrial wastewater, report granule formation times of 30–150 days (e.g. Meunier *et al.* 2016; Caluwé *et al.* 2017; Stes *et al.* 2020).

The full-scale morphology showed more diversity with the presence of larger granular-like structures embedded in a matrix of loose sludge particles. Besides the short time of the study, other factors may have played a role. Most importantly, the full-scale installation suffered periodical disturbances, both in the wastewater treatment plant, especially the DAF, as in the truck cleaning operation. Without going into too much detail, these disturbances lead (1) to significant overloading of the SBR, (2) insufficient COD storage in the anaerobic phase, and (3) extended low DO periods during the aerobic phase. Previous research shows that the presence of suspended solids, e.g. due to a failing DAF, could prevent or be detrimental for (full) granulation (de Kreuk *et al.* 2010; Martins *et al.* 2011; De Vleeschauwer *et al.* 2021). Also leaking of soluble, potentially rapidly degradable, COD into the aerobic phase likewise hinders granulation, especially if aeration is insufficient (Pronk *et al.* 2015b; Haaksman *et al.* 2020). In addition, the absence of a substrate build up, due to the slow mixed feeding, may represent an insufficient driver for granulation (Franca *et al.* 2018; Stes *et al.* 2019). As discussed in the introduction, most literature references on AGS formation in industrial settings applied a pulse feeding, to mimic the plug flow regime applied in the full-scale Nereda plants (Pronk *et al.* 2015a; Franca *et al.* 2018). In the presented full-scale study, the sludge loading rate during the feeding phase, and the associated substrate build-up, may be increased by shortening the feeding time, within the practical limitations of the existing pumps, and taking the impact of feeding acidic wastewater into account. Finally, the absence of a hydraulic selection pressure prevents the selective washout of the floccular sludge fraction, resulting in a hybrid sludge consisting of both floccular and granular structures (Stes *et al.* 2020; De Vleeschauwer *et al.* 2021). Further research should explore these hypotheses in more detail, to optimize the granulation in the full-scale treatment plant.

CONCLUSIONS

The introduction of an anaerobic (feeding) step in the SBR operation alleviated the filamentous sludge bulking problem in the industrial wastewater treatment plant. Slow, unaerated feeding of the wastewater led to the metabolic selection of substrate-storing GAO, the elimination of filamentous genera, and resulted in the formation of well-settling dense sludge, although full granulation was not achieved in the time frame of the study. The implementation of the densification strategy is easy and straightforward: the main investment to be done is the installation of a submerged mixer to keep the sludge in suspension during the unaerated phase, while an anaerobic (feeding) step can easily be implemented in the PLC system of the plant. These results are of great importance for the applicability of the proposed strategy in full-scale industrial SBR systems.

DATA AVAILABILITY STATEMENT

All relevant data are included in the paper or its Supplementary Information.

REFERENCES

Albertsen, M., Karst, S. M., Ziegler, A. S., Kirkegaard, R. H. & Nielsen, P. H. 2015 Back to basics – the influence of DNA extraction and primer choice on phylogenetic analysis of activated sludge communities. *PLoS ONE* **10**, e0132783. https://doi.org/10.1371/journal.pone.0132783.

Andersen, K. S., Kirkegaard, R. H., Karst, S. M. & Albertsen, M. 2018 Ampvis2: an R package to analyse and visualise 16S rRNA amplicon data (preprint). *Bioinformatics*. https://doi.org/10.1101/299537.

APHA 1998 *Standard Methods for the Examination of Water and Wastewater*, 20th edn. American Public Health Association, American Water Works Association and Water Environmental Federation, Washington, DC, USA.

Burow, L. C., Kong, Y., Nielsen, J. L., Blackall, L. L. & Nielsen, P. H. 2007 Abundance and ecophysiology of Defluviicoccus spp., glycogen-accumulating organisms in full-scale wastewater treatment processes. *Microbiology* **153**, 178–185. https://doi.org/10.1099/mic.0.2006/001032-0.

Caluwé, M., Dobbeleers, T., D'aes, J., Miele, S., Akkermans, V., Daens, D., Geuens, L., Kiekens, F., Blust, R. & Dries, J. 2017 Formation of aerobic granular sludge during the treatment of petrochemical wastewater. *Bioresource Technology* **238**, 559–567. https://doi.org/10.1016/j.biortech.2017.04.068.

Cornelissen, R., Van Dyck, T., Dries, J., Ockier, P., Smets, I., Van den Broeck, R., Van Hulle, S. & Feyaerts, M. 2018 Application of online instrumentation in industrial wastewater treatment plants – a survey in Flanders, Belgium. *Water Science and Technology* **78**, 957–967. https://doi.org/10.2166/wst.2018.375.

de Kreuk, M. K. & van Loosdrecht, M. C. M. 2004 Selection of slow growing organisms as a means for improving aerobic granular sludge stability. *Water Science and Technology* **49**, 9–17. https://doi.org/10.2166/wst.2004.0792.

de Kreuk, M. K., Kishida, N. & van Loosdrecht, M. C. M. 2007 Aerobic granular sludge – state of the art. *Water Science and Technology* **55**, 75–81. https://doi.org/10.2166/wst.2007.244.

de Kreuk, M.K., Kishida, N., Tsuneda, S. & van Loosdrecht, M. C. M. 2010 Behavior of polymeric substrates in an aerobic granular sludge system. *Water Research* **44**, 5929–5938. https://doi.org/10.1016/j.watres.2010.07.033.

De Vleeschauwer, F., Caluwé, M., Dobbeleers, T., Stes, H., Dockx, L., Kiekens, F., Copot, C. & Dries, J. 2021 A dynamically controlled anaerobic/aerobic granular sludge reactor efficiently treats brewery/bottling wastewater. *Water Science and Technology*. wst2021470. https://doi.org/10.2166/wst.2021.470.

Dueholm, M. S., Nierychlo, M., Andersen, K. S., Rudkjøbing, V., Knutsson, S., the MiDAS Global ConsortiumAlbertsen, M. & Nielsen, P. H. 2021 MiDAS 4: a global catalogue of full-length 16S rRNA gene sequences and taxonomy for studies of bacterial communities in wastewater treatment plants (preprint). *Microbiology*. https://doi.org/10.1101/2021.07.06.451231.

Edgar, R. C. 2013 UPARSE: highly accurate OTU sequences from microbial amplicon reads. *Nature Methods* **10**, 996–998. https://doi.org/10.1038/nmeth.2604.

Franca, R. D. G., Pinheiro, H. M., van Loosdrecht, M. C. M. & Lourenço, N. D. 2018 Stability of aerobic granules during long-term bioreactor operation. *Biotechnology Advances* **36**, 228–246. https://doi.org/10.1016/j.biotechadv.2017.11.005.

Guo, F. & Zhang, T. 2013 Biases during DNA extraction of activated sludge samples revealed by high throughput sequencing. *Applied Microbiology and Biotechnology* **97**, 4607–4616. https://doi.org/10.1007/s00253-012-4244-4.

Haaksman, V. A., Mirghorayshi, M., van Loosdrecht, M. C. M. & Pronk, M. 2020 Impact of aerobic availability of readily biodegradable COD on morphological stability of aerobic granular sludge. *Water Research* **187**, 116402. https://doi.org/10.1016/j.watres.2020.116402.

Henriet, O., Meunier, C., Henry, P. & Mahillon, J. 2017 Filamentous bulking caused by Thiothrix species is efficiently controlled in full-scale wastewater treatment plants by implementing a sludge densification strategy. *Scientific Reports* **7**, 1430. https://doi.org/10.1038/s41598-017-01481-1.

Iorhemen, O. T. & Liu, Y. 2021 Effect of feeding strategy and organic loading rate on the formation and stability of aerobic granular sludge. *Journal of Water Process Engineering* **39**, 101709. https://doi.org/10.1016/j.jwpe.2020.101709.

Jenkins, D., Richard, M. G. & Daigger, G. T. 2003 *Manual on the Causes and Control of Activated Sludge Bulking, Foaming, and Other Solids Separation Problems*, 0 edn. CRC Press. https://doi.org/10.1201/9780203503157.

Martins, A. M. P., Heijnen, J. J. & van Loosdrecht, M. C. M. 2003 Effect of feeding pattern and storage on the sludge settleability under aerobic conditions. *Water Research* **37**, 2555–2570. https://doi.org/10.1016/S0043-1354(03)00070-8.

Martins, A. M. P., Heijnen, J. J. & van Loosdrecht, M. C. M. 2004a Bulking sludge in biological nutrient removal systems. *Biotechnology and Bioengineering* **86**, 125–135. https://doi.org/10.1002/bit.20029.

Martins, A. M. P., Pagilla, K., Heijnen, J. J. & van Loosdrecht, M. C. M. 2004b Filamentous bulking sludge – a critical review. *Water Research* **38**, 793–817. https://doi.org/10.1016/j.watres.2003.11.005.

Martins, A. M. P., Karahan, Ö. & van Loosdrecht, M. C. M. 2011 Effect of polymeric substrate on sludge settleability. *Water Research* **45**, 263–273. https://doi.org/10.1016/j.watres.2010.07.055.

Meunier, C., Henriet, O., Schoonbroodt, B., Boeur, J.-M., Mahillon, J. & Henry, P. 2016 Influence of feeding pattern and hydraulic selection pressure to control filamentous bulking in biological treatment of dairy wastewaters. *Bioresource Technology* **221**, 300–309. https://doi.org/10.1016/j.biortech.2016.09.052.

Nielsen, P. H., de Muro, M. A. & Nielsen, J. L. 2000 Studies on the in situ physiology of Thiothrix spp. present in activated sludge. *Environmental Microbiology* **2**, 389–398. https://doi.org/10.1046/j.1462-2920.2000.00120.x.

Nittami, T., Kasakura, R., Kobayashi, T., Suzuki, K., Koshiba, Y., Fukuda, J., Takeda, M., Tobino, T., Kurisu, F., Rice, D., Petrovski, S. & Seviour, R. J. 2020 Exploring the operating factors controlling Kouleothrix (type 1851), the dominant filamentous bacterial population, in a full-scale A2O plant. *Scientific Reports* **10**, 6809. https://doi.org/10.1038/s41598-020-63534-2.

Pronk, M., Abbas, B., Al-zuhairy, S. H. K., Kraan, R., Kleerebezem, R. & van Loosdrecht, M. C. M. 2015a Effect and behaviour of different substrates in relation to the formation of aerobic granular sludge. *Applied Microbiology and Biotechnology* **99**, 5257–5268. https://doi.org/10.1007/s00253-014-6358-3.

Pronk, M., de Kreuk, M. K., de Bruin, B., Kamminga, P., Kleerebezem, R. & van Loosdrecht, M. C. M. 2015b Full-scale performance of the aerobic granular sludge process for sewage treatment. *Water Research* **84**, 207–217. https://doi.org/10.1016/j.watres.2015.07.011.

Rossetti, S., Tandoi, V. & Wanner, J. 2017 *Activated Sludge Separation Problems: Theory, Control Measures, Practical Experiences*, 2nd edn. Water Intell Online, p. 16, 9781780408644. https://doi.org/10.2166/9781780408644.

Stes, H., Aerts, S., Caluwe, M., D'aes, J., De Vleesschauwer, F., Dobbeleers, T., De Langhe, P., Kiekens, F. & Dries, J. 2019 Influence of mixed feeding rate in a conventional SBR on biological P-removal and granule stability while treating different industrial effluents. *Water Science and Technology* **79**, 645–655. https://doi.org/10.2166/wst.2019.081.

Stes, H., Caluwé, M., Dockx, L., Cornelissen, R., De Langhe, P., Smets, I. & Dries, J. 2020 Cultivation of aerobic granular sludge for the treatment of food-processing wastewater and the impact on membrane filtration properties. *Water Science and Technology* **83**, 39–51. https://doi.org/10.2166/wst.2020.531.

van Loosdrecht, M. C. M., Nielsen, P. H., Lopez-Vazquez, C. M. & Brdjanovic, D. 2016 *Experimental Methods in Wastewater Treatment*. Water Intelligence Online, p. 15, 9781780404752–9781780404752. https://doi.org/10.2166/9781780404752.

Yao, J., Liu, J., Zhang, Y., Xu, S., Hong, Y. & Chen, Y. 2019 Adding an anaerobic step can rapidly inhibit sludge bulking in SBR reactor. *Scientific Reports* **9**, 10843. https://doi.org/10.1038/s41598-019-47304-3.

First received 13 December 2021; accepted in revised form 5 April 2022. Available online 15 April 2022

doi: 10.2166/wst.2022.113

Ammonium recovery from agro-industrial digestate using bioelectrochemical systems

A. Carucci [a,b,*], G. Erby[b], G. Puggioni[a], D. Spiga[a], F. Frugoni[c] and S. Milia[b]

[a] Department of Civil-Environmental Engineering and Architecture (DICAAR), University of Cagliari, Via Marengo 2, 09123, Cagliari, Italy
[b] National Research Council of Italy (CNR), Institute of Environmental Geology and Geoengineering (IGAG), Via Marengo 2, 09123, Cagliari, Italy
[c] Agrofortis srl, via Monte Suello 18, 25128, Brescia, Italy
*Corresponding author. E-mail: carucci@unica.it

ABSTRACT

Growing food and biomass production at the global scale has determined a corresponding increase in the demand for and use of nutrients. In this study, the possibility of recovering nitrogen from agro-industrial digestate using bioelectrochemical systems was investigated: two microbial electrolysis cells (MECs) were fed with synthetic and real digestate (2.5 gNH_4^+-N L^{-1}). Carbon felt and granular graphite were used as anodes in MEC-1 and MEC-2, respectively. As to synthetic wastewater, the optimal nitrogen load (NL) for MEC-1 and -2 was 1.25 and 0.75 gNH_4^+-N d^{-1}, respectively. MEC-1 showed better performance in terms of NH_4^+-N removal efficiency (39 ± 2.5%) and recovery rate (up to 70 gNH_4^+-N $m^{-2}d^{-1}$), compared to MEC-2 (33 ± 4.7% and up to 30 gN $m^{-2}d^{-1}$, respectively). At the optimal hydraulic retention time, lower NH_4^+-N removal efficiencies and recovery rates were observed when real digestate was fed to MEC-1 (29 ± 6.6% and 60 ± 13 gNH_4^+-N $m^{-2}d^{-1}$, respectively) and MEC-2 (21 ± 7.9% and 10 ± 3.6 gNH_4^+-N $m^{-2}d^{-1}$, respectively), likely due to the higher complexity of the influent. The average energy requirements were 3.6–3.7 kWh $kgN_{removed}^{-1}$, comparable with values previously reported in the literature and lower than conventional ammonia recovery processes. Results are promising and may reduce the need for costly and polluting processes for nitrogen synthesis.

Key words: circular economy, electricity-driven resource recovery, microbial electrochemical technologies, nutrients, wastewater treatment

HIGHLIGHTS

- Bio-electrochemical nitrogen recovery from agro-industrial digestate was achieved.
- Carbon felt anode allowed higher nitrogen recovery than granular graphite.
- Reducing the HRT had an opposite effect on nitrogen removal, depending on the anode material.
- Nitrogen removal rates were negatively affected by real digestate complexity.
- Specific energy consumption was competitive with conventional technologies for nitrogen recovery.

INTRODUCTION

The growing food and biomass production at the global scale has determined a corresponding increase in the demand and use of nutrients (N, P, and K). According to the Food and Agriculture Organization of the United Nations (FAO), global demand for nutrients will increase above 200 million tonnes in 2022; nitrogen will represent approximately 60% (i.e., 153 Mt) of such demand, and it will be primarily used for the synthesis of fertilisers (112 Mt). This scenario indicates the need to adopt innovative strategies to implement reactive nitrogen compounds production, especially in more sustainable and less energy-intensive ways. Ammonia synthesis mainly relies on the Haber-Bosch process, which requires up to 60 MJ kgN^{-1} (depending on which fossil fuel is used to generate syngas) and represents about 2% of the total energy use at global scale (Ledezma *et al.* 2015; Monetti *et al.* 2019), thus having a significant negative impact on the environment. Moreover, the progressive nitrogen accumulation due to inefficient agricultural practices and inadequate treatment of liquid streams is responsible for eutrophication, with the consequent worsening of water quality and human health problems (Nancharaiah *et al.* 2016). Within this framework, a circular economy-based approach should promote the development of new strategies to minimise the release of nutrients into the environment and shift toward a more competitive, resource-efficient bio-economy where the waste economic sector plays an important role (Lin *et al.* 2016). In this sense, the recovery of nutrients from wastewater using technical- and cost-effective processes should be encouraged. For ammonium-rich wastewater streams (>0.5 gN L^{-1}), such as manure, digestate, urine, black water, landfill leachate, and sludge reject water, nitrogen recovery is possible by conventional processes such as ammonia stripping, struvite precipitation (i.e., $MgNH_4PO_4{\cdot}6H_2O$), and ion exchange. Such techniques are energy-intensive and often require massive chemical dosing (Kuntke *et al.* 2018). The high operating costs and the relative cheapness of competing ammonia products derived from the Haber–Bosch process make the economic feasibility of conventional nitrogen recovery technologies still low.

An emerging approach for recovering ammonia is through bioelectrochemical systems (BES) (Monetti *et al.* 2019; Sharma & Mutnuri 2019; Zhang 2020): the oxidation of COD at the anode is catalysed by microorganisms and generates electrons, which migrate from the anode to the cathode through an external electrical circuit. This flow of electrons drives positively charged ammonium ions' migration across a cation exchange membrane (CEM) to the cathodic compartment, where several grams of nitrogen per litre can be accumulated (Kelly & He 2014; Nancharaiah *et al.* 2016) and subsequently recovered by stripping, chemisorption, or forward osmosis. Depending on the counter-reaction occurring at the cathode, either energy can be harvested (i.e., microbial fuel cells, MFCs), or energy needs to be supplied (i.e., microbial electrolysis cells, MECs) (Kuntke *et al.* 2018). Interactions between several parameters, including ammonium ion concentration in the treated wastewater, stripping removal efficiency, catholyte pH and conductivity, current density, membrane and electrode type, and concentrations of other competing ions, need to be investigated to optimise the process (Nancharaiah *et al.* 2016; Zhang 2020).

Although BES were proved as promising for ammonium nitrogen recovery from high ammonium concentration streams (Wu & Modin 2013; Gildemyn *et al.* 2015; Zhang & Angelidaki 2015; Zou *et al.* 2017), studies about nitrogen recovery from agro-industrial wastes are still limited (Cerrillo *et al.* 2016, 2018).

In this study, two MECs equipped with carbon felt and granular graphite as anodes were fed with a synthetic medium mimicking agro-industrial digestate with high ammonium concentrations. Different hydraulic retention times and nitrogen loads were tested to maximise system performance in terms of ammonium recovery. Carbon felt and granular graphite are relatively cheap electrodes to be used in bioelectrochemical systems, and their application was subsequently investigated with real agro-industrial digestate with the perspective of process scale-up. No modifications were made to the electrodes (e.g., chemical cleaning as described by Gildemyn *et al.* 2015) to keep the costs at their minimum. Results were compared in terms of, among the others, nitrogen removal efficiencies, nitrogen removal and recovery rates, and specific power consumption to find the most appropriate material for the treatment of real digestate. Results were promising and provided a comprehensive set of valuable information for the possible combination of BES and anaerobic digestion for agro-waste valorisation within a circular economy-based approach.

MATERIALS AND METHODS

Experimental setup

The experiments were carried out using two double-chamber cells made of Perspex (MEC-1 and MEC-2). Two frames with an internal dimension of 8 × 8 × 2 cm were linked with screws. The hydraulic seal was ensured by linking each cell frame with a rubber sheet, appropriately cut to fit the Perspex frame. Anodic and cathodic chambers, both with a volume of 130 cm^3, were physically separated by a CEM. The MEC-1 anode was a 64 cm^2 carbon felt electrode (thickness 1.12 cm, degree of purity 99.9%, AlfaAesar, Germany), connected to a stainless-steel mesh current collector; the MEC-2 anode consisted of graphite granules (d = 2–3 mm) that filled the anode chamber for about 1/3 of the total volume, and a graphite bar was used as the current collector. For both MEC-1 and MEC-2, the cathode material was a stainless-steel mesh (AISI 316, dimensions 8 × 8 cm, 0.1 mm thickness), with a projected surface of 64 cm^2. The anode chamber was equipped with a reference electrode (Ag/AgCl, +0.197 V vs SHE, mod. MF2052, BioAnalytical Systems, USA) for both cells. The anode (working electrode), cathode (counter electrode), and reference electrode were connected to a multi-channel potentiostat (Ivium Technologies, Ivium-N-stat, NL) for process control. The cathode compartment was coupled to a stripping and an absorption column, both made of Perspex, with an inner diameter of 8 and 70 cm height, half-filled with Raschig rings (L × A 4 × 4 mm, Carlo Erba, Italy). The catholyte was forced to flow continuously from the cathode chamber to the stripping column. The airflow for the ammonia stripping was ensured with a membrane pump, connected in a closed loop with the two columns. A schematic setup overview is shown in Figure 1. The absorption column was filled with 1 M sulphuric acid solution.

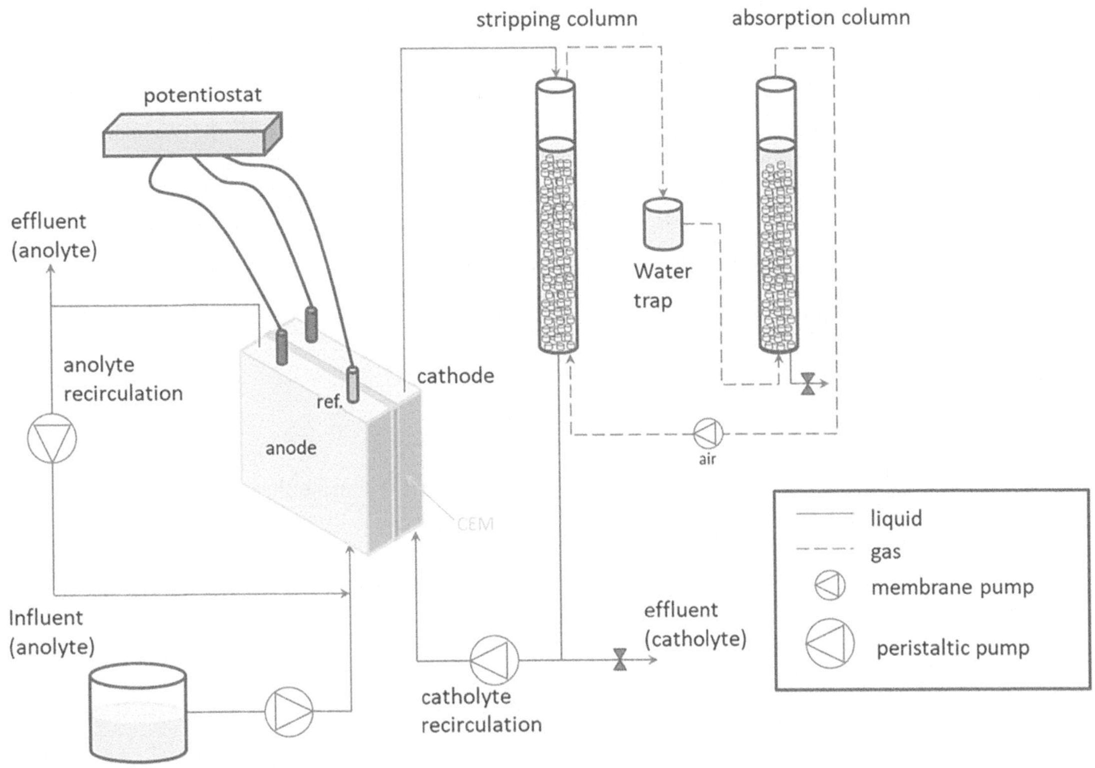

Figure 1 | Schematic representation of the BES setup.

Influent and media composition

The synthetic anolyte (Medium A) composition was prepared according to Gildemyn *et al.* (2015). The concentrations of ammonium (dosed as NH_4HCO_3) and acetate were 2.5 gNH_4^+-N L^{-1} and 3.0 g L^{-1}, respectively. The real anolyte consisted of the digestate produced by an anaerobic digester treating corn silage, livestock manure, and other agro-industrial residues. The 3-stage continuous wet digestion plant consists of three anaerobic mixed reactors working in series for a total digestion volume of approximately 4,200 m^3. The first reactor (i.e., primary digester) works in the low range of thermophilic process temperature. It is fed with biomass, while the second stage (i.e., post-digester) works in the high thermophilic range to boost the methanogenic conversions. The digestate is subsequently stored in the final tank, where the undigested residues are further processed in mesophilic conditions. The daily feedstock is composed of approximately 7 t of dedicated crop silage (corn and weed), 2 t of bran, 1 t of sheep manure and 15 m^3 of bovine slurry, partially replaced by olive pomace during the olive harvesting season. When available, small amounts of other agro-industrial by-products could also be fed during the year. The digestate was sieved at 125 μm, and the main composition was: NH_4^+-N, up to 2,500 mg L^{-1}; $COD_{filtered}$, up to 4,600 mg L^{-1}; total solids, up to 0.95%. The catholyte consisted of 0.1 M NaCl solution (Medium B).

BES operation

The cells were initially batch-fed using Medium A and operated as MFCs to monitor the growth and activity of the electroactive biomass (results not shown). The anodic chamber of each cell was inoculated with a mixture (50:50 v:v) consisting of the activated sludge drawn from the wastewater treatment plant of Cagliari (Italy) and the effluent sludge from the 3-stage anaerobic digester treating agro-industrial residues. After the process start-up, both cells were switched to MEC mode. The working electrode (i.e., the anode) was poised at −200 mV vs the reference electrode (Ag/AgCl) using the potentiostat. The anodic compartments were fed continuously with intense recirculation, and the cathodic chambers were batch-operated. Different hydraulic retention times (HRT) and nitrogen loads (NL: 0.75, 1.0, and 1.25 gNH_4^+-N d^{-1}) were tested in MEC1 and 2 fed with synthetic anolyte, and the optimal ones were chosen to treat the real agro-industrial digestate.

Analytical methods

Samples were collected from the anolyte (influent and effluent), the catholyte, and the absorption column. Anolyte samples were analysed in terms of chemical oxygen demand (COD), pH, acetate, and ammonia concentrations. Catholyte and absorption column samples were analysed in terms of ammonia concentration. Acetate concentration was measured through high-performance liquid chromatography (P680, Dionex, USA) equipped with a UV lamp detector and an organic acid column. COD was determined using a spectrophotometer (DR-2800, Hach, USA) after filtration of samples (1.2 μm). The anolyte conductivity was measured with an HQ30d meter (Hach, USA), equipped with an intellical CDC-401 probe (Hach, USA). NH_4^+-Nitrogen concentration was determined by spectrophotometric analysis (U-2000, Hitachi, Japan) at a wavelength of 420 nm. The pH was measured for all samples through a pH meter (GLP 22, Crison, Spain). X-ray diffractometry (XRD) analysis was performed on ammonium sulfate crystals recovered from the absorption column.

Calculations

The concentrations of NH_4^+-N measured in the influent anolyte ($C_{An,inf}$, mg L^{-1}) and effluent ($C_{An,eff}$, mg L^{-1}) were used for the calculation of the removal efficiency (RE, %), according to Equation (1):

$$RE = \frac{C_{An,inf} - C_{An,eff}}{C_{An,inf}} \times 100 \tag{1}$$

The nitrogen flux across the cation exchange membrane (i.e., the removal rate) was calculated according to Equation (2):

$$J_N = \frac{(C_{An,inf} - C_{An,eff}) \cdot Q}{A_m} \tag{2}$$

where J_N = flow of NH_4^+-N through the membrane (g $m^{-2}d^{-1}$); $C_{An,inf}$ = concentration of NH_4^+-N in the influent anolyte (g L^{-1}); $C_{An,eff}$ = concentration of NH_4^+-N in the effluent anolyte (g L^{-1}); Q = influent anolyte flowrate (L d^{-1}); A_m = surface of the membrane (m^2).

The recovery rate of NH_4^+-N ($R^+_{NH4\text{-}N}$, g $m^{-2}d^{-1}$) in the absorption column was calculated according to Equation (3):

$$R_{NH_4^+-N} = \frac{C_{abs.col.}}{A_m \cdot t} \tag{3}$$

where $C_{abs.col.}$ = NH_4^+-N accumulated in the absorption column (g L^{-1}); A_m = surface of the membrane (m^2); t = operating time (d).

The current density (j, A m^{-2}) was calculated according to Equation (4):

$$j = \frac{I}{A_m} \tag{4}$$

where I = current produced (A); A_m = surface of the membrane (m^2).

The specific energy consumption (SEC, kWh $kgN_{removed}^{-1}$) for ammonium transfer through the membrane was calculated according to Equation (5):

$$SEC = \frac{j \cdot A_m \cdot \Delta V \cdot 24}{(C_{An,inf} - C_{An,eff}) \cdot Q} \tag{5}$$

where j = current density (mA m^{-2}); A_m = surface of the membrane (m^2); ΔV = potential difference between anode and cathode (V); $C_{An,inf}$ = NH_4^+-N concentration in the influent anolyte (g L^{-1}); $C_{An,eff}$ = NH_4^+-N concentration in the effluent anolyte (g L^{-1}); Q = flowrate of influent anolyte (L d^{-1}).

Finally, the current efficiency (CE, %) was calculated according to Gildemyn *et al.* (2015).

RESULTS AND DISCUSSION

BES performance treating synthetic agro-industrial digestate

The operating conditions applied in the experimental phases with synthetic digestate (Medium A) are summarised in Table 1.

Process performances in terms of acetate and ammonium removal efficiency, current efficiency, and SEC for MEC-1 and MEC-2 are summarised in Table 2. Acetate removal efficiency did not change significantly in MEC-1 from Phase I onward, while a progressive decrease was observed in MEC-2, indicating the worsening of process performance with lower HRTs.

Figure 2 shows BES performance in terms of ammonium removal efficiency and removal rates. The highest values of NH_4^+-N removal rates were 78 $\pm$ 3.8 gN $m^{-2}d^{-1}$ for MEC-1 (Phase III) and 33 $\pm$ 4.7 gN $m^{-2}d^{-1}$ for MEC-2 (Phase I), respectively. As for MEC-1, a slight decrease in NH_4^+-N removal efficiency (from 48 $\pm$ 5.2 to 39 $\pm$ 2.5%) was observed, as the NL was increased from 0.75 to 1.25 gNH_4^+-N d^{-1}. However, the resulting NH_4^+-N removal rates increased from 49 $\pm$ 5.0 to 78 $\pm$ 3.8 gNH_4^+-N $m^{-2}d^{-1}$

Table 1 | Experimental phases using synthetic wastewater

Phase	Acetate [g L^{-1}]	NH_4^+-N [g L^{-1}]	NL [gNH_4^+-N d^{-1}]	Anolyte conductivity [mS cm^{-1}]	Anolyte pH [–]	HRT (MEC-1) [h]	HRT (MEC-2) [h]
I	3.0	2.5	0.75	22.9 $\pm$ 0.5	8.2 $\pm$ 0.2	12	8
II	3.0	2.5	1	22.9 $\pm$ 0.5	8.2 $\pm$ 0.2	8	6
III	3.0	2.5	1.25	22.9 $\pm$ 0.5	8.2 $\pm$ 0.2	6	4

Table 2 | Overall BES performance with synthetic influent

Phase	Acetate RE [%]		Ammonium RE [%]		Current efficiency [%]		SEC [kWh kgN^{-1}]	
	MEC-1	MEC-2	MEC-1	MEC-2	MEC-1	MEC-2	MEC-1	MEC-2
I	50 $\pm$ 10	36 $\pm$ 11	48 $\pm$ 5.2	33 $\pm$ 4.7	52 $\pm$ 6.6	48 $\pm$ 11	5.7 $\pm$ 0.8	4.9 $\pm$ 1.6
II	51 $\pm$ 10	34 $\pm$ 11	45 $\pm$ 4.9	21 $\pm$ 1.9	62 $\pm$ 8.5	50 $\pm$ 6.4	3.6 $\pm$ 0.8	5.2 $\pm$ 0.6
III	49 $\pm$ 3	24 $\pm$ 11	39 $\pm$ 2.5	14 $\pm$ 1.6	69 $\pm$ 2.8	63 $\pm$ 17	3.6 $\pm$ 0.1	3.9 $\pm$ 0.2

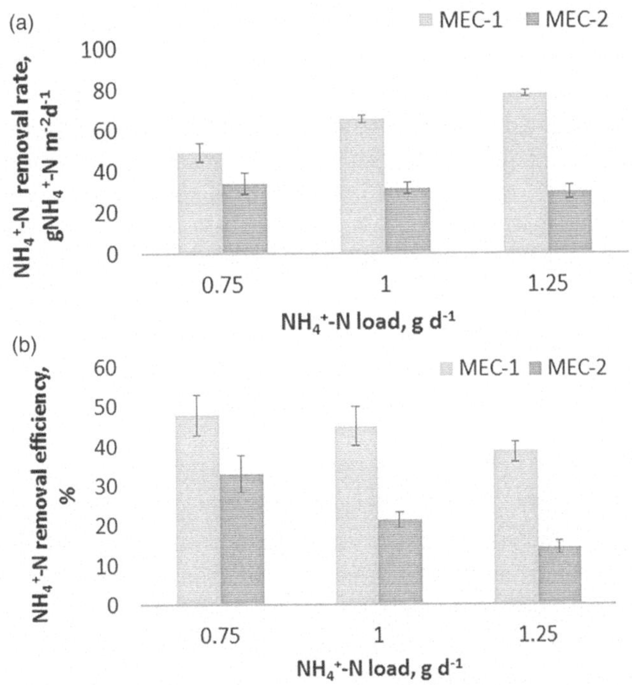

Figure 2 | NH_4^+-N removal rate (a) and removal efficiency (b) in MEC-1 and 2, with different applied NL.

due to the higher amount of influent treated per day. Correspondingly, the average SEC progressively decreased from 5.7 to 3.6 kWh kgN^{-1}. As for MEC-2, the NH_4^+-N removal efficiency decreased significantly (from 33 ± 4.7 to 14 ± 1.6%) as the NL increased. Such decrease was not counterbalanced by the increasing amount of influent treated per day, therefore, the observed nitrogen removal rate slightly decreased from 34 ± 5.2 to 30 ± 3.3 gNH_4^+-N $m^{-2}d^{-1}$. The average SEC was in the range 3.9–5.2 kWh kgN^{-1}. For both MECs, open circuit tests were used as controls and showed negligible NH_4^+-N removal (below 4%), suggesting that bioelectrochemically driven NH_4^+-N removal was the main process involved.

The highest nitrogen recovery rates were consistently observed during Phase III (up to 70 gNH_4^+-N $m^{-2}d^{-1}$, corresponding to 3.5 $kgNH_4^+$-N $m^{-3}d^{-1}$) in MEC-1 and during Phase I in MEC-2 (up to 30 gNH_4^+-N $m^{-2}d^{-1}$, corresponding to 2.3 $kgNH_4^+$-N $m^{-3}d^{-1}$).

The difference in overall process performance between MEC-1 and 2 may be ascribed to the different anode materials. As previously reported by Vilà-Rovira *et al.* (2015), the biofilm growth between granular graphite particles may block the fluid flow path and produce clogging, negatively affecting the electrogenic microbial activity. Ceballos-Escalera *et al.* (2021) indicated the proper control of fluid dynamics in granular graphite-based electrodes as a key for minimising the effects of mass transfer limitations on catalytic rates and the transfer of charge-balancing ions outwards from the electroactive biofilm. In our study, the physical continuity of the carbon felt electrode in MEC-1 may have caused less hydraulic and electrical interferences than discontinuous granular graphite, with consequent better performance.

BES performance treating real agro-industrial digestate

MEC-1 and 2 were operated at their optimal HRT, as determined using the synthetic anolyte (i.e., 6 and 8 h, respectively). The average COD removal efficiencies were 32.7 ± 8.1 and 22 ± 6.6% in MEC-1 and 2, respectively, suggesting that most of the organic matter in the influent was slowly or not degradable. Low COD removal efficiencies (12 ± 1.8%, corresponding to 37% of the biodegradable organic matter) due to substrate complexity were also reported by Cerrillo *et al.* (2016) in an MEC fed with anaerobically digested pig slurry and operated in steady-state conditions at high HRT (30 h).

A progressive decrease in nitrogen removal rates and efficiencies was observed in MEC-1 (Figure 3_top) and MEC-2 (Figure 3_bottom) as real digestate was fed. This decrease was more evident in MEC-2 and ascribed to progressive clogging of intergranular pores due to the presence of solids in the influent. For this reason, cleaning and unclogging of the anodic compartment was performed on day 15. However, the consequent increase in process performance was only temporary, and nitrogen removal decreased as solids accumulated again. As reported by Bolognesi *et al.* (2021), the accumulation of solid particles in an MFC anodic compartment filled with granular graphite influences the contact between substrates and electrodes, and significantly modifies the distribution of the influent in the cell compartment, thus reducing process performance. However, MEC-1 and 2 showed similar performances in terms of NH_4^+-N removal efficiency (29 ± 6.6% and 21 ± 7.9%, respectively), while average nitrogen removal rates were 59 ± 15 and 19 ± 7.6 gN $m^{-2}d^{-1}$, due to the difference in NL applied. Such values were lower than those achieved with synthetic digestate, likely due to the much higher complexity of the real digestate (e.g., lower availability of readily degradable organic substrates, progressive clogging due to high solids content), as previously described also by Zeppilli *et al.* (2017) with the anaerobically digested organic fraction of municipal solid wastes (OFMSW). As reported by Cerrillo *et al.* (2016), the daily supply of mixed volatile fatty acids (namely acetate, propionate, and butyrate) led to an increase in ammonium removal efficiencies from 12 up to 60% in a MEC fed with pre-filtered and diluted pig slurry digestate, due to the higher availability of readily degradable organic substrate and the corresponding higher currents involved.

The average CE in our study was 70 ± 7.9 and 33 ± 24% in MEC-1 and -2, respectively. Again, open circuit tests showed that bio-electrochemically driven NH_4^+-N removal was the main process involved.

Both in MEC-1 and 2, ammonium was successfully recovered as $(NH_4)_2SO_4$, as confirmed by XRD analysis performed on crystals collected from the absorption system (Figure 4).

The NH_4^+-N recovery rate achieved in MEC-1 was much higher than MEC-2 (60 ± 13 vs 10 ± 3.6 gNH_4^+-N $m^{-2}d^{-1}$, corresponding to 3.0 ± 0.7 and 0.8 ± 0.3 $kgNH_4^+$-N $m^{-3}d^{-1}$, respectively), due to the higher NL applied. Such values were lower than those achieved with the synthetic anolyte, in accordance with the nitrogen removal rates.

Nevertheless, the SEC observed in MEC-1 and -2 (3.6 ± 0.4 kWh kgN^{-1} and 3.7 ± 1.1 kWh kgN^{-1}, respectively) were still comparable with other published results (Wu & Modin 2013; Cerrillo *et al.* 2018) and lower than those reported for conventional ammonia removal techniques, such as the nitrification-denitrification process (13 kWh kgN^{-1}), the Sharon-Anammox process (5 kWh kgN^{-1}), and ammonia stripping (9 kWh kgN^{-1}) (Cerrillo *et al.* 2018). As reported by Wu & Modin (2013), electric energy requirements as high as 3.8–8.6 kWh kgN^{-1} still led to a positive energy balance of 6.5–1.7 kWh kgN^{-1}

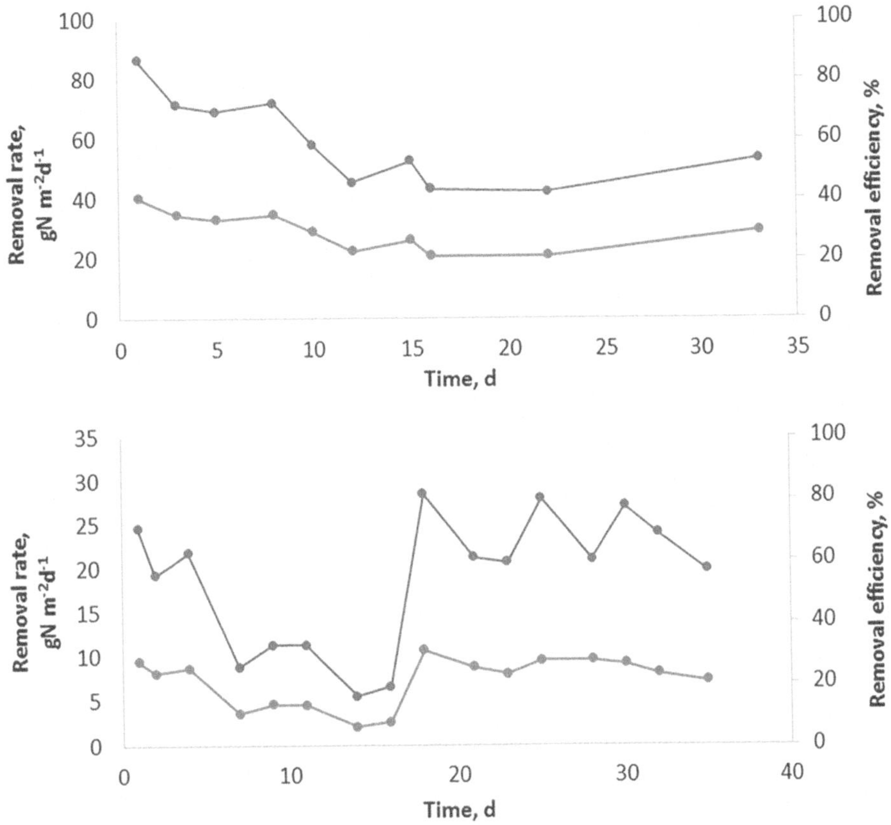

Figure 3 | Trend of NH_4^+-N removal efficiency (red) and removal rates (blue) observed in MEC-1 (top) and MEC-2 (bottom) fed with real digestate.

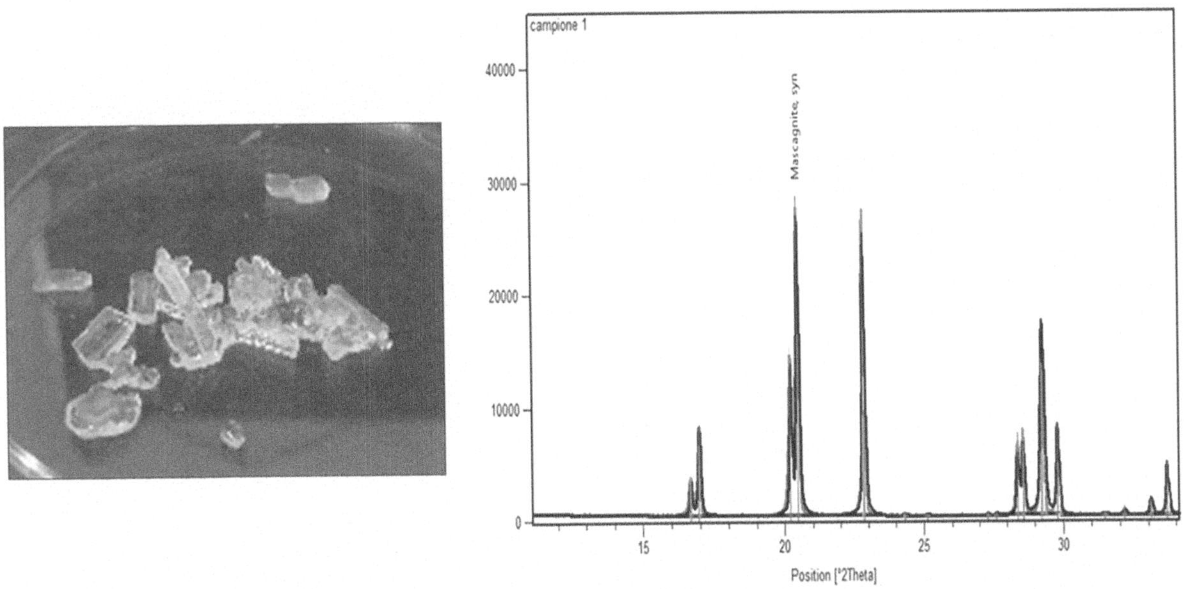

Visible	Ref.Code	Score	Compound Name	Displ.[°2Th]	Scale Fac.	Chem. Formula
*	00-040-0	**71**	Mascagnite, syn	0.000	1.013	$(NH_4)_2SO_4$
*	00-044-1	**80**	Mascagnite, syn	0.000	0.891	$(NH_4)_2SO_4$

Figure 4 | Typical XRD analysis on ammonium sulfate crystals collected from the absorption line of MEC-1 (fed with real digestate). Red line represents PDF card #00-044-1 (Mascagnite).

Table 3 | Comparison among different studies concerning ammonium recovery from digestate using bioelectrochemical and electrochemical systems

	Influent type	Reactor Type	Anode	Cathode	Influent NH_4-N	Nitrogen removal rate ($gN\ m^{-2}d^{-1}$; *$kgN\ m^{-3}d^{-1}$*)	Nitrogen recovery rate ($gN\ m^{-2}d^{-1}$; *$kgN\ m^{-3}d^{-1}$*)	SEC ($kWh\ kgN^{-1}$)
This study	Synthetic agro-industrial digestate	2-chamber MEC	Graphite felt	Stainless steel mesh	2,500	78 ± 3.8; *3.9 ± 0.2*	up to 70; *up to 3.5*	3.6 ± 0.1
	Real agro-industrial digestate	2-chamber MEC	Graphite felt	Stainless steel mesh	2,447 ± 121	59 ± 15; *3.0 ± 0.8*	60 ± 13; *3.0 ± 0.6*	3.6 ± 0.4
Cerrillo *et al.* (2018)	Filtered digested pig slurry	2-chamber MEC	Carbon felt	Granular graphite	1,590–2,970	14.5	–	3
Cerrillo *et al.* (2016)	Filtered digested pig slurry	2-chamber MEC	Carbon felt	Stainless steel mesh	2,190	3.7 ± 0.5	–	–
Wu & Modin (2013)	Real reject water	2-chamber MEC	Carbon felt	Steel wire	1,000	–	–	3.8–20.5
Zeppilli *et al.* (2017)	Digested OFMSW	2-chamber MEC	Graphite granules	Graphite granules	932 ± 95	*0.04 ± 0.01*	–	–
	OFMSW fermentate-digestate mix.	2-chamber MEC	Graphite granules	Graphite granules	818 ± 51	*0.23 ± 0.03*	–	–
Desloover *et al.* (2012)	Digestate	2-chamber electrochemical cell	Ti-MMO	Stainless steel mesh	2,120	90	–	13–26

Numbers in italics are expressed as $kgN\ m^{-3}\ d^{-1}$.

compared with the Haber-Bosch process in a 2-chambered MEC fed with real reject wastewater (1,000 mgNH_4^+-N L^{-1}) at the cathode and synthetic municipal wastewater at the anode, and operated in galvanostatic mode (5–10 mA). Although calculations do not take into account the energy consumption related to the air stripping, it should be considered that other much more energy-efficient methods can be used for capturing ammonia, such as diffusion through a membrane (Rothrock *et al.* 2010; Wu & Modin 2013).

Some practical considerations

A comparison between the results of this study with agro-industrial digestate and previous studies concerning ammonium recovery from digestate using (B)ES is summarised in Table 3. Though the results of our study were promising, it is interesting to notice that further improvement of MEC performance can still be achieved. As reported by Zeppilli *et al.* (2017), a significant increase in volumetric nitrogen removal rates (from 0.04 ± 0.01 to 0.23 ± 0.03 kgN $m^{-3}d^{-1}$) in a methane-producing MEC fed with the effluent of a two-phase anaerobic digester was achieved, as the feeding was switched from anaerobic digestate to a fermentate-digestate mixture (ratio of 1:10). The high concentration of volatile fatty acids in the fermentate (80% of soluble COD) led to increasing currents (from 23 ± 4 to 60 ± 4 mA) and enhanced NH_4^+-N contribution to ionic transport from the anodic to the cathodic compartment (up to 20%). A similar approach can be used in our study by mixing digestate with VFA-rich fermentate taken from the primary digester, thus enhancing ammonium recovery. However, such a strategy would lower biogas production potential and reduce the plant's overall energy efficiency. MEC application for ammonium recovery looks particularly suitable if applied to 1-stage digestion plants that usually have a low conversion efficiency and work with a typical 30-day hydraulic retention time. Biomasses are considered costless for such systems, while major concerns are usually about nitrogen-induced methanogenic suppression (Ramm *et al.* 2020) and slurry/digestate distribution cost. The readily degradable organic acids content in filtered digestate should be sufficient to sustain nitrogen recovery using MEC. A fraction of the filtration residues, still rich in complex organic compounds, could also be recirculated into the digester to virtually increase its HRT and partially recover the biogas production potential loss, even if further field studies should be performed to evaluate the energetic balance of this last step.

Another option, designed for large utility-scale plants and prone to reduce transportation costs, is based on manure processing directly in the production site, to obtain a manure-thickened fraction, which can be sold to the biogas plant, and a filtered fraction which can be treated by MEC and, thanks to the reduced nitrogen content, could be applied to the nearest field, dramatically reducing energy consumption and pollution related to transportation. Manure or digestate filtration represents the step that mainly impacts process energetic yield; therefore, the ideal application of the process described in our study is on already pre-filtered matrices, according to standard industry practices. Bolognesi *et al.* (2021) observed that colloidal and small solid particles gradually accumulated in the anodic chamber of a microbial fuel cell filled with granular graphite and fed with pre-screened mature leachate and dairy wastewater. The progressive accumulation of solid particles limited the contact between substrate and electrode surface, reduced the net free volume in time with consequent reduction of HRT, and likely caused partial fouling of the cation exchange membrane, negatively affecting ion transfer efficiency between anodic and cathodic compartments. In this sense, implementing a more particle-selective influent screening was suggested to minimise clogging and avoid the worsening of process performance. If this is not possible, the recovery of nitrogen and other potential advantages like reduced transportation and distribution costs (filtered residues can be distributed by fertirrigation) should compensate for the extra costs related to filtration.

Since the effluent of both MECs still had a relatively high nitrogen concentration, the possible integration with a completely autotrophic post-treatment system (e.g., one- and two-step anammox) may represent an attractive solution. The integration of these processes would allow obtaining, in addition to the economic advantage generated by the partial recovery of nitrogen (MEC), also a significant reduction in the operating costs associated with its completely autotrophic removal downstream.

CONCLUSION

Bioelectrochemically driven ammonium recovery was successfully achieved in two microbial electrolysis cells fed with agro-industrial digestate. The carbon felt anode allowed higher and more stable nitrogen removal and recovery rates than granular graphite due to better hydraulic distribution and electrical continuity. Although the complexity of real digestate in terms of organic matter and total solids content led to worse nitrogen removal and recovery rates compared to the synthetic influent, the results were promising. They proved BES as a viable, cost-effective option for recovering nitrogen from agro-industrial digestate instead of conventional processes aimed at nitrogen recovery or synthesis.

ACKNOWLEDGEMENTS

This study was funded by Sardegna Ricerche and Autonomous Region of Sardinia (Regional Law n.7, 7/8/2007 – Bando Capitale Umano ad Alta Qualificazione – Annualità 2015). Project ID: CRP_102.

DATA AVAILABILITY STATEMENT

All relevant data are included in the paper or its Supplementary Information.

REFERENCES

Bolognesi, S., Cecconet, D., Callegari, A. & Capodaglio, G. 2021 Bioelectrochemical treatment of municipal solid waste landfill mature leachate and dairy wastewater as co-substrates. *Environmental Science and Pollution Research* **28**, 24639–24649. doi/10.1007/s11356-020-10167-7.

Ceballos-Escalera, A., Pous, N., Chiluiza-Ramos, P., Kort, B., Harnisch, F., Bañeras, L., Dolors Balaguer, M. & Puig, S. 2021 Electro-bioremediation of nitrate and arsenite polluted groundwater. *Water Research* **190**, 116748. doi: 10.1016/j.watres.2020.116748.

Cerrillo, M., Viñas, M. & Bonmatí, A. 2016 Removal of volatile fatty acids and ammonia recovery from unstable anaerobic digesters with a microbial electrolysis cell. *Bioresource Technology* **219**, 348–356. doi: 10.1016/j.biortech.2016.07.103.

Cerrillo, M., Viñas, M. & Bonmatí, A. 2018 Anaerobic digestion and electromethanogenic microbial electrolysis cell integrated system: increased stability and recovery of ammonia and methane. *Renewable Energy* **120**, 178–189. doi: 10.1016/j.renene.2017.12.062.

Desloover, J., Woldeyohannis, A. A., Verstraete, W., Boon, N. & Rabaey, K. 2012 Electrochemical resource recovery from digestate to prevent ammonia toxicity during anaerobic digestion. *Environmental Science and Technology* **46** (21), 12209–12216. doi: 10.1021/es3028154.

Gildemyn, S., Luther, A. K., Andersen, S. J., Desloover, J. & Rabaey, K. 2015 Electrochemically and bioelectrochemically induced ammonium recovery. *Journal of Visualized Experiments* **95**, 1–12. doi: 10.3791/52405.

Kelly, P. T. & He, Z. 2014 Nutrients removal and recovery in bioelectrochemical systems: a review. *Bioresource Technology* **153**, 351–360. doi: 10.1016/j.biortech.2013.12.046.

Kuntke, P., Sleutels, T. H. J. A., Rodríguez Arredondo, M., Georg, S., Barbosa, S. G., ter Heijne, A., Hamelers, H. V. M. & Buisman, C. J. N. 2018 (Bio)electrochemical ammonia recovery: progress and perspectives. *Applied Microbiology and Biotechnology* **102**, 3865–3878. doi: 10.1007/s00253-018-8888-6.

Ledezma, P., Kuntke, P., Buisman, C. J. N., Keller, J. & Freguia, S. 2015 Source-separated urine opens golden opportunities for microbial electrochemical technologies. *Trends in Biotechnology* **33** (4), 214–220. doi: 10.1016/j.tibtech.2015.01.007.

Lin, Y., Guo, M., Shah, N. & Stuckey, D. C. 2016 Economic and environmental evaluation of nitrogen removal and recovery methods from wastewater. *Bioresource Technology* **215**, 227–238. doi: 10.1016/j.biortech.2016.03.064.

Monetti, J., Ledezma, P., Virdis, B. & Freguia, S. 2019 Nutrient recovery by bio-electroconcentration is limited by wastewater conductivity. *ACS Omega American Chemical Society* **4**, 2152–2159. doi: 10.1021/acsomega.8b02737.

Nancharaiah, Y. V., Venkata Mohan, S. & Lens, P. N. L. 2016 Recent advances in nutrient removal and recovery in biological and bioelectrochemical systems. *Bioresource Technology* **215**, 173–185. doi: 10.1016/j.biortech.2016.03.129.

Ramm, P., Abendroth, C., Latorre-Pérez, A., Herrmann, C., Sebök, S., Geißler, A., Vilanova, C., Porcar, M., Dornack, C., Bürger, C., Schwarz, H. & Luschnig, O. 2020 Ammonia removal during leach-bed acidification leads to optimized organic acid production from chicken manure. *Renewable Energy* **146**, 1021–1030. doi:10.1016/j.renene.2019.07.021.

Rothrock Jr., M. J., Szögi, A. A. & Vanotti, M. B. 2010 Recovery of ammonia from poultry litter using gas-permeable membranes. *Transactions of the ASABE* **53**, 1267–1275. doi: 10.13031/2013.32591.

Sharma, P. & Mutnuri, S. 2019 Nutrient recovery and microbial diversity in human urine fed microbial fuel cell. *Water Science and Technology* **79** (4), 718–730. doi: 10.2166/wst.2019.089.

Vilà-Rovira, A., Puig, S., Dolors Balagues, M. & Colprim, J. 2015 Anode hydrodynamics in bioelectrochemical systems. *RSC Advances* **5** (96), 78994–79000. doi: 10.1039/c5ra11995b.

Wu, X. & Modin, O. 2013 Ammonium recovery from reject water combined with hydrogen production in a bioelectrochemical reactor. *Bioresource Technology* **146**, 530–536. doi: 10.1016/j.biortech.2013.07.130.

Zeppilli, M., Pavesi, D., Gottardo, M., Micolucci, F., Villano, M. & Majone, M. 2017 Using effluents from two-phase anaerobic digestion to feed a methane-producing microbial electrolysis. *Chemical Engineering Journal* **328**, 428–433. doi: 10.1016/j.cej.2017.07.057.

Zhang, G. 2020 Bioelectrochemical Ammonium Production – Nitrogen Removal and Recovery in BES. *Bioelectrosynthesis: Principles and Technologies for Value-Added Products*, Chapter 8, 217-229. https://doi.org/10.1002/9783527343829.ch8

Zhang, Y. & Angelidaki, I. 2015 Submersible microbial desalination cell for simultaneous ammonia recovery and electricity production from anaerobic reactors containing high levels of ammonia. *Bioresource Technology* **177**, 233–239. https://doi.org/10.1016/j. biotech.2014.11.079.

Zou, S., Qin, M., Moreau, Y. & He, Z. 2017 Nutrient-energy-water recovery from synthetic sidestream centrate using a microbial electrolysis cell-forward osmosis hybrid system. *Journal of Cleaner Production* **154**, 16–25. https://doi.org/10.1016/j.jclepro.2017.03.199.

First received 16 December 2021; accepted in revised form 22 March 2022. Available online 4 April 2022

doi: 10.2166/wst.2022.095

An essential tool for WRRF modelling: a realistic and complete influent generator for flow rate and water quality based on data-driven methods

Feiyi Li [a,b,*] and Peter A. Vanrolleghem [a,b]
[a] modelEAU, Université Laval, 1065, Avenue de la Médecine, Québec, QC, G1 V 0A6, Canada
[b] CentrEau, Québec Water Research Center, 1065 avenue de la Médecine, Québec, QC, G1 V 0A6, Canada
*Corresponding author. E-mail: feiyi.li.1@ulaval.ca

FL, 0000-0003-4278-730X; PAV, 0000-0003-1695-1313

ABSTRACT

Modelling, automation, and control are widely used for water resource recovery facility (WRRF) optimization. An influent generator (IG) is a model, aiming to provide the flowrate and pollutant concentration dynamics at the inlet of a WRRF for a range of modelling applications. In this study, a new data-driven IG model is proposed, only using routine data and weather information, and without need for any additional data collection. The model is constructed by an artificial neural network (ANN) and completed with a multivariate regression to generate time series for certain pollutants. The model is able to generate flowrate and quality data (TSS, COD, and nutrients) at different time scales and resolutions (daily or hourly), depending on various user objectives. The model performance is analyzed by a series of statistical criteria. It is shown that the model can generate a very reliable dataset for different model applications.

Key words: data mining, digitalization, machine learning, multivariate regression, wastewater modelling, water quality

HIGHLIGHTS

- A data-driven influent generator (IG) to create a long-term time series for flow and pollutant concentration at hourly and daily frequencies.
- The IG application only needs WRRF routine data and weather information, without need for any additional data collection.
- The IG also generates nutrient concentration time series.
- A random walk stochastic process is added to better approximate observed temporal variability.

GRAPHICAL ABSTRACT

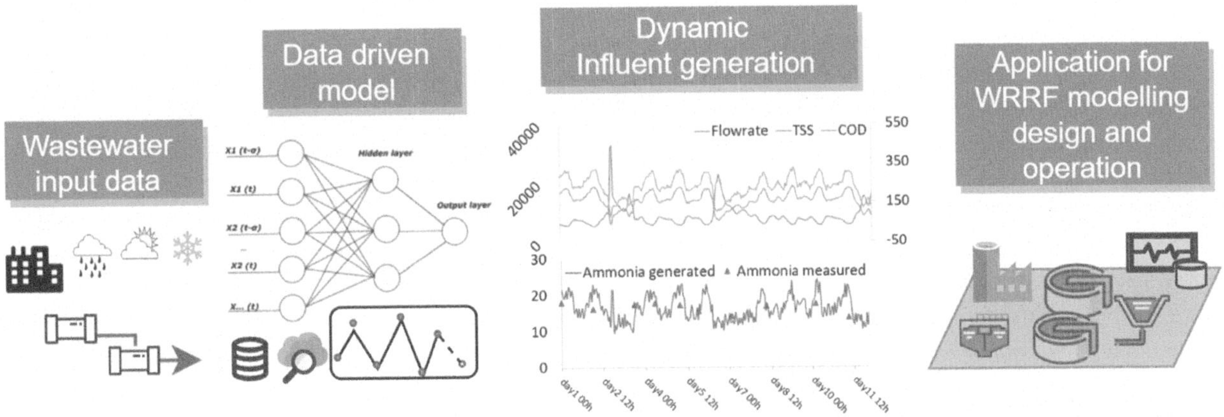

1. INTRODUCTION AND OBJECTIVES

Nowadays, modelling is widely used for water resource recovery facility (WRRF) design, upgrade and controller evaluation (Sweeney & Kabouris 2013). However, because of a lack of adequate (dynamic) input datasets, most of the models are still used under steady state conditions. For example, for WRRF design, engineers usually make an initial sizing by using design guidelines and safety factors or some statistical evaluations to come up with the values for the design inputs, and the consequence of overly conservative safety factors is oversizing the WRRF (Talebizadeh 2015). However, steady-state simulations are not able to represent the temporal variability present in reality. Thus, a reliable dynamic influent generator (IG) is essential as key input for WRRF modelling (Gernaey *et al.* 2011; Martin & Vanrolleghem 2014).

An Influent Generator (IG) is a model that generates realistic dynamic (or static) influent scenarios at the inlet of a WRRF. The outputs of an IG include flowrates and pollutant concentrations at different time scales (long term, short term) and resolutions (daily, hourly), according to different user objectives.

An IG model can be applied in very diverse domains, such as the integrated urban wastewater systems (IUWS) to globally improve the wastewater treatment process (Benedetti *et al.* 2013), and WRRF design and upgrade to face the growing amounts of wastewater produced by increasing urbanization and population. IGs can help to provide a fast and accurate estimation for WRRF designers and a credible input time series to anticipate the treatment performance under different operating conditions.

An IG model can also be used for influent database quality evaluation, optimization and completion (Martin & Vanrolleghem 2014). Since data observation and collection become more and more important in wastewater management, an IG model enables quality evaluation of a dataset collected from online measurements in different ways, e.g. by identifying errors due to clogged sensors, detecting extreme measurement values (outliers), etc. An IG can also complete missing data (gap filling, i.e. temporary measurement failure) and interpolate a low frequency time series into a denser time series.

Different researchers (see below) have been developing IGs based on different modelling principles. The current IGs can be divided into two types: data-driven IGs and phenomenological IGs, also called 'black' box and 'grey' box, respectively. Data-driven IGs focus on finding the relation between their inputs and outputs, without any knowledge of the internals of the system. Therefore, the performance of data-driven IG depends on the provided dataset. Varying degrees of model complexity have been studied, such as harmonic function models, regression equations, and also artificial neural networks (ANN). The Fourier series IG based on harmonic function has been used to develop a simple and reliable generator of diurnal variations for dry weather influent data, and has been used in different studies (Langergraber *et al.* 2008; Mannina *et al.* 2011). Ahnert *et al.* (2016) developed a statistical method for the generation of a continuous time series of influent quality by only using routine data based on the Weibull distribution. Troutman *et al.* (2017) developed an automated toolchain based on Gaussian processes to predict the dynamics of combined sewer system flow. Recently, machine learning is also being studied for flow forecasting over a short time horizon, with model structures such as artificial neural networks (ANN) (El-Din & Smith 2002), multiple linear regression (Zhu & Anderson 2016) or nonlinear autoregressive exogenous models (NARX) (Banihabib *et al.* 2019).

Unlike data-driven approaches, the phenomenological model is built by integrating some important processes, which influence the generation of the influent. This type of model is usually constructed around different submodels, such as dry weather generation, sewer system transport, soil model including infiltration of groundwater etc. (Gernaey *et al.* 2011). For example, the dry weather flowrate has a diurnal pattern with two peaks corresponding to the morning and evening urban activities, e.g. characterized by Butler *et al.* (1995). This diurnal pattern can reflect different lifestyles and can be described by a harmonic function. For storm flow, the dilution and first flush during a rain event can be integrated in the model (Bechmann *et al.* 1999; Talebizadeh *et al.* 2016; Langeveld *et al.* 2017). Gernaey *et al.* (2011) developed and Flores-Alsina *et al.* (2014) extended a more conceptual phenomenological model, by focusing on flowrate scenarios and submodels of the urban drainage system.

There are advantages and disadvantages in both types of models (Price & Vojinovic 2011). Compared with a data-driven model, a phenomenological model contains more details of the influent generation process, which leads to an explicit result and good extrapolation power beyond the calibration range. Normally, such model consists of submodels with parameters related to physical processes. One of the shortcomings of such IG comes from the need of calibration, because of the series of parameters and submodels, it is difficult to calibrate all parameters needed, such as the catchment information, sewer system etc., which leads to less flexibility when applying to a different case study. In contrast, a data-driven model focuses only on the relation between input and output, instead of providing more understanding of the system. Thus, it is easier to calibrate and use a data-driven model. However, the high dependence on data demands a more complete dataset, in order to reach adequate model quality.

Currently, most influent generators in the literature (either phenomenological or data-driven) suffer from the following two issues. The first issue is that it is difficult to balance the complexity and precision of the IG. Usually, a higher complexity leads to better performance, but it makes the modelling more time consuming and more intense in terms of calculation. The other issue is the ability to generate an adaptable influent profile that can be easily adjusted to different user objectives. For example, the design process demands a long-term series but allows for low time resolution, while a process control application needs a higher time resolution profile.

To solve the issues with the available IG models, in this study, a data-driven IG is proposed that can generate dynamic WRRF influents of different time scales and resolutions. The model aims to generate reliable influent properties and more complete time series for the influent dynamics, including daily and hourly flowrate, daily and hourly total suspended solids (TSS), and daily chemical oxygen demand (COD) and nutrient concentrations for the case study at hand. With the available daily concentration results the IG is constructed such that it is able to obtain hourly concentration dynamics by applying an observed daily pattern or by developing a more detailed model in future studies.

The IG proposed in this work is intended to generate influent time series only based on weather information, without any further collection of flowrate or concentration data. This allows obtaining long-term simulation inputs and since it does not need a complicated dataset, it reduces the investment of labor and money for measurement campaigns. The proposed IG model is not built around a physical submodel, resulting in lower modelling efforts than with a phenomenological model, so that it can provide a satisfactory compromise between model complexity and prediction quality.

2. MATERIALS AND METHODS

This section describes the case study and the preparation of the data set, followed by an overview of the modelling approach. First, the basic ANN model is presented followed by the stochastic process model that is added to increase the generated variability. Subsequently, the multivariate regression that allows calculating the nutrient time series is introduced. This section finishes with a definition of the criteria that will be used to assess the quality of the proposed IG.

2.1. Case studies and dataset preparation

The modelling approach was developed and tested on two urban catchments, Quebec City (Canada) (Tik & Vanrolleghem 2017) and Bordeaux, Clos de Hilde (France) (Ledergerber *et al.* 2020). These two case studies are both combined sewer systems, with a similar number of population equivalents, 300,000 PE and 200,000 PE, respectively. However, the two catchments provide different extents (range, data frequency) of available data.

In this research, the Quebec City case study is developed around routine influent data that are collected regularly at the entry of the WRRF, including flowrate, COD, TSS. The nutrient concentrations (ammonia and phosphorus) are sampled and measured only once or twice per week since the plant is only required to remove organic matter. Data were available for 8 years (2011–2018). For the Bordeaux case study, which is also a carbon removal WRRF, hourly flowrate and TSS concentrations were available for three months in 2017 and 2018.

2.2. Modelling approach

The proposed IG was developed with MATLAB R2019b (www.mathworks.com) using the following machine learning methodology. In short, the dataflow through the model is according to the flowchart shown in Figure 1, detailed in the following sections. Figure 1 shows that the raw data series is pre-treated. The pre-treatment aims to remove outliers detected by a univariate method and the faulty data are replaced by estimated values obtained by the Gaussian kernel smoother (Alferes & Vanrolleghem 2016). Then the cleaned flowrate data are Fourier transformed to obtain a yearly-seasonal-daily pattern for the baseflow P(t). The input dataset of the ANN consists of this P(t) together with weather data (temperature T and rainfall intensity R, provided by the WRRF's pluviometry and treated at hourly frequency), creating input vectors with a number of lags τ, which represents the weather data in the previous τ time steps. A stochastic generator is subsequently applied on the ANN's output, to get a time series, which better mimics the reality in terms of variability (Q_{sim}). Finally, the nutrient concentrations ($C_{nutrisim}$) are obtained by applying a multivariate regression with input of the ANN's results and the weather data time series.

2.2.1. ANN model

The artificial neural network takes inspiration from the biological learning process. As a powerful data-driven tool, it can simulate nonlinear systems and is increasingly used in water engineering, urban hydrology and catchment modelling (Maier & Dandy 2000; Rajurkar *et al.* 2004; Fu *et al.* 2010).

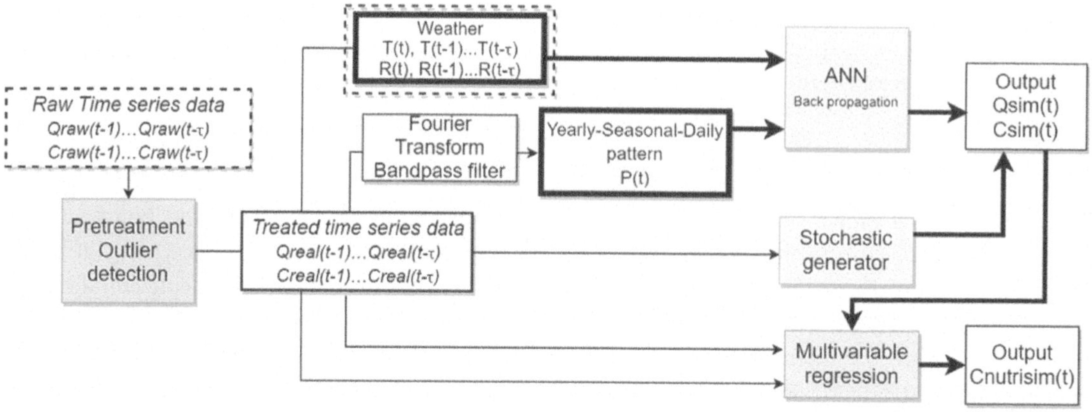

Figure 1 | Description of the IG model and interaction between ANN, stochastic generator, and multivariate regression submodels. The dashed line boxes represent the real data needed for training and the bold line boxes represent the model input for new influent series generation.

Input data selection is important before using any ANN model. For a combined sewer system, the dry weather flow (DWF) component varies with the pattern of urban behaviour and is completed with hydrological processes (snowmelt, groundwater) and wet weather flow (WWF) also including direct storm water inflow, and rainfall-dependent infiltration and inflow (RDII) (Wright *et al.* 2001). Thus, the input of the ANN-based IG consists of a basic domestic flowrate pattern, weather information, including temperature and rainfall measurements with a number of lags τ time steps, to represent an internal autoregression.

The DWF pattern is obtained by applying a Chebyshev bandpass filter after Fourier transform (Heideman *et al.* 1984). The Fourier transform enables converting the signal from the time domain into the frequency domain, and the Chebyshev bandpass filter allows only the signal with selected frequencies to pass (Schlichthärle 2011). By removing high-frequency contributions such as measurement noise, the dry weather flow pattern can be obtained. In this study, the bandpass filter is focusing on extracting two major signals: the seasonal/yearly effect and the daily effect. To this end, the frequencies that were retained are: 4 year^{-1} and 2 year^{-1} for the seasonality and 1 day^{-1} and 2 day^{-1} for the diurnal phenomena.

As activation function in the ANN's hidden layers, the sigmoid function was selected to capture the non-linearities. The output layer uses a simple linear activation function. In order to minimize the loss function, the Levenberg–Marquardt backpropagation algorithm (Levenberg 1944) was used for training, because it is fast and easy to use and has been a first-choice in supervised search algorithms. In this study, the performance evaluation is focused on measuring the residual between the model result and target value, therefore, the root mean square of errors (RMSE) was used, one of the most used loss functions in regression problems in ANN learning:

$$RMSE = \sqrt{\frac{1}{N}\sum_{i=1}^{N}(y_{oi} - y_{si})^2} \tag{1}$$

where, y_{oi} and y_{si} represent observed data and simulated data, respectively.

The dataset was divided into a training set, a cross-validation set and a test set (70, 15 and 15%, respectively). To determine the architecture of the ANN (hyper-parameters: number of neurons and layers), a series of ANNs with different numbers of layers and neurons were trained. For each training, the performance of the model was recorded and compared. The number of layers and neurons giving the best performance for the cross-validation dataset, was selected as final architecture. To avoid local minima during training, a series of iterations of the training procedure was applied. The final test of the ANN model was performed with the test set.

2.2.2. Stochastic process

After obtaining the ANN result, it was observed that the model output exhibited less variability than the measurements. However, it is important to have a probability distribution of the generated time series similar to reality, i.e. this allows especially to better design WRRF parameters such as the expected load and hydraulic capacity and certain percentiles of their distribution etc. Therefore, a stochastic process was added in order to optimize the IG model adequacy. The aim was to simulate a more random time series, with a statistical characterization that is more reflecting the distribution of the measured reality.

Theorical analysis shows that adding an autocorrelation model residual can improve model reliability (Villez *et al.* 2020). Therefore, a stochastic process was added that is modelled by a k-order random walk based on the error between the ANN output and the measurements:

$$R_t = \sum_{1}^{k} \varphi_k * R_{t-k} + \varepsilon \tag{2}$$

where R_t is the error value at time t, corresponding to the difference between the ANN model and the actual data, φ_k are the model's k coefficients and ε is a white noise sequence with a Gaussian distribution.

2.2.3. Multivariate regression

Experiments show that nutrients (such as ammonia) are highly correlated with the flowrate and COD concentrations (Petersen *et al.* 2002). Thus, it is proposed that nutrient data can be generated by a multivariate regression using the COD and flowrate results of the ANN together with weather information as additional input variable. The adopted multivariate polynomial regression aims to calculate the nutrient concentration time series by using its relationship with other variables:

$$\begin{bmatrix} y_1 \\ y_2 \\ \cdots \\ y_m \end{bmatrix} = \begin{bmatrix} 1 & x_{11} & x_{21} & x_{k1} & & x_{11}^2 & & x_{k1}^k \\ 1 & x_{12} & x_{22} & x_{k2} & \cdots\cdots & x_{12}^2 & \cdots\cdots & x_{k2}^k \\ \cdots & \cdots & \cdots & \cdots & & \cdots & & \cdots \\ 1 & x_{1m} & x_{2m} & x_{km} & & x_{1m}^2 & & x_{km}^k \end{bmatrix} \begin{bmatrix} \theta_1 \\ \theta_2 \\ \cdots \\ \theta_{..} \end{bmatrix} + \begin{bmatrix} \varepsilon_1 \\ \varepsilon_2 \\ \cdots \\ \varepsilon_{...} \end{bmatrix} \tag{3}$$

where y_i are the target variables, x_{ij} are the multivariable data, k refers to the kth-order of the polynomial, θn are the regression coefficients, and ε is white noise. To avoid overfitting during higher order regression (when k increases), the Lasso (Least Absolute Shrinkage and Selection Operator) regression regularization method was used to determine the optimal model complexity and regression parameters (Tibshirani 1996):

$$\theta^{lasso} = argmin\left\{ \sum_{i=1}^{m} \left(y_i - \theta_0 - \sum_{j=1}^{n} x_{ij}\theta_j \right)^2 + \lambda \sum_{j=1}^{n} |\theta_j| \right\} \tag{4}$$

where λ is a pre-specified parameter, determining the amount of regularization and θ_j are coefficients of regression. Thus, the Lasso method solves the minimization problem with respect to the MSE, under the condition of minimizing the model parameters.

2.3. Criteria and error analysis

The model was optimized for MSE during the training procedure. In addition, the model performance was also evaluated by the mean absolute percentage error (MAPE) and the Nash-Sutcliffe efficiency (NSE), expressed as

$$NSE = 1 - \frac{\sum_{t=1}^{T} (Q_s^t - Q_o^t)^2}{\sum_{t=1}^{T} (Q_o^t - \overline{Q_o})^2} \tag{5}$$

where, Q_s is the simulated data and Q_o is the observed data for the variable under consideration. The optimal NSE value equals one, corresponding to a perfect match of modelled and observed data, and zero indicates that the model is equally good as a mean value model. A negative value means that the model is performing worse than the mean value of the observations (Hauduc *et al.* 2015).

In addition, a frequency histogram of the generated time series, the probability density function (PDF) and the cumulative density function (CDF) were created to compare the statistical characteristics of the measured and the simulated time series. The statistical characterization allows evaluating whether the model output can well represent the real dataset. In addition, the Kullback–Leibler divergency (KL divergency) (Kullback & Leibler 1951) was calculated, in order to measure the difference between the probability distributions of model output and measurements, which is expressed as

$$D_{kl}(Q||P) = \sum_{x \in \chi} P(x) \ln\left(\frac{Q(x)}{P(x)}\right) \quad (6)$$

where, P(x) and Q(x) represent the probability distribution of the model output and the real measurement, respectively, and $P(x) > 0$ and $Q(x) > 0$ for any x in χ, which is the distribution space.

3. RESULTS AND DISCUSSION

In this section, first, the result of each of the IG's submodels is presented, including the daily and seasonal pattern, and the ANN modelling results of flowrate and pollutant concentrations for both Quebec City and Bordeaux. Next, attention is focused on the usefulness of adding a stochastic process to increase the variability of the generated time series. Subsequently, the result of the nutrient concentration calculations is shown. Then, the model performance is analyzed with two other criteria. Finally, this section ends with a detailed evaluation of the proposed IG from an overall perspective.

3.1. Model and submodel results

3.1.1. Daily and seasonal pattern

A Chebyshev bandpass filter (2nd order Chebyshev Type I) was applied after the Fourier transform of each time series to get the seasonal and daily DWF pattern of the urban activity. Figure 2 shows the pattern for the Quebec City case study. The flowrate pattern exhibits a clear increase during the snowmelt from March to the end of April, while the COD and TSS

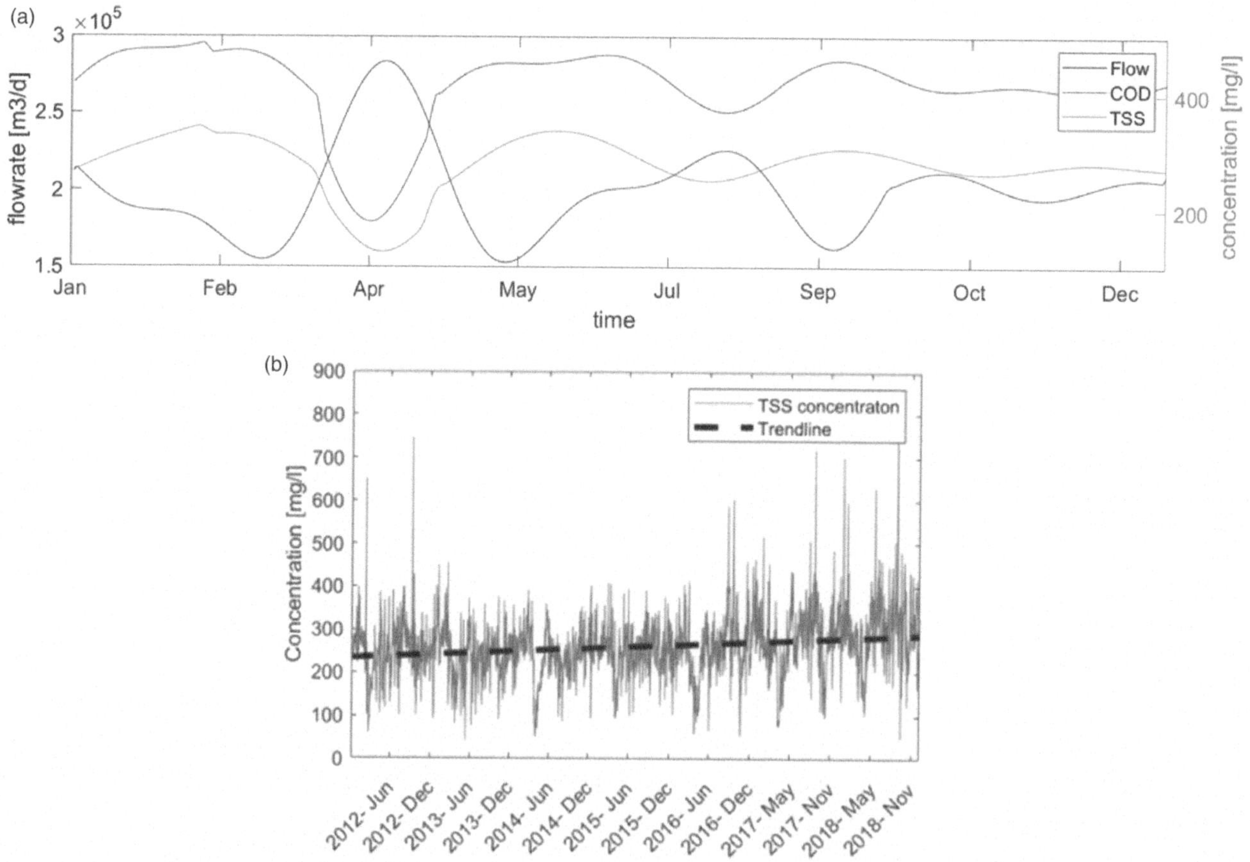

Figure 2 | (a) Yearly pattern for flowrate, COD and TSS based on daily data for the Quebec City case study of 2011–2018, (b) yearly trend in TSS concentration from 2011 to 2018.

patterns demonstrate dilution by the snowmelt. The TSS daily concentration shows a slight trend from 2011 to 2018. There are two explanations for this phenomenon: (i) an increase of the urbanization brings increasing TSS discharges and (ii) thanks to the improvements made to the sewer system, the infiltration inflow has decreased, which leads to a higher TSS concentration.

Similarly, the 2nd order Chebyshev Type I bandpass filter is applied for the Bordeaux case study, in order to extract the daily pattern. Figure 3 illustrates that the daily pattern has two peaks corresponding to the increased urban activity in the morning and evening, as reported elsewhere in the literature (Butler *et al.* 1995). However, the peak and form are different for weekday and weekend days, because of the different weekend behaviour.

3.1.2. ANN modelling result for Quebec City

The Quebec City case study aimed at generating the flowrate, COD and TSS time series using an ANN model. The selected ANN has one input layer, one hidden layer and one output layer. The hidden layer contains four neurons for flowrate and five neurons for COD and TSS. Figure 4 shows the daily flowrate, TSS and COD concentration in the test set. The result clearly demonstrates the snowmelt effect at the end of winter, with an increase of the flowrate and dilution of TSS and COD. The impact of each rain event is adequately described by the IG model.

3.1.3. ANN modelling result for Bordeaux

The ANN model also demonstrated good results for Bordeaux, as shown in Figure 5 for the hourly flowrate (a) and hourly TSS (b). The flowrate is generated by using a daily dry weather flow pattern and hourly rain data as input. The daily dry weather flow pattern profile and the increase of inflow by storm water can both be seen for each rain event. The TSS results demonstrate that the data-driven methodology is also able to describe water quality at high frequency. The TSS concentration follows a daily concentration pattern and it is diluted by the inflow of storm water.

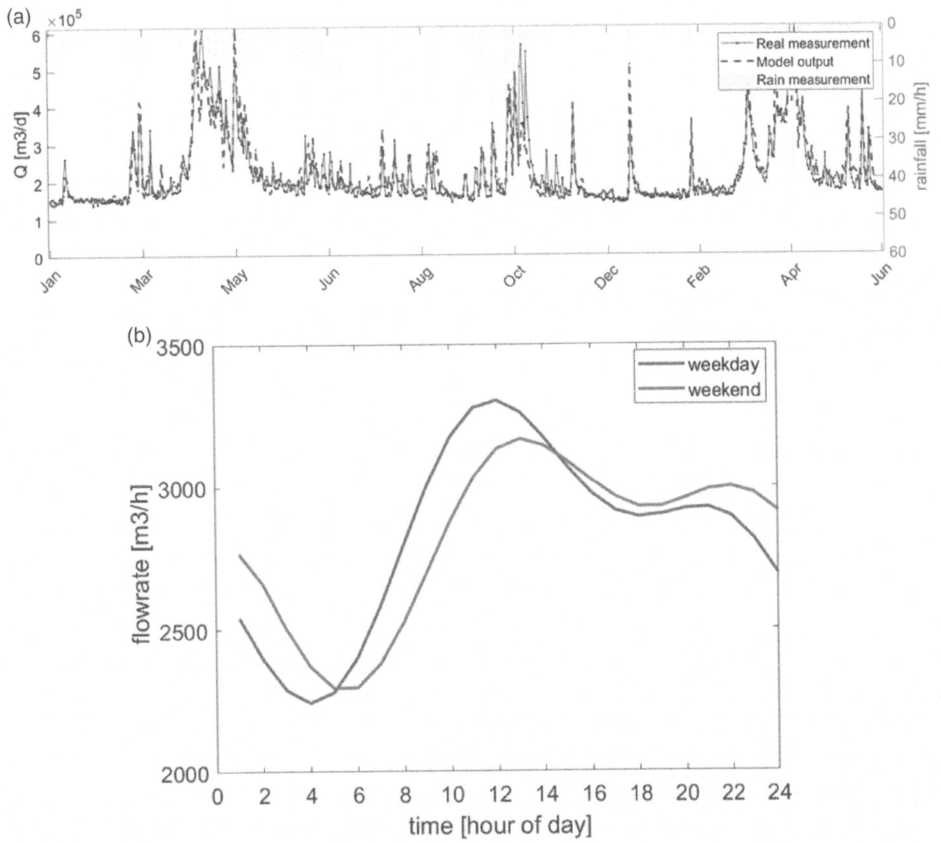

Figure 3 | Daily pattern for flowrate in Bordeaux: hourly flowrate for weekday and weekend days based on the hourly data for May to August, 2018.

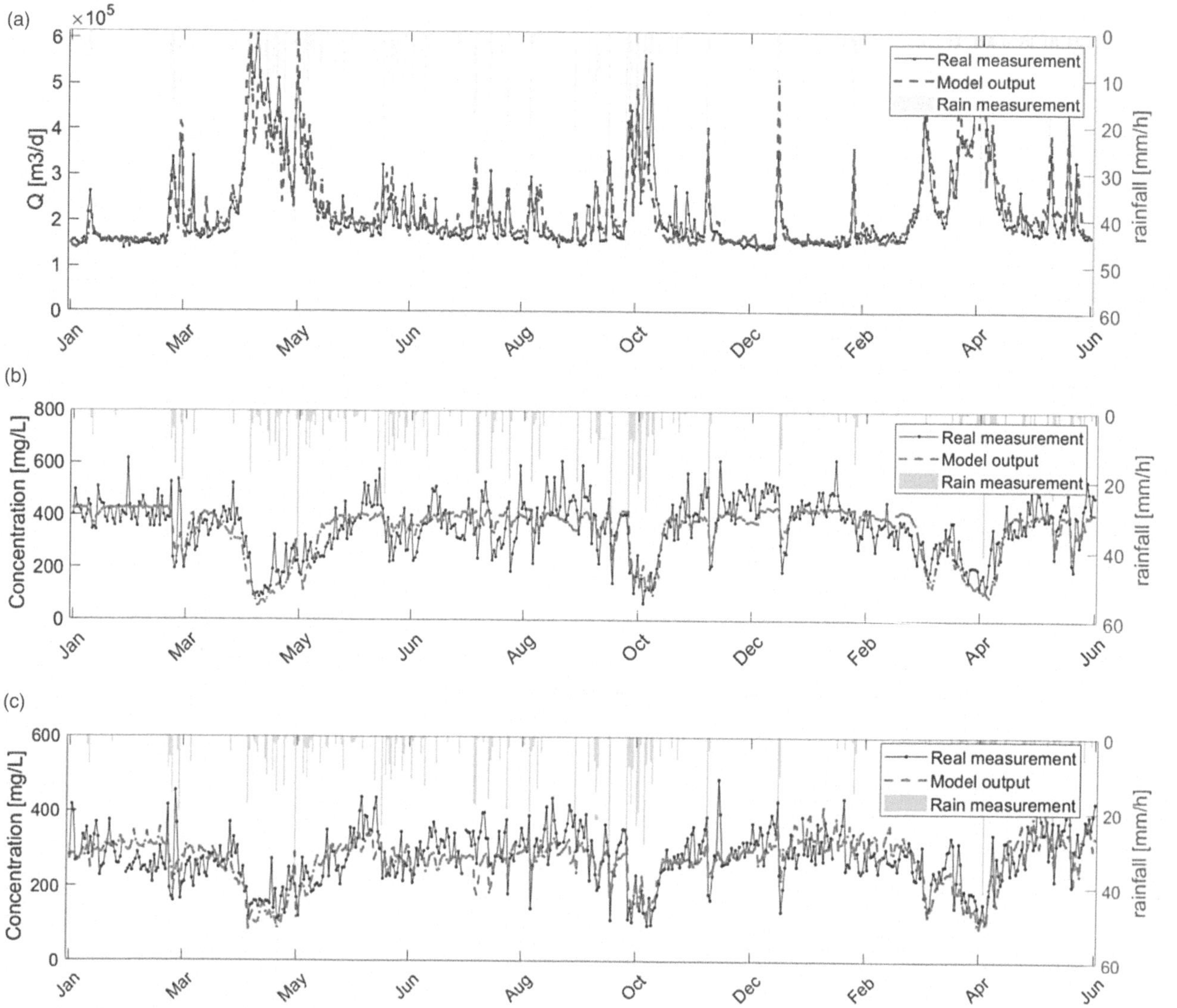

Figure 4 | IG model output of the test set for the Quebec City case study: (a) daily flowrate, (b) COD concentration, (c) TSS concentration. Measurement were collected in 2017 and 2018.

The ability of the IG model to generate concentration time series can also be used for dataset gap filling, especially under wet weather conditions when sensor clogging occurs more frequently. Moreover, as shown in Figure 6, the model can be improved by also feeding it with daily average laboratory measurements as input. Although the model output is still less dynamic than what the sensor data exhibit, this extra input enables the model to better represent the variability of reality (by representing the differences of wastewater discharge on different days). For instance, thanks to this add-on, the lower concentrations on 13 and 14 June could be captured even though there was no wet weather effect. Finally, note that a longer time series will help to find a yearly or seasonal pattern so that the TSS concentration results would be further improved.

3.1.4. Better representing variability by adding a stochastic process

As mentioned before in the Materials and Methods section Stochastic process, the obtained ANN model was extended with a random walk model after having analyzed the error between the ANN model output and the measurement data. The stochastic process extension allows the influent generator to statistically better approach the variability in the real measurements.

The random walk model is a special case of the autoregressive (AR) model. In order to define the order of the autoregression model, orders 1–5 time lags were evaluated for the Quebec City case study. As the autocorrelation coefficient plot and

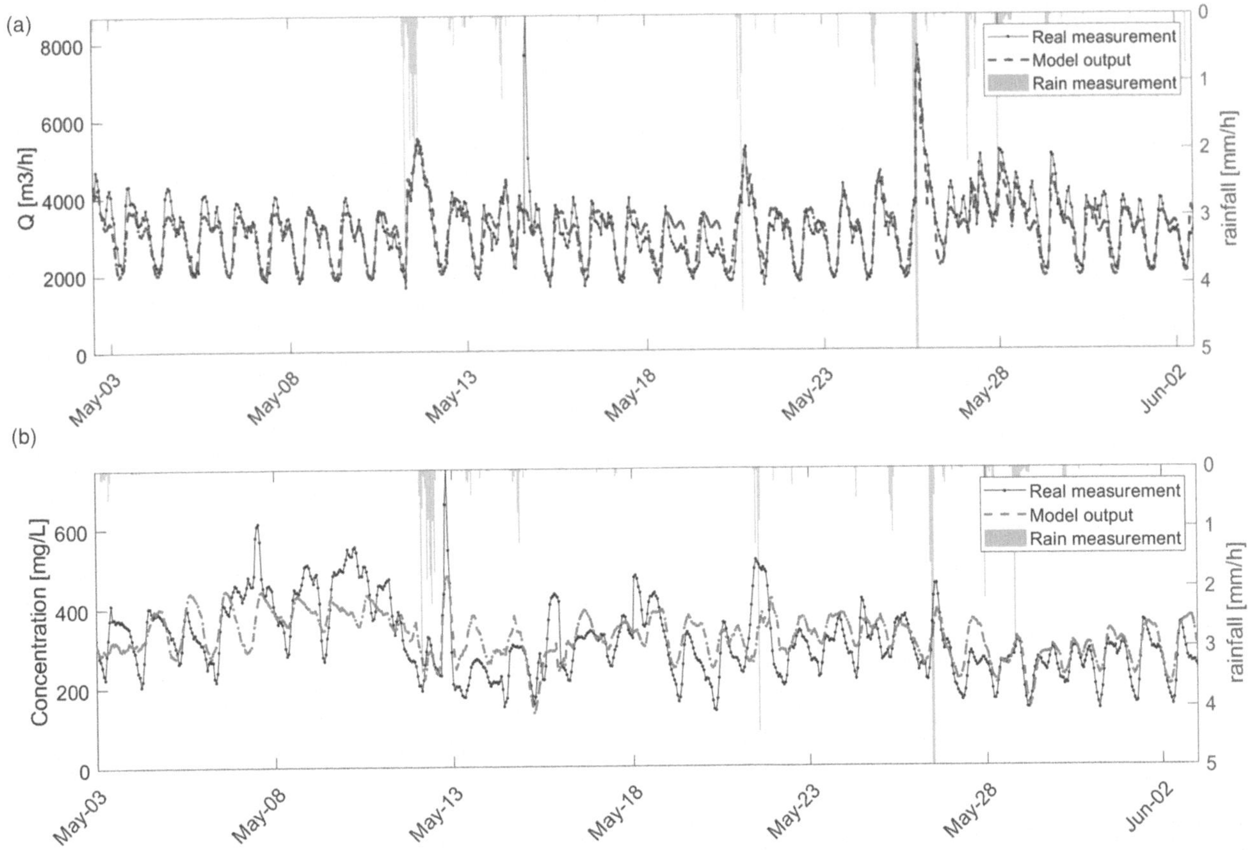

Figure 5 | Hourly flowrate (a) and TSS concentration (b) generation by the IG model for the Bordeaux case study.

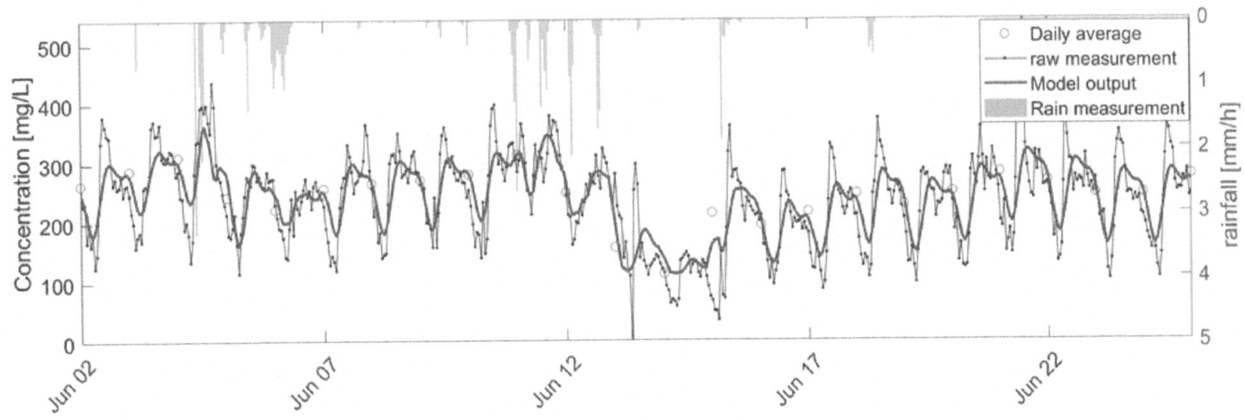

Figure 6 | Hourly TSS generation improved by adding daily average measurements.

the MAE results of Figure 7 show, the best stochastic model is obtained at 4 time lags autoregression because this gave the best compromise between model complexity and precision.

Figure 8 shows the results of COD and TSS generation with addition of the stochastic process. Compared with the ANN-only model, the stochastic process allows to better mimic the influent random variation. A more detailed analysis is provided in the performance analysis section.

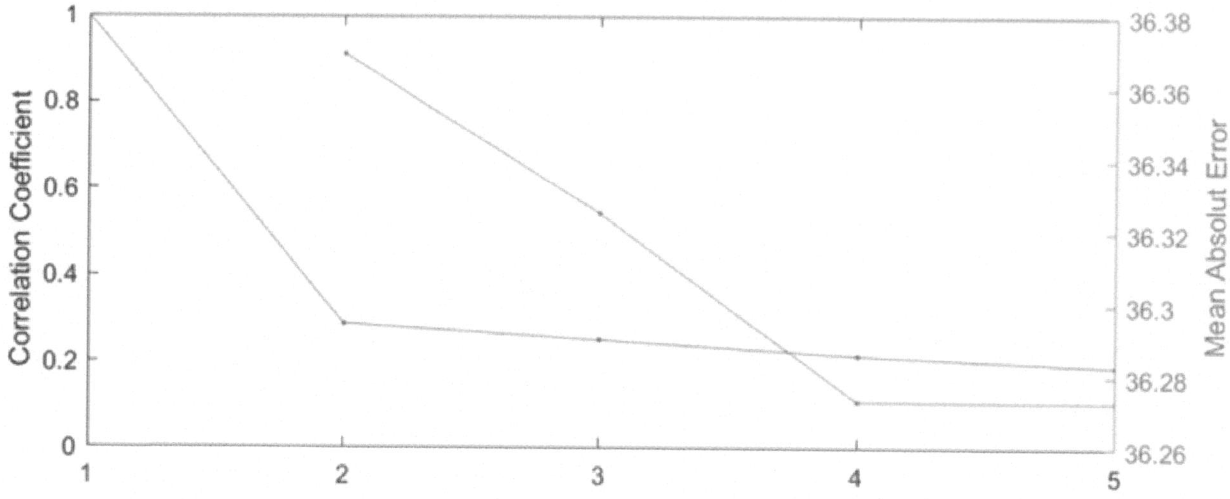

Figure 7 | The autocorrelation coefficient plot and the MAE for each time lag for the Quebec City case study.

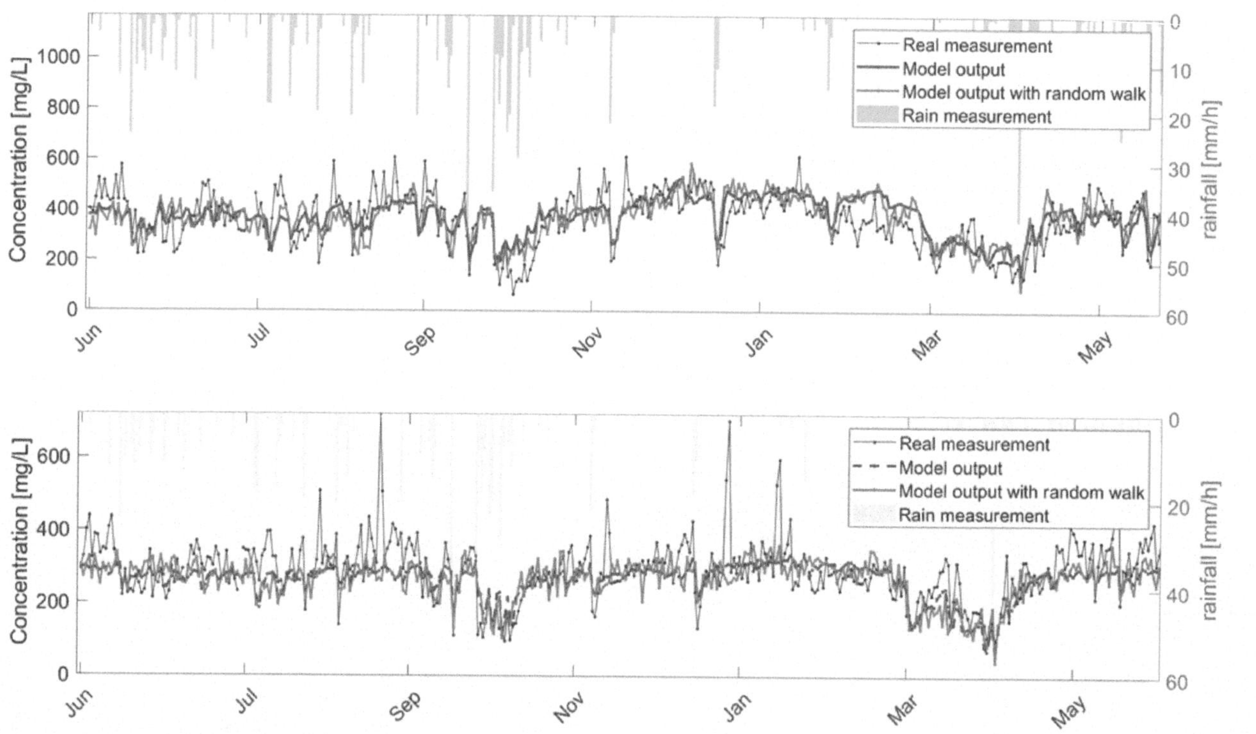

Figure 8 | COD (top) and TSS (bottom) concentrations generated for the Quebec City case study by the ANN with and without stochastic process extension.

3.1.5. Nutrient concentration generation by multivariate regression

For the reasons explained before, in practice, the influent nutrient concentrations (ammonia and phosphorus) in Quebec City's COD removing plant are measured only once or twice a week. This makes it difficult to estimate the current treatment result from the limited nutrient data or to create an influent time series in view of upgrading the plant for nutrient removal. Thanks to the high correlation between the nutrient concentrations and the flowrate, it was found that the concentrations could be generated quite well by multivariate regression, see Equation (2). Based on the same principles of regression modelling as used in section 3.1.4, regression orders between 1 and 5 were tested and the best order of the regression was found to be 3, giving the best performance (RMSE) on both training set and validation set.

Figure 9 shows the model performance for the test set, demonstrating that the regression is able to adequately describe the ammonia concentrations on the basis of only flowrate and weather data.

In the same way, Figure 10 presents the generated phosphorus concentration time series. Thus, the multivariate model enables generating a high-frequency time series (daily simulation) from a low frequency measurement time series (weekly measurement).

3.2. Performance analysis and discussion

The performance of the final data-driven influent generation is summarized in Table 1. The criteria are all calculated based on the test sets. Table 1 demonstrates that the model is able to generate time series with an error of around 10% for flowrate and 13–20% for water quality. The NSE is around 0.5–0.7 indicating that the model can match the observed dataset well. For water quality, the model is able to provide a good generation, too. However, it is worth remembering that the observed data consist of imperfect lab and sensor measurements and that these errors thus indirectly influence the quality of the water quality generation. Considering the high uncertainty and measurement errors of raw wastewater data (Montgomery & Sanders 1986; Bertrand-Krajewski *et al.* 2007), the RMSE and MAPE obtained can be considered to be in the same order of magnitude as real-life measurements. These results thus indicate that the model performance is sufficiently good.

Figure 11 compares the observed flow data with the model generated data in different ways. First the quantile–quantile (q–q) plot, see Figure 11(a), shows the good correspondence of the generated data with the observation data (with coefficient of determination $r^2 = 0.88$). The CDF in Figure 11(b) compares the statistical characteristics of the observed and generated dataset. The time series was also split into two subsets, one for winter (from January to May), and one for summer (from July to December) in order to demonstrate that the model's performance is not different for different seasons, see Figure 11(c)

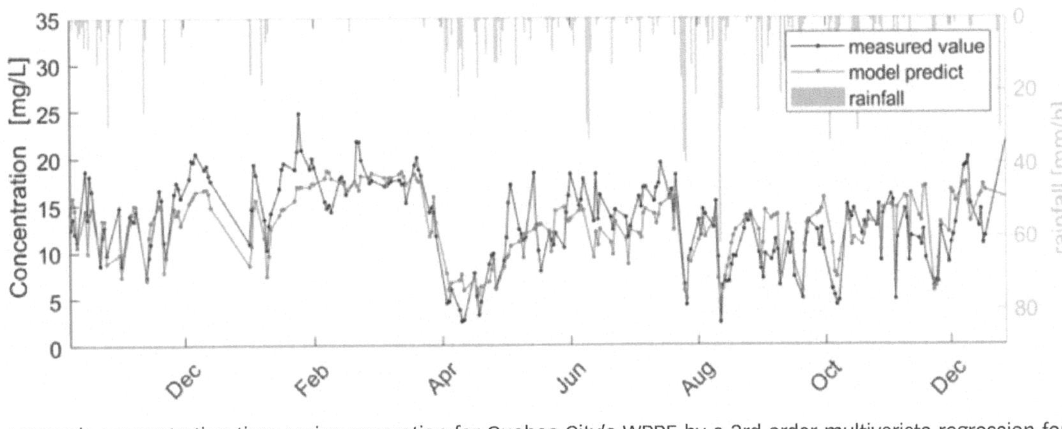

Figure 9 | Ammonia concentration time series generation for Quebec City's WRRF by a 3rd order multivariate regression for the test set.

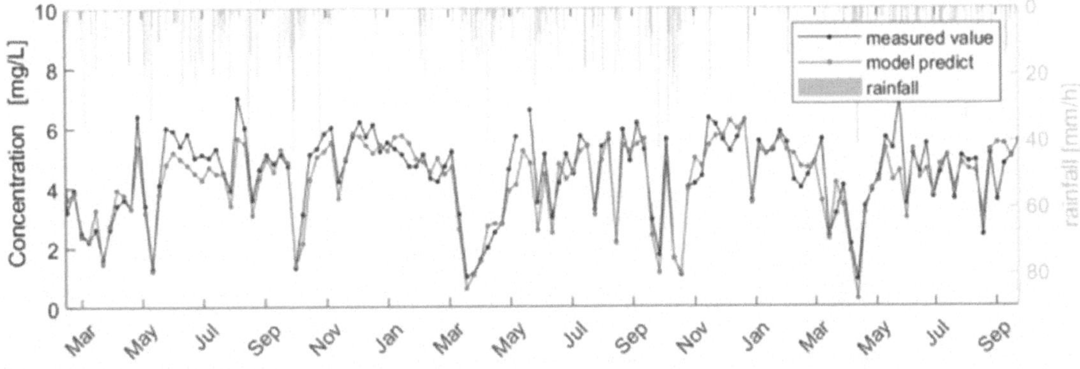

Figure 10 | Phosphorus concentration time series generation for Quebec City's WRRF, by a 3rd order multivariate regression for the test set.

Table 1 | Model performance for test set, evaluated by MAPE, RMSE and NSE

Variable	Average	RMSE	MAPE	NSE
Case study for Quebec City: daily data				
Flowrate	200,000 m^3/d	38,000	11.8%	0.74
COD	400 mg/L	52	19.8%	0.43
TSS	250 mg/L	30	16.7%	0.59
Ammonia	12.5 mg/L	2.6	13.7%	0.68
Phosphorus	4.3 mg/L	0.8	13.4%	0.71
Case study for Bordeaux: hourly data				
Flowrate	12,000 m^3/h	1,200	13.5%	0.61
TSS	300 mg/L	42	17.5%	0.70

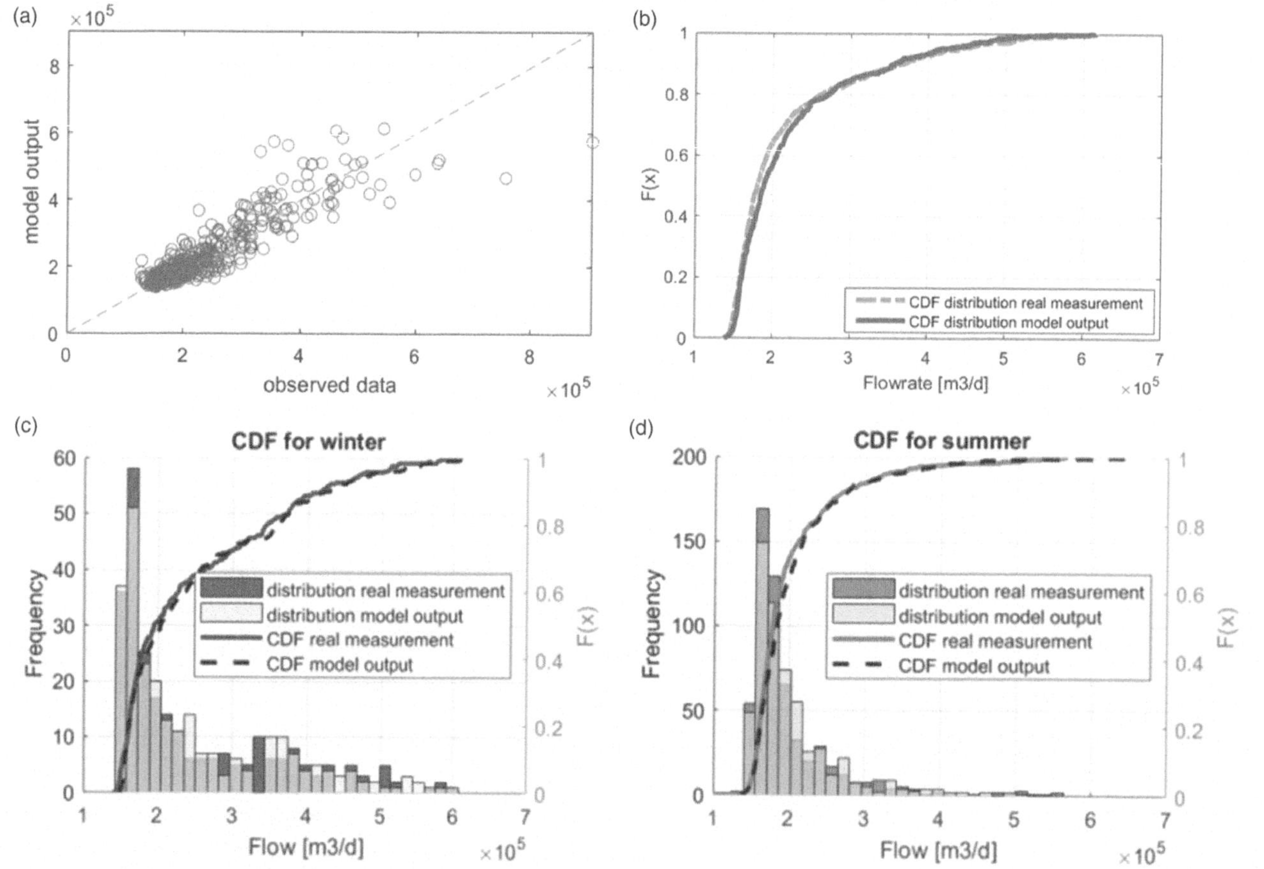

Figure 11 | Model performance analysis for flow: (a) quantile–quantile plot for observed data and model output, (b) CDF for the complete test set. (c) PDF and CDF for flow data in winter and (d) in summer.

and 11(d). Although the winter and summer flow distributions are different, i.e. the winter flow distribution is more dispersed than the summer one: in winter 85% of the flow rate is below $3.8*10^5\ m^3/d$ while in summer 85% of flow is below $2.4*10^5\ m^3/d$, both the seasonal PDF and CDF analysis show that the distribution of the model outputs is similar to the real data distribution.

Figure 12 illustrates the COD concentration generation by the ANN model without (a) and with stochastic process extension (b). It can be concluded that the ANN model with the stochastic process can better mimic the observed variability. The

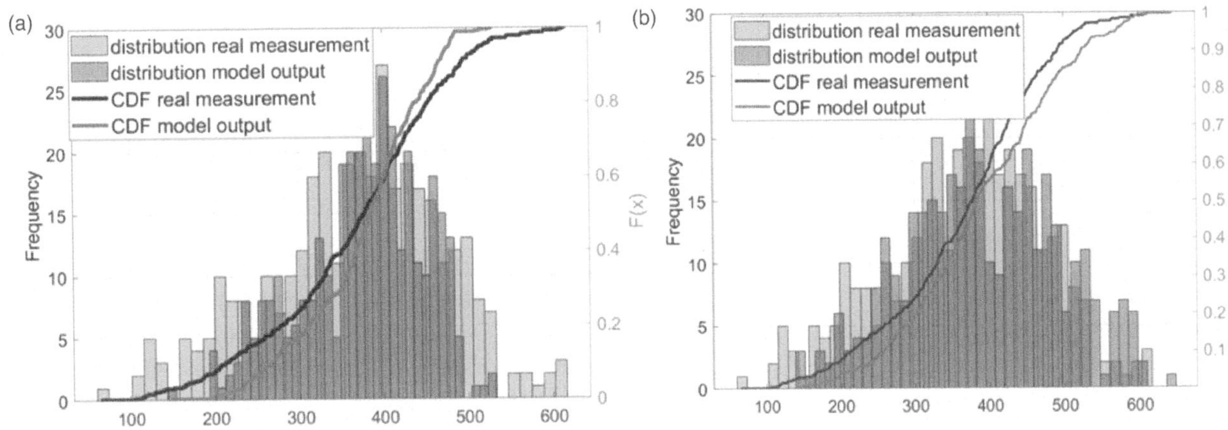

Figure 12 | PDF and CDF of the COD concentrations for the Quebec City test set without stochastic process (a) and with stochastic process (b).

stochastic process enables the model output distribution to be wider and closer to the distribution of the real measurement time series.

This improvement is confirmed by the KL divergence values in Table 2. A smaller KL value represents a better similarity between two distributions: i.e. the KL divergence value with random walk is smaller than the KL value for the ANN model output without the extension, which indicates it is closer to the distribution of the observations.

3.3. Discussion and evaluation

In general, the error analysis (Table 1) shows the high precision of the IG model outputs. Figure 11 demonstrates a good match between the model results and the measured data for the test set, which means the model can successfully generate flowrate and pollutant concentration time series. However, some discrepancies can be noticed for the hourly generation occurring at the beginning of wet weather flows (Figure 5). This might be caused by overflows generated in the sewer system or depression storage effects, which are not included in the model. On the other hand, high-frequency water quality time series generation is based on sensor measurements, and it must be recognised that sensor clogging and anormalies often occur, especially during rain events (Bertrand-Krajewski *et al.* 2007).

The snowmelt period in the generated data is also shifted in time for some of the years that were studied (Figures 4 and 5). This timing error is due to the model considering snowmelt water infiltration as an average amount and at an average time in the year. This issue may be dealt with by modifying the yearly pattern for different years, using for instance temperature data.

The proposed IG model provides a tool for fast influent profile generation at different resolutions and it benefits not only the carbon removal but also the nutrient removal process. Moreover, instead of being a deterministic model, the proposed IG creates a stochastic process, which may be helpful for further WRRF modelling uncertainty and scenario analysis. Compared with available phenomenological models presented in the literature review, the advantage of using the proposed data-driven model is that there is no need to gather any information on the catchment, nor does it need calibration of physical parameters. Therefore, It is suitable for cachtments with unknown physical details.

Finally, to confirm its utility and advantages, the proposed model was compared with a very simple data-driven method: the generated TSS concentration was calculated simply as the average TSS load (calculated for the whole available data series), divided by measured flowrate data. This reflects the fact that a city typically generates a stable daily pollution load. The results

Table 2 | KL divergence calculation for COD and TSS concentrations generated with and without stochastic process extension for the Quebec City test set

KL divergence	COD	TSS
ANN model (a)	0.573	0.204
ANN model with stochastic process extension (b)	0.115	0.058

of Appendix 1. show that the obtained ANN model is considerably more precise, using the same input and the same size of training set.

This simple data-driven model faces the same problem as any ANN model, i.e. it is risky to extrapolate if the new time series is too much different from the training set. To counter this limitation, a more complete database is required to make sure the model has been exposed to 'more experiences' and can thus capture more of the system properties, such as more diverse rain data, wider and higher temporal distribution of water quality measurements, etc.

4. CONCLUSION

This work developed an urban wastewater influent generator model for flowrate and pollutant concentration generation. The proposed IG model is data-driven and includes an ANN, a multivariate regression and a random walk process. Its performance was analyzed by different criteria: MAPE, NSE and statistical characteristics evaluating variability (KL_divergency). The IG model showed a high generation precision for the two case studies: 10% error for flowrate and 15–20% for the concentrations. Given the fact that the performance of a data-driven model depends on the quality and coverage of the training data availability, a more complete dataset (in temporal or spatial sense) can help further improve the model performance.

The proposed IG is balanced in terms of modelling efforts and precision. Compared with a simple data-driven IG model based on average pollutant load and flow (result of Appendix 1), the proposed IG demonstrated a higher accuracy. On the other hand, compared with phenomenological IG models, this model requires less modelling efforts, especially in calibrating the parameters for which prior expert knowledge is needed. Another significant improvement of the proposed IG is that stochastic variability is included in the model, which enables the generated time series to better represent the reality in terms of probability distribution of the generated data. Last but not least, the application of the proposed IG requires only routine WRRF data, thus not requiring any additional investments in data collection. Once the IG model is calibrated, the generation of new time series only requires weather data and is not relying on historical influent data. This ensures the prediction result will be stable over a long time horizon and will not accumulate errors over time.

In conclusion, the proposed IG generates a dynamic and complete data set (flowrate and pollutant concentrations, both organics and nutrients) with good performance, and it is able to generate a time series at different time resolutions (daily and hourly). Further research will be focusing on the optimization of the IG in order to better support WRRF modelling studies.

ACKNOWLEDGEMENTS

Sincere thanks are extended to SUEZ SGAC and NSERC (Natural Sciences and Engineering Research Council of Canada) for funding the research project. Québec City and Bordeaux (SUEZ Le Lyre) are thanked for its contribution with all relevant data. Peter Vanrolleghem holds the Canada Research Chair on Water Quality Modelling.

DATA AVAILABILITY STATEMENT

All relevant data are included in the paper or its Supplementary Information.

REFERENCES

Ahnert, M., Marx, C., Krebs, P. & Kuehn, V. 2016 A black-box model for generation of site-specific WWTP influent quality data based on plant routine data. *Water Sci. Technol.* **74**, 2978–2986.

Alferes, J. & Vanrolleghem, P. A. 2016 Efficient automated quality assessment: dealing with faulty on-line water quality sensors. *AI Commun.* **29**, 701–709.

Banihabib, M., Bandari, R. & Peralta, R. 2019 Auto-regressive neural-network models for long lead-time forecasting of daily flow. *Water Resour. Manage.* **33**, 159–172.

Bechmann, H., Nielsen, M. K., Madsen, H. & Kjølstad Poulsen, N. 1999 Grey-box modelling of pollutant loads from a sewer system. *Urban Water* **1**, 71–78.

Benedetti, L., Langeveld, J., Comeau, A., Corominas, L., Daigger, G., Martin, C., Mikkelsen, P. S., Vezzaro, L., Weijers, S. & Vanrolleghem, P. A. 2013 Modelling and monitoring of integrated urban wastewater systems: review on status and perspectives. *Water Sci. Technol.* **68**, 1203–1215.

Bertrand-Krajewski, J. L., Winkler, S., Saracevic, E., Torres, A. & Schaar, H. 2007 Comparison of and uncertainties in raw sewage COD measurements by laboratory techniques and field UV-visible spectrometry. *Water Sci. Technol.* **56**, 17–25.

Butler, D., Friedler, E. & Gatt, K. 1995 Characterising the quantity and quality of domestic wastewater inflows. *Water Sci. Technol.* **31**, 13–24.

El-Din, A. G. & Smith, D. W. 2002 A neural network model to predict the wastewater inflow incorporating rainfall events. *Water Res.* **36**, 1115–1126.

Flores-Alsina, X., Saagi, R., Lindblom, E., Thirsing, C., Thornberg, D., Gernaey, K. V. & Jeppsson, U. 2014 Calibration and validation of a phenomenological influent pollutant disturbance scenario generator using full-scale data. *Water Res.* **51**, 172–185.

Fu, G., Makropoulos, C. & Butler, D. 2010 Simulation of urban wastewater systems using artificial neural networks: embedding urban areas in integrated catchment modelling. *J. Hydroinf.* **12**, 140–149.

Gernaey, K. V., Flores-Alsina, X., Rosen, C., Benedetti, L. & Jeppsson, U. 2011 Dynamic influent pollutant disturbance scenario generation using a phenomenological modelling approach. *Environ. Modell. Software* **26**, 1255–1267.

Hauduc, H., Neumann, M. B., Muschalla, D., Gamerith, V., Gillot, S. & Vanrolleghem, P. A. 2015 Efficiency criteria for environmental model quality assessment: areview and its application to wastewater treatment. *Environ. Modell. Software* **68**, 196–204.

Heideman, M., Johnson, D. & Burrus, C. S. 1984 Gauss and the history of the Fast Fourier Transform. *IEEE Signal Process Mag.* **1**, 14–21.

Kullback, S. & Leibler, R. A. 1951 On information and sufficiency. *Ann. Math. Stat. TA – TT* **22**, 79–86.

Langergraber, G., Alex, J., Weissenbacher, N., Woerner, D., Ahnert, M., Frehmann, T., Halft, N., Hobus, I., Plattes, M., Spering, V. & Winkler, S. 2008 Generation of diurnal variation for influent data for dynamic simulation. *Water Sci. Technol.* **57**, 1483–1486.

Langeveld, J., Van Daal, P., Schilperoort, R., Nopens, I., Flameling, T. & Weijers, S. 2017 Empirical sewer water quality model for generating influent data for WWTP modelling. *Water* **9**, 1–18.

Ledergerber, J. M., Maruéjouls, T. & Vanrolleghem, P. A. 2020 No-regret selection of effective control handles for integrated urban wastewater systems management under parameter and input uncertainty. *Water Sci. Technol.* **81**, 1749–1756.

Levenberg, K. 1944 A method for the solution of certain non-linear problems in least squares. *Q. Appl. Math.* **2**, 164–168.

Maier, H. R. & Dandy, G. C. 2000 Neural networks for the prediction and forecasting of water resources variables: a review of modelling issues and applications. *Environ. Modell. Software* **15**, 101–124.

Mannina, G., Cosenza, A., Vanrolleghem, P. A. & Viviani, G. 2011 A practical protocol for calibration of nutrient removal wastewater treatment models. *J. Hydroinf.* **13**, 575–595.

Martin, C. & Vanrolleghem, P. A. 2014 Analysing, completing, and generating influent data for WWTP modelling: a critical review. *Environ. Modell. Software* **60**, 188–201.

Montgomery, R. H. & Sanders, T. G. 1986 Uncertainty in water quality data. In: *Statistical Aspects of Water Quality Monitoring* (El-Shaarawi, A. H. & Kwiatkowski, R. E. B. T.-D., eds). Elsevier, Amsterdam, The Netherlands, pp. 17–29.

Petersen, B., Gernaey, K., Henze, M. & Vanrolleghem, P. A. 2002 Evaluation of an ASM1 model calibration procedure on a municipal-industrial wastewater treatment plant. *J. Hydroinf.* **4**, 15–38.

Price, R. K. & Vojinovic, Z. 2011 *Urban Hydroinformatics: Data, Models and Decision Support for Integrated Urban Water Management.* IWA Publishing, London, UK.

Rajurkar, M. P., Kothyari, U. C. & Chaube, U. C. 2004 Modeling of the daily rainfall-runoff relationship with artificial neural network. *J. Hydrol.* **285**, 96–113.

Schlichthärle, D. 2011 *Analog Filters – Digital Filters: Basics and Design.* Springer Berlin Heidelberg, Berlin, Heidelberg.

Sweeney, M. & Kabouris, J. 2013 Modeling, instrumentation, automation, and optimization of wastewater treatment facilities. *WATER Environ. Res.* **85**, 1322–1338.

Talebizadeh, M. 2015 *Probabilistic Design of Wastewater Treatment Plants.* PhD Thesis, génie des eaux, Universite Laval, Quebec, QC, Canada.

Talebizadeh, M., Belia, E. & Vanrolleghem, P. A. 2016 Influent generator for probabilistic modeling of nutrient removal wastewater treatment plants. *Environ. Modell. Software* **77**, 32–49.

Tibshirani, R. 1996 Regression shrinkage and selection via the Lasso. *J. R. Stat. Soc. Ser. B* **58**, 267–288.

Tik, S. & Vanrolleghem, P. A. 2017 Chemically enhancing primary clarifiers: model-based development of a dosing controller and full-scale implementation. *Water Sci. Technol.* **75**, 1185–1193.

Troutman, S. C., Schambach, N., Love, N. G. & Kerkez, B. 2017 An automated toolchain for the data-driven and dynamical modeling of combined sewer systems. *Water Res.* **126**, 88–100.

Villez, K., Del Giudice, D., Neumann, M. B. & Rieckermann, J. 2020 Accounting for erroneous model structures in biokinetic process models. *Reliab. Eng. Syst. Saf.* **203**, 107075.

Wright, L. T., Dent, S., Mosley, C., Kadota, P. & Djebbar, Y. 2001 Comparing rainfall dependent inflow and infiltration simulation methods. *J. Water Manage. Modell.* **9**, 235–257.

Zhu, J. & Anderson, P. R. 2016 Assessment of a soft sensor approach for determining influent conditions at the MWRDGC calumet WRP. *J. Environ. Eng.* **142**, 04016023.

First received 8 November 2021; accepted in revised form 7 March 2022. Available online 18 March 2022

doi: 10.2166/wst.2022.096

Going beyond conventional wastewater treatment plants within circular bioeconomy concept – a sustainability assessment study

Hadis Marami[a], Panagiotis Tsapekos[b], Benyamin Khoshnevisan [c], Jeanette Agertved Madsen [d], Jacob Kragh Andersen[d,*], Shahin Rafiee[a] and Irini Angelidaki [b]

[a] Department of Mechanical Engineering of Agricultural Machinery, Faculty of Agricultural Engineering and Technology, College of Agriculture and Natural Resources, University of Tehran, Tehran, Iran
[b] Department of Chemical and Biochemical Engineering, Technical University of Denmark, Kgs Lyngby DK-2800, Denmark
[c] Department of Chemical Engineering, Biotechnology and Environmental Technology, University of Southern Denmark, Odense, Denmark
[d] EnviDan A/S, Fuglebækvej 1A, Kastrup 2770, Denmark
*Corresponding author. E-mail: JKA@envidan.dk

BK, 0000-0003-0236-5970

ABSTRACT

Wastewater treatment plants (WWTP) have extensive energy processes that undermine their economic and environmental performance. In this context, the integration of wastewater treatment with other biochemical processes such as co-digestion of sludge with organic wastes, and production of value-added products at their downstream processes will shift conventional WWTPs into biorefinery platforms with better sustainability performance. The sustainability of such a biorefinery platform has been investigated herein using an economic and life cycle assessment approach. This WWTP-based biorefinery treats wastewater from Copenhagen municipality, co-digests the source-sorted organic fraction of municipal solid waste and sludge, and upgrades biogas into biomethane using a hydrogen-assisted upgrading method. Apart from bioenergy, this biorefinery also produces microbial protein (MP) using recovered nutrients from WWTP's reject water. The net environmental savings achieved in two damage categories, i.e., -1.07×10^{-2} species.yr/FU in ecosystem quality and -1.68×10^{6} USD/FU in resource scarcity damage categories along with high potential windows for the further environmental profile improvements make this biorefinery platform so encouraging. Despite being promising in terms of environmental performance, the high capital expenditure and low gross profit have undermined the economic performance of the proposed biorefinery. Technological improvements, process optimization, and encouraging incentives/subsidies are still needed to make this platform economically feasible.

Key words: biological biogas upgrading, biorefinery, circular bioeconomy, microbial protein, nutrient recovery, wastewater treatment

HIGHLIGHTS

- A biorefinery for microbial protein and bioenergy production has been developed.
- Wastewater treatment and anaerobic digestion of sludge incorporated in biorefinery.
- Microbial production had lower environmental profile than soybean meal.
- High CAPEX undermined the economic feasibility of MP production.

GRAPHICAL ABSTRACT

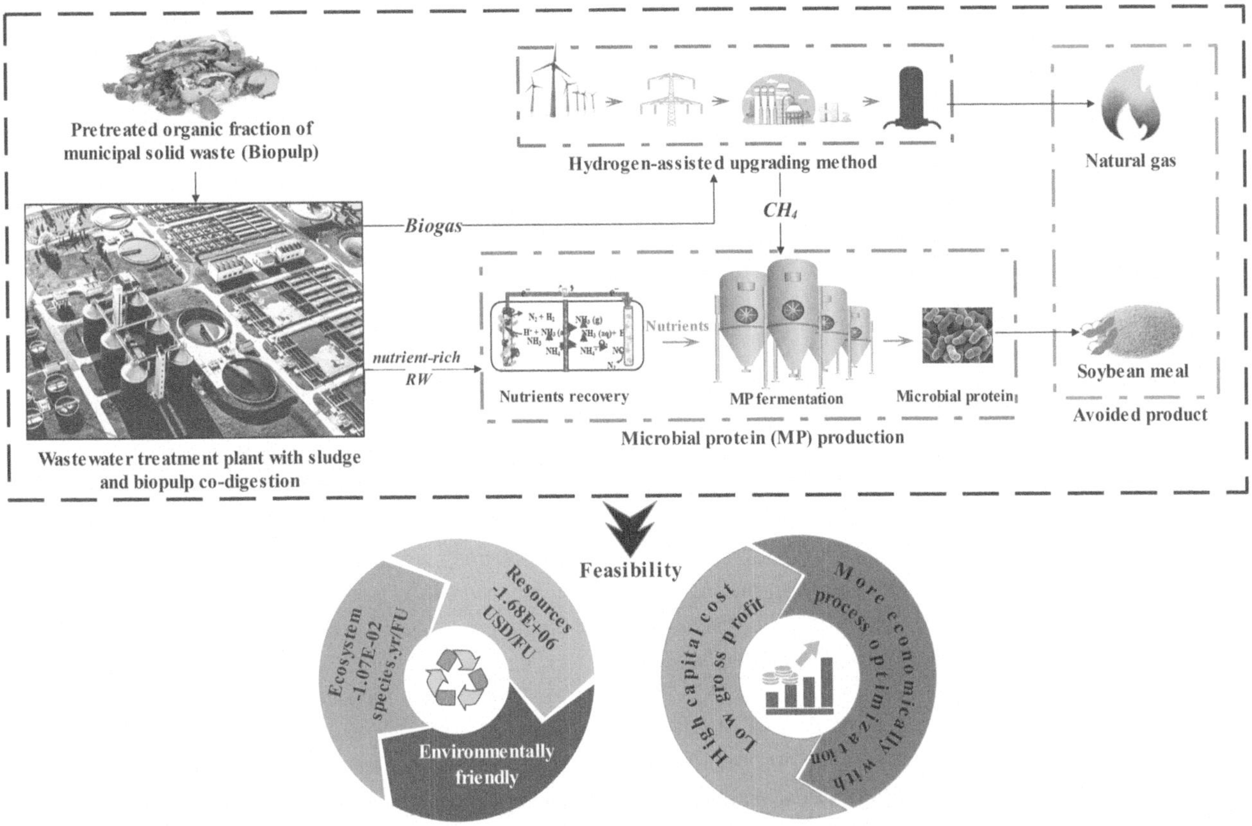

ABBREVIATIONS

LCA life cycle assessment
LCI life cycle inventory
CLCA consequential life cycle assessment
FU functional unit
WWTP wastewater treatment plant
RW reject water
NH_4-N ammonium nitrogen
TAN total ammonium nitrogen
MOB methane-oxidizing bacteria
AD anaerobic digestion
CSTR continuously stirred tank reactor
HRT hydraulic retention time
MP microbial protein
dAMS diluted ammonium medium salt
N nitrogen
P phosphorus
BES bio-electrochemical system
MFC microbial fuel cells
CH_4 methane
CO_2 carbon dioxide
N_2O nitrous oxide
O_2 oxygen
DALY disability-adjusted life year
USD United States dollar
LHV low heat value

NPV	net present value
ROI	return on investment
M€	million €
IRR	internal rate of return
CAPEX	capital expenditure
OPEX	operational expenditure
FOPEX	fixed operational expenditure
VOPEX	variable operational expenditure
GWP	global warming potentials

1. INTRODUCTION

Extensive energy demand (i.e., electricity and heat) is an inevitable challenge that wastewater treatment plants (WWTPs) are facing (Tian *et al.* 2020). Energy from external sources is supplied to meet WWTPs' demand, which brings about serious challenges among them undermining environmental and economic performance (Gu *et al.* 2017). Most of the large WWTPs worldwide are now supplemented with anaerobic digesters to supply their internal energy demand by biogas production via anaerobic digestion (AD) of sewage sludge. However, AD cannot always compensate for the high energy demand in WWTPs. In some countries such as Denmark, utility companies are front runners and have worked intensely to decrease energy consumption optimize overall processes. Their attempts have resulted in a positive energy balance (150% or more) (WATER VISION DENMARK 2021; Energy & Resources 2022). However, to produce even more energy and contribute to the future circular bioeconomy, the AD of other organic wastes from external sources is in focus. Despite the great potential of integrating the wastewater treatment process and AD to enhance the energy flow within WWTPs, the low biodegradability of sludge results in lower energy recovery compared with agro-industrial biogas plants fed with biowaste with high biogas yield. For instance, the organic fraction of municipal solid waste (OFMSW) has a biochemical methane yield of 474 $mLCH_4/gVS$ (Hanum *et al.* 2019; Zhang & Li 2019) while that of sludge is 376 $mLCH_4/gVS$ (Khoshnevisan *et al.* 2018a; Zha *et al.* 2020). In this context, the anaerobic co-digestion of sludge and OFMSW is an attractive option to deal with the above-mentioned problems (Zha *et al.* 2020). The co-digestion of sludge and OFMSW has shown synergistic effects leading to increased biomethane potential of the feedstocks (Zha *et al.* 2020). It is worth mentioning that there are considerable debates in the literature on full-scale anaerobic co-digestion, the relative importance of performance conditions, cost-effectiveness, and technical requirements (Cavinato *et al.* 2013; Wehner *et al.* 2021). Regarding energy recovery, environmental, and financial benefits, full-scale anaerobic co-digestion had very encouraging results (Nghiem *et al.* 2017), even more than doubling energy recovery (Koch *et al.* 2016). Most of the published literature reached similar results when other wastes were co-digested with sludge in a full-scale application (Park *et al.* 2011; Di Maria *et al.* 2015; Wong *et al.* 2018).

While the co-digestion of sludge and OFMSW increases the energy flows within WWTPs, the reject water (RW), i.e., the liquid fraction of digestate also contains considerable amounts of nutrients, more specifically nitrogen (N) and phosphorous (P), which can be further valorized into value-added products (Tsapekos *et al.* 2021). Overall, the nutrients in the RW are supplied from two sources: wastewater and biopulp. The source of N and P in wastewater is greywater (i.e., any water that is consumed at home, including washing, bathing, and kitchen, except toilet water), urine, and faeces (Vinnerås *et al.* 2006; Morales Pereira 2014). Also, biopulp is mainly composed of source-separated organic household waste, green waste, superstore waste, and leftover food from restaurants and cafeterias (Alvarado-Morales *et al.* 2021), which make up the rest of the nutrients in the RW. The RW after the dewatering process reportedly contains a high amount of recoverable ammonium nitrogen, i.e., ~15–20% of the input flow (Horan 1994; Gustavsson *et al.* 2010), the removal of which plays a key role in making WWTPs environmentally friendly (Guo *et al.* 2010). The co-digestion of sludge with OFMSW increases the nutrient content of RW compared to that of conventional WWTPs as OFMSW is rich in nutrients (Khoshnevisan *et al.* 2018b). Therefore, the valorization of nutrient-rich RW into value-added products can contribute to circular bioeconomy and play the same role as a non-waste-derived primary source without environmental damage. In this way we can apply the end-of-waste status to RW and cease to be waste if it meets the end-of-waste criteria in accordance with the Waste Framework Directive (Delgado *et al.* 2009). Achieving end-of-waste criteria for waste materials can provide potential environmental and economic benefits (Zorpas 2016).

The valorization of N-rich effluents into microbial protein (MP), i.e., a protein-rich biomass that can be used as supplementary animal feed, has provoked considerable attention (Matassa *et al.* 2016b; Tsapekos *et al.* 2019; Khoshnevisan *et al.*

2020a). The growth of the world's population and increased living standards have raised concerns over food insecurity and an imbalanced diet. To meet the high demand for stable food sources, proteinaceous food from vegetal and animal sources has expanded rapidly over the past decades, leading to significant environmental damages (Herrero *et al.* 2016; Khoshnevisan *et al.* 2021). The environmental degradation caused by feed production has undermined the sustainability of animal protein sources (Du *et al.* 2018; Uwizeye *et al.* 2020). In this context, the production of land-independent MP, i.e., microorganisms which are rich in protein and amino acid (i.e., 10% to 80% based on culture media, growth conditions, and microorganism type (Jones *et al.* 2020)), can substitute conventional proteinaceous feed with high environmental impacts such as soybean meal and fishmeal (Matassa *et al.* 2016a; Tsapekos *et al.* 2019).

Methane-oxidizing bacteria (MOB), a bacterial-based MP, have shown great potential to assimilate ammonium-nitrogen (NH_4-N) using methane as their energy/carbon source (Strong *et al.* 2015; Khoshnevisan *et al.* 2019). Initial attempts have led to the commercialization of 1st generation MOB-based MP such as UniProtein and FeedKind, which are used as a protein feed supplement for livestock and aquaculture. Unibio® and Calysta® companies use synthetic nitrogen and natural gas as their N and C sources, respectively, which do not seem to be long-term sustainable solutions (Sillman *et al.* 2020; Verbeeck *et al.* 2020). This challenge has directed investigators towards the production of 2nd generation MP by using waste effluents as N source and biomethane as C source (Matassa *et al.* 2015a, 2015b). Previous studies have demonstrated that the upcycling of N-rich waste effluent to high-quality MP would decrease the induced environmental impacts of conventional proteinaceous feed such as fishmeal and soybean meal (Khoshnevisan *et al.* 2020b; Matassa *et al.* 2020). Furthermore, biogas or biomethane (biologically upgraded biogas) have been shown as promising alternatives to natural gas for the production of MOB-based MP (Khoshnevisan *et al.* 2019; Acosta *et al.* 2020).

The previous studies have provided valuable insights on the possibility of valorizing different types of waste streams into MP such as municipal solid waste (Khoshnevisan *et al.* 2020b), pig manure (Acosta *et al.* 2020; Verbeeck *et al.* 2021), urine (Khoshnevisan *et al.* 2020a), household waste (Tsapekos *et al.* 2020), and sewage sludge (Zha *et al.* 2021). The literature review showed that the potential of WWTP-based biorefinery platforms for MP and bioenergy production by the treatment of municipal wastewater and OFMSW has not been scrutinized yet. A recent study was conducted by Marami *et al.* (2022) to investigate various pathways for nutrients extraction from wastewater and deploying methane in different forms (i.e., biogas and upgraded biomethane) for MP production. They found that the lack of sufficient nutrients in wastewater would hinder the sustainability of integrated WWTPs and MP production facilities. Accordingly, the main objective of this study was to investigate the environmental and economic performance of a biorefinery platform that produces MP and bioenergy. This platform is established on a WWTP to valorize N and carbon flows from wastewater treatment and anaerobic co-digestion of sludge as well as OFMSW. To do so, a real WWTP in Copenhagen, Denmark, was selected as our case study to model N, C, and energy flows within the WWTP and implement life cycle assessment (LCA), economic assessment (net present value (NPV), gross profit, return on investment (ROI), and internal rate of return (IRR)) to evaluate the sustainability of the proposed biorefinery.

2. MATERIALS AND METHOD

2.1. LCA

In the following sub-sections, the detailed description of how LCA is to be implemented will be presented in terms of goal and scope of the study, life cycle inventory (LCI), life cycle impact assessment (LCIA), and interpretation of results. Guidelines and recommendations made by ISO 14040 (Finkbeiner *et al.* 2006) and International Reference Life Cycle Data System (ILCD) handbook (JRC 2010) were followed to correctly implement this LCA study.

2.1.1. Goal and scope definition

The main objective of this study was to assess the environmental impacts of a biorefinery platform established on a WWTP aimed at valorizing N and carbon flows into an animal grade proteinaceous feed supplement (known as microbial protein) and bioenergy. To do so, Avedøre municipal WWTP, which is one of the WWTP facilities managed by BIOFOS company (https://biofos.dk/), located in the south of Copenhagen (56.2639° N, 9.5018° E), Denmark, was selected as our case study (plant description is shown in SI-1). Hence, the geographical scope of this study was Denmark and the temporal scope was 2030. The production of 2nd generation MP using N- and carbon-rich waste streams relies on various technologies some of which are still at pilot scale demonstrations and have not been commercialized yet. Hence, scaled-up MP production and the penetration of the final product into the animal feed market as a substitution for conventional protein feeds such as soybean meal need a minimum time frame of 10 years. Because the mass-production of MP and its use as supplementary animal feed may affect the current animal feed market and will have consequences outside the system boundary of this study (it will be later discussed in section 2.1.2.6), consequential LCA was applied herein to include the above-mentioned impacts within the system boundary opted for this study. The multifunctionality problem was dealt with system expansion with substitution as recommended by ISO guidelines (Finkbeiner *et al.* 2006).

A cradle to gate system boundary was used covering all the processes within the WWTP, nutrient recovery, and MP production facility as well as electrolytic hydrogen production using off-peak surplus electricity needed for biological biogas upgrading. The system boundary of this study along with the mass and energy balances is shown in Figure 1 (see details in section 2.1.2). Because this biorefinery is proposed on a real WWTP as our case study, the inlet raw sewage stream to the Avedøre WWTP in 2030, i.e., 27.3 million m^3/year, was used as the functional unit (FU). However, a second FU, i.e., 1 metric ton of finished pure protein, was also opted later to allow comparing the environmental impacts of 2nd generation MP with other proteinaceous feed sources.

2.1.2. LCI

The data used for LCA modeling were collected from various sources: (1) data needed to perform material and energy flows within WWTP were obtained from Avedøre WWTP as well as HCS company supplying biopulp (i.e., source sorted OFMSW pretreated by biopulp technology, more details in the supplementary file 'SI-2'); (2) data related to MP production and biological biogas upgrading were collected from authors' earlier studies conducted in this context; (3) background data related to the production of auxiliaries, energy carriers, transportation, etc. were obtained from ecoinvent database (version V3.4); and finally (4) other required information that could not be gathered from the above-mentioned sources were obtained through literature review, technical reports, and commercialized protein suppliers. A summary of all inventory data is shown in Table 1 and detailed information is provided in the following sections.

2.1.2.1. Substrate. The RW from Avedøre WWTP (i.e., liquid fraction from dewatering process of digested sludge, Figure 1) and biopulp collected from HCS Company were considered as the primary N and P sources for MP production in this study. The water and electricity demands for the biopulping process, producing a biopulp with a TS of 12–16%, reportedly are 400 kg water/t biopulp and 26.6 kWh_{el}/t biopulp, respectively (Khoshnevisan *et al.* 2018b).

The WWTP model was created based on the annual average data and the projected data for 2030. The comprehensive characteristics of biopulp have been already detailed in the earlier study of Khoshnevisan *et al.* (2018b) and the key components are shown in Table 2. The amount of biopulp that is co-digested with sludge in AD reactors was determined based on the optimal volumetric ratio of sludge/biopulp, i.e., 60:40 (VS), which creates the synergistic effect and maximizes the biomethane potential (Zha *et al.* 2020). Previous studies have reported a methane yield of 428 $mLCH_4$/gVS, for mono-digestion of biopulp under continuous mode (Khoshnevisan *et al.* 2020b). The data from our case study demonstrated an average methane yield of 265 $mLCH_4/gVS_{sludge}$ under continuous mode. Zha *et al.* (2020) achieved a methane yield of up to 430 $mLCH_4$/gVS under the co-digestion of biopulp and sewage sludge at the above-mentioned volumetric ratio. The characteristics of various streams within Avedøre WWTP are shown in Table 2.

2.1.2.2. Nutrient recovery. N and methane are two key elements for the fermentation of methane oxidizing bacteria referred to as MP. However, other nutrients are also needed to prepare a balanced cultivation medium. P, after N, is the second important element used in the cultivation medium of methane-oxidizing bacteria. Accordingly, the recovery of both P and N – in the form of ammonium – from RW was targeted using a two-step bio-electrochemical system (BES). The successful nutrient recovery from AD effluents and RW via BES for further valorization into MP has been reported in the authors' earlier studies (Yang *et al.* 2021) and its sustainability has been reported by Marami *et al.* (2022). In BES, active microorganisms transfer ions between electrodes and stimulate denitrification at the cathode compartment (Arredondo *et al.* 2015). BES is available in two types: microbial fuel cell (MFC) and microbial electrolysis cell (MEC), both of them are working based on living microorganisms (Zou & He 2018). For NH_4-N and P recovery from the RW, MFC was considered due to its negative energy balance (Kuntke *et al.* 2018; Zou & He 2018). In other words, within MFCs electricity generation takes place simultaneously with the NH_4-N and P recovery (Kuntke *et al.* 2012; Rodríguez Arredondo *et al.* 2015). Having considered a two-stage NH_4-N recovery, 150 t NH_4-N per year with a concentration of 297 mg/l can be achieved (total recovery efficiency of 61%). P recovery was performed in a two-chamber-MFC as previously reported by Almatouq & Babatunde (2018). By the control of cell design and the operation condition such as

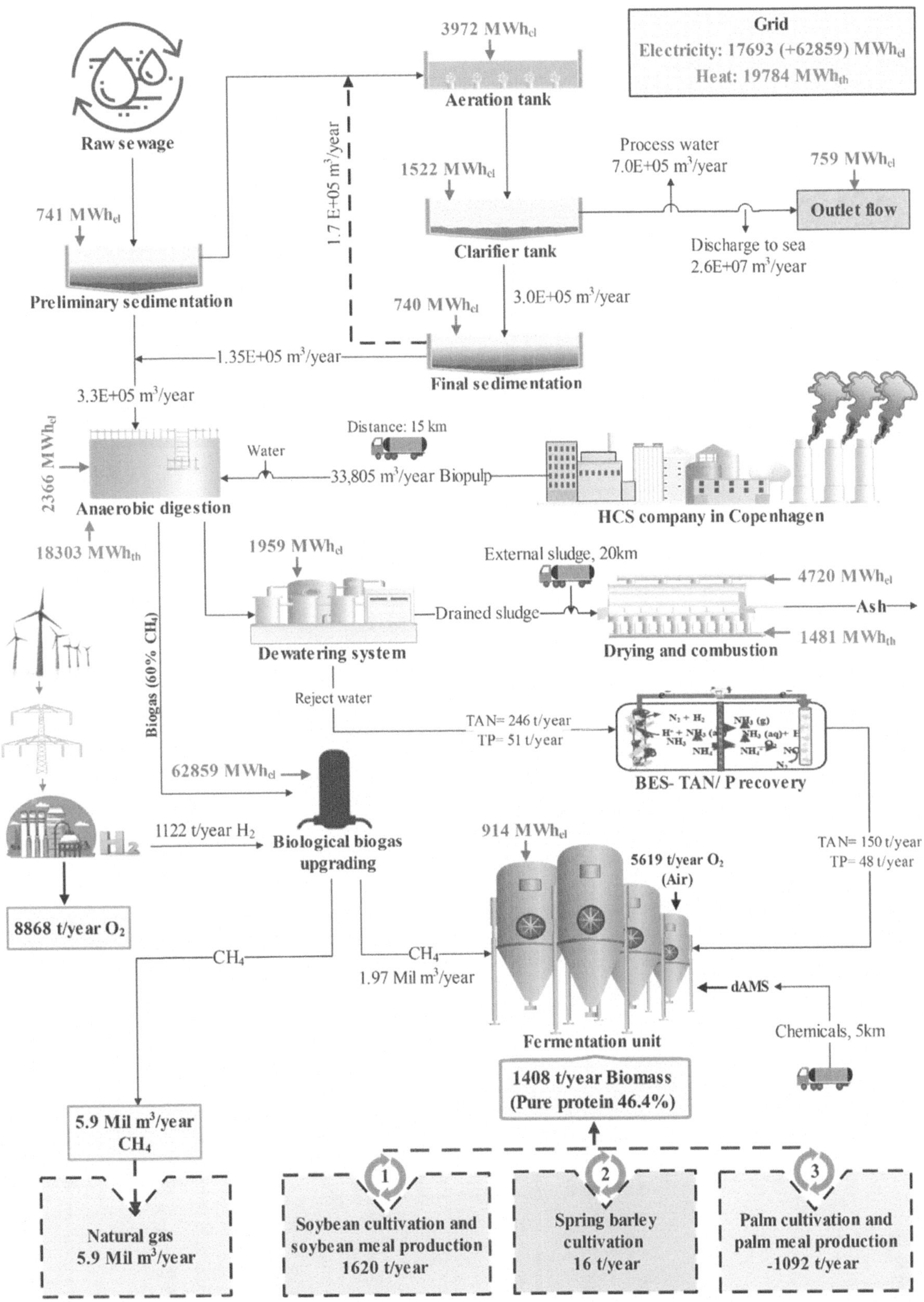

Figure 1 | Process flow diagram, mass flow, and energy flow through the WWTP. Marginal products/ energies and final products are shown with dashed boxes and yellow boxes, respectively. Numbers 1, 2, and 3 represent the soybean-palm oil-barley loops (see details in subsection 2.1.2.6).

Table 1 | Overview of life cycle inventory data

Unit process	Parameter	Amount	Unit	Ref.	Description
WWTP inventory data					
Preliminary sedimentation	Power consumption	741	MWh_{el}	Provided by WWTP	Energy intensity for pumping: 0.027 kWh/m^3 WW
	N_2O diffusion	139	kg	Sun *et al.* (2015); Bao *et al.* (2016); Hwang *et al.* (2016)	0.002–0.005 g N_2O /m^3 WW
	CH_4 diffusion	919	kg	Liu *et al.* (2014); Masuda *et al.* (2015)	0.034 g CH_4/m^3 WW
Aeration tank	Power consumption	3,972	MWh_{el}	Provided by WWTP	0.145 kWh/m^3 WW
	N_2O diffusion	164	t	Hwang *et al.* (2016); Nguyen *et al.* (2019)	1.5 g N_2O /m^3 WW
	CH_4 diffusion[a]	39	t	Liu *et al.* (2014); Bao *et al.* (2016); Nguyen *et al.* (2019)	1.425 g CH_4/m^3 WW
	CO_2 diffusion[a]	9,609	t	Bao *et al.* (2016); Nguyen *et al.* (2019)	350.82 g CO_2/m^3 WW
Clarifying tank	Power consumption	1,522	MWh_{el}	Provided by WWTP	Based on energy balance 2016
	N_2O diffusion	37	t	Kyung *et al.* (2015)	0.027 kg N_2O/kg TN
Pumping station	Power consumption	759	MWh_{el}	Provided by WWTP	0.028 kWh/m^3 WW
	N_2 diffusion	94	t	–	Emission to water based on mass balance 2016.
Final sedimentation	Power consumption	740	MWh_{el}	Provided by WWTP	Based on energy balance 2016.
	N_2O diffusion	21	kg	Sun *et al.* (2015); Bao *et al.* (2016); Hwang *et al.* (2016)	0.002–0.005 g N_2O /m^3 WW
	CH_4 diffusion	10	kg	Liu *et al.* (2014); Masuda *et al.* (2015)	0.034 g CH_4/m^3 WW
Anaerobic digestion	Power consumption	2,366	MWh_{el}	Pöschl *et al.* (2010)	3% of produced biogas
	Heat consumed	18,303	MWh_{th}	Based on energy balance 2016.	39% of MWh produced CH_4
	CH_4 diffusion[a]	91	t	Naroznova *et al.* (2016); Khoshnevisan *et al.* (2018b)	2.7% of the total CH_4 produced
	Water used	33,805	m^3	Khoshnevisan *et al.* (2020b)	Dilution of biopulp with a ratio of 1: 1
Sludge dewatering	Power consumption	1,959	MWh_{el}	Calculated	Energy intensity for pumping and centrifuge: 0.027 and 3.55 kWh/m^3
Drying-combustion	Power consumption	4,720	MWh_{el}	Provided by WWTP	Energy intensity for drying and combustion: 23.59 and 86.82 kWh/m^3
	Heat consumed	13,485	MWh_{th}	Provided by WWTP	Was supplied internally
Biological biogas upgrading inventory data					
Biological biogas upgrading	Inlet biogas	7.89	Mm^3	–	–
	Power consumption	237	MWh_{el}		10% consumed electricity in digester
Electrolysis	Water used	11,348	m^3	Kuckshinrichs *et al.* (2017)	10.11 kg water/kgH_2
	Power consumption	62,859	MWh_{el}	Ivy (2004)	56 kWh_{el}/kgH_2
MP production inventory data					
BES-P recovery	Na_2HPO_4 consumed	0.43	t	Xavier *et al.* (2014); Almatouq & Babatunde (2018)	Based on the stoichiometric relationships
	$Mg(OH)_2$ consumed	74	t	Xavier *et al.* (2014); Almatouq & Babatunde (2018)	Based on the stoichiometric relationships

(Continued.)

Table 1 | Continued

Unit process	Parameter	Amount	Unit	Ref.	Description
BES-TAN recovery	Na_2HPO_4 consumed	0.43	t	–	Calculated
Fermentation process	Power consumption	495	MWh_{el}	Calculated	Air blower: 0.025 kWh/m^3
	Chemical consumed	Table 3	t	Khoshnevisan *et al.* (2018b), Unibio®	–
	Biomethane consumed	1.97	Mm^3	Provided by DTU	CH_4/TAN: 9.4
Purification process	Power consumption	419	MWh_{el}	Calculated	GNLW554EP-VFD: 0.83 kWh/m^3
Protein substitution	Soybean meal	1,620	t	Calculated	Soybean meal loop (Section 2.1.2.6)
	Palm oil	−1,092	t	Calculated	Soybean meal loop (Section 2.1.2.6)
	Barley grain	16	t	Calculated	Soybean meal loop (Section 2.1.2.6)
Heat substitution	Biomass-wood chips/ wood pellet	7,617	MWh_{th}	Calculated	–
Oxygen substitution	Oxygen	8,868	t	Calculated	Based on stoichiometric relationships
Biomethane substitution	Natural gas	5.91	Mm^3	Calculated	–
Biopulp production inventory data					
Biopulp unit	Power consumption	851	MWh_{el}	Khoshnevisan *et al.* (2018b)	26.6 kWh/t biopulp
	Water used	12,846	m^3	Khoshnevisan *et al.* (2018b)	400 kg water/t biopulp
Transportation inventory data					
Transportation	Chemical materials	2,589	tkm/ year	Calculated	Distance: 5 km, it was supposed to be shipped from the markets of central Copenhagen.
	Drying-combustion unit	49,240	tkm/ year	Calculated	Distance: 20 km, it was supposed to be shipped from Lynetten plant[b].
	Biopulp	468,876	tkm/ year	Calculated	Distance: 15 km, it was supposed to be shipped from HCS company.

Mm^3, Million m^3.

[a]Biogenic.

[b]Denmark (55.696° N, 12.637° E).

Table 2 | Characteristics of the feedstock used in this study

Characteristics	Raw sewage	Inlet sludge to AD	Reject water	Biopulp	Unit
Flow	27,295,120	325,875	380,997	33,805	m^3/year
COD	16,205	10,485	247	8,381	t/year
BOD	6,441	–	267	–	t/year
SS	8,250	8,817	149	–	t/year
TN	1,300	315	38	158	t/year
TP	193	218	380,997	17	t/year

pH and temperature control, recovery efficiency can be promoted. Considering struvite as a form of recovered P, as shown in Equation (1), a molar ratio of 1: 1: 1 is acquired with ammonium (NH_4^+), phosphate (PO_4^{3-}), and magnesium (Mg^{2+}) for struvite settlement. In this regard, pumping magnesium hydroxide solution i.e., $Mg(OH_2)$, to the cathode compartment as a source of Mg^{2+} would lead to a 95% P precipitation efficiency (Xavier *et al.* 2014; Almatouq & Babatunde 2018). It is worth mentioning that in both NH_4-N and P recovery, Na_2HPO_4 was used for pH adjustment in the cathode chamber. It should be pointed out that the nutrients needed for the growth of active microorganisms in the anodic chamber were supplied by the trace elements available in biopulp.

$$NH_4^+ + Mg^{2+} + PO_4^{-3} + 6H_2O \rightarrow MgNH_4PO_4 \cdot 6H_2O \quad (1)$$

2.1.2.3. Biological upgrading of biogas. The number of studies on methanation strategies for biogas upgrading has been increasing. Elyasi *et al.* (2021) studied the sustainability of the biological biogas upgrading using off-peak surplus electricity in Denmark. Their findings confirmed that from economic and environmental points of view, biological upgrading of biogas can be a good competitor to conventional water scrubbing technology. During the high penetration of renewable electricity into the Danish power system, MOB-based MP production integrated with biological biogas upgrading as a downstream process has been introduced as a sustainable pathway (Khoshnevisan *et al.* 2022). Accordingly, the use of biological biogas upgrading as a sub-category of power-to-gas technology is considered in this study.

The implementation of biological biogas upgrading improves the efficiency of WWTPs by converting the CO_2 content of biogas to biomethane by the addition of hydrogen (i.e., $4H_2 + CO_2 \rightarrow CH_4 + 2H_2O$). Taking the thermodynamic relationships and mass balance into account, 0.18 kg of hydrogen is required to upgrade 1 kg CO_2 to biomethane (Mehmeti *et al.* 2018). To produce 1 kg of hydrogen, 10.11 kg H_2O water is consumed (Kuckshinrichs *et al.* 2017). Considering the stoichiometric relationships, the amount of water consumed by the electrolysis unit and subsequently the amounts of H_2 and O_2 generated were estimated. Therefore, the amount of electricity consumption was calculated at 56 kWh/kg H_2 (Ivy 2004). Accordingly, in the biological biogas upgrading facility, standing at upstream of the MP production facility, 1,122 t/year hydrogen was required to react with 6,175 t/year CO_2 to generate 5,626 t/year biomethane. Furthermore, 62,859 MWh of electricity was annually consumed by electrolyzers to break down water into H_2 and O_2, which was supplied by excess wind power. About 25% of the total upgraded biomethane was consumed for MP fermentation, i.e., 1,974 million m^3/year. The remaining biomethane (i.e., 5,913 million m^3/year), after supplying the requirements of the fermentation unit, was assumed to be injected into the natural gas grid.

It should be pointed out that excess wind power is assumed to be utilized by electrolyzers to produce hydrogen for biological biogas upgrading. Surplus power refers to the amount of electricity produced in excess and has no market demand leading to wind turbines curtailment and zero/negative electricity price in the Nord Pool electricity market (Elyasi *et al.* 2021). The curtailment of wind turbines increases the environmental impacts of wind electricity production (caused by the manufacturing of fixed and moving parts of wind turbines) within their life cycle. On the contrary, avoiding curtailment by using surplus electricity for other purposes would decrease the impact per kWh electricity that is produced within the life cycle of wind turbines. On the other hand the surplus electricity can be further valorized into high-added-value products or stored in the existing gas pipelines infrastructures.

2.1.2.4. MP fermentation. Mixed methanotrophic culture is used for the fermentation of the MP. The highest concentration of extracted TAN without any dilution was considered for the preparation of the culture medium, i.e., 297 mg/l. Moreover, industrial data has shown a water circulation rate of 80% in the 1st generation MP fermentation process (https://www.unibio.dk/end-product/protein/), which decreases the freshwater requirement. The carbon/energy source for methane oxidizing bacteria was supplied from biologically upgraded biogas as shown promising results for MP production (Khoshnevisan *et al.* 2019; Tsapekosa *et al.* 2019). The earlier studies showed that a CH_4/TAN ratio of 9.4 would maximize MP yield, hence, a total of 1,408 t CH_4/year is consumed within the fermentation process.

The pretreated sludge and biopulp were digested in a continuously stirred tank reactor (CSTR). The fermentation of MP took place at ambient temperature; hence, this stage did not have any heat demand. The power demand of the fermenters was estimated based on the amount of air/oxygen required to be pumped in the culture medium (i.e., the energy demand of pumps). The oxygen and methane ratio of 2:1 would provide an optimal growth condition for MP production (Khoshnevisan *et al.* 2019; Tsapekos *et al.* 2019). Accordingly, an air blower with a capacity of 43.5–134 m^3/min and electricity consumption of 0.025 kWh/m^3 (model: GM130 L/DN300) was considered to supply air/oxygen to the fermenters.

The output flow from BES lacks micro/macro nutrients required for the balanced growth of the methanotrophic culture. Diluted ammonium medium salt (dAMS) is the standard medium used as the growth medium for the cultivation of methanotrophs. As N and P are supplied from the RW, the original protocol is altered and the modified-dAMS is used for the fermentation of MP. The modified-dAMS were prepared according to the recommendations of previous studies (Khoshnevisan *et al.* 2019). This protocol has been modified based on the authors' experiences and the instructions made by experts from Unibio®. Table 3 provides an overview of the chemicals needed for the preparation of modified-dAMS. As the cultivation medium has been patented by Unibio®, detailed information is not allowed to be shown herein.

2.1.2.5. System expansion and marginal technologies. Consequential LCA is characterized by using marginal data/technology instead of average technology. Weidema *et al.* (1999) have recommended a 5-step procedure to identify marginal technologies, which was employed herein. As the temporal scope of this study is 2030, offshore wind electricity was used as marginal technology. The choice of offshore wind electricity as marginal technology was in line with projections made by Denmark's Energy and Climate Outlook for 2030 (Danish Energy Agency 2019). Regarding the marginal source of heat, the renewable energy sources in central heating installations have the highest proportion of Denmark's heat supply and are expected to grow 80% by 2030 (Danish Energy Agency 2019). Biomass, with a 10% increase in annual consumption, is projected to be the best alternative to the most common district heating sources in Denmark, such as natural gas, heat pumps, and coal (Danish Energy Agency 2019). Accordingly, biomass-wood chip and biomass-wood pellet were chosen as marginal qualified sources for the centralized heating system. Regarding the biomethane produced, natural gas was considered to be substituted, which is in line with the Danish energy outlook as shown in the recent studies (Elyasi *et al.* 2021; Faragò *et al.* 2021). Considering conventional protein sources for animal feeds, the most preferred proteinaceous feeds that are imported to Denmark are rapeseed, sunflower, and soybean meal. Among them, soybean meal accounts for up to 64% of the overall imported protein-based feeds (Gylling & Hermansen 2018; Index Mundi data 2020); mostly from Germany, Argentina, and Brazil (OEC 2018). The statistical analysis demonstrated that the demand for soybean/soybean meal in Denmark will rise in the next decade (KNOEMA 2016; Index Mundi data 2020); hence, soybean meal was chosen as marginal technology herein.

Table 3 | Modified-dAMS instruction for MP production

Chemicals	Amount (t/year)
NH_4-N from RW	149.82
Potassium and P source	417.1
Magnesium source	12.1
Calcium source	128.8
Iron source	5.5
Trace metals	0.39

2.1.2.6. Soybean meal loop. As mentioned earlier, it is expected that 2nd generation MP substitutes for soybean meal in the feed market. However, this substation will have important consequences on the soybean oil market as a decrease in soybean meal production is accompanied by a drop in soybean production, which should be compensated by palm oil production (Khoshnevisan *et al.* 2020b). Accordingly, any changes in soybean demand will have subsequent changes in demand/supply of other products, i.e., soybean oil, palm oil, palm kernel meal, and barley. For a detailed description please read Dalgaard *et al.* (2007). So, a continuous chain of demands/supplies (loop) will be created, due to the replacement of soybean meal by MP and consequently soybean oil by palm oil (Schmidt & Weidema 2008), which must be addressed within the framework of this study. The accumulation of protein and energy of each product was considered, as a major criterion, to determine the amounts of marginal feed substituted. The soybean meal loop has been shown in Figure 1 using a 'loop' sign, which implies the quantity of avoided soybean meal and spring barley as well as the stimulated palm oil (Dalgaard *et al.* 2007).

2.1.3. LCIA

The impact assessment method, ReCiPe 2016 v1.1 (hierarchies perspective) (Huijbregts *et al.* 2017) at the endpoint and mid-point levels were used to investigate the environmental impacts of the developed biorefinery. This impact assessment method has 18 midpoint impact categories (i.e., global warming, ionizing radiation, water consumption, land use, ozone formation (terrestrial ecosystems), ozone formation (human health), stratospheric ozone depletion, human non-carcinogenic toxicity, human carcinogenic toxicity, fine particulate matter formation, freshwater eutrophication, freshwater ecotoxicity, marine eco-toxicity, terrestrial ecotoxicity, mineral and fossil resource scarcity, terrestrial acidification, and marine eutrophication) and three endpoint areas of protection (i.e., human health, resource availability, and ecosystem quality).

2.1.4. Economic assessment

An overview of all the steps, included in the economic assessment of the proposed biorefinery is demonstrated in Figure 2. The sum of capital expenditure (CAPEX) and operating expenditure (OPEX) has been made up of the total cost of the project. In this respect, operating expenditure OPEX can be classified into fixed OPEX and variable OPEX, both of which are included in our economic assessment.

The CAPEX of MP production as a downstream process of WWTP covered all the costs associated with equipment, building and construction, pipework, equipment installation, power generation facility, etc. Most equipment prices were collected from authentic sources such as literature, technical reports, and supplier companies. The prices of equipment were updated using the scale factor and the Chemical Engineering Plant Cost Index (CEPCI) (see Equations (2) and (3)). In Equation (2), 'P' stands for 'scale factor', which was considered to be 0.7 as suggested in the literature (Lynd *et al.* 2005). The parameters 'Model cost' and 'Model scale' indicate expenditure and size in the reporting year. In this Equation, 'Scaling size' stands for the size of equipment in the current study. The cost of equipment in the current year should be then updated for the target year 2030 using Equation (3). The CEPCI of the reference year ($CEPCI_a$) was taken from annual reports published in the chemical engineering magazines (*Chemical Engineering* 2020). As there are no projections of CEPCIs for 2030,

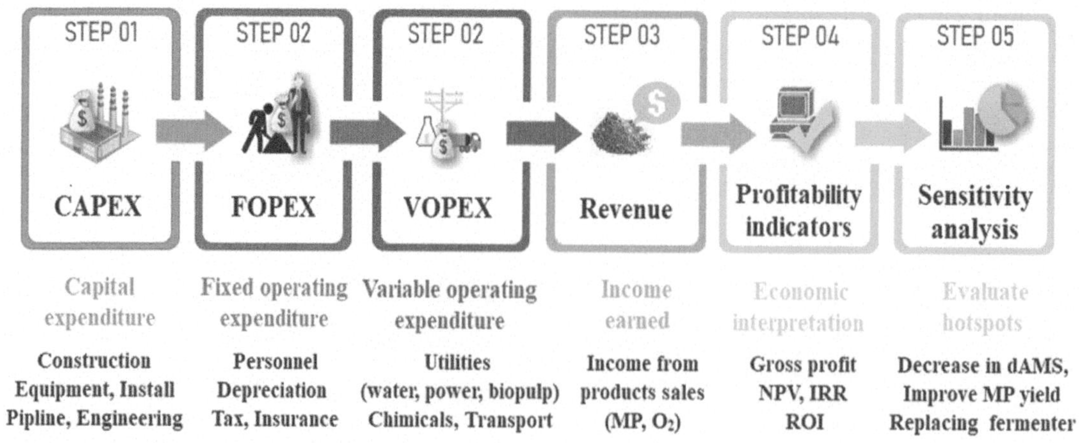

Figure 2 | Economic assessment steps in current work.

historical data from 2000 to 2019 were employed to predict CEPCI. Accordingly, a CEPCI of 679 was used herein. Detailed information in this context is shown in the supplementary file 'SI-3'.

$$\mathrm{Cost_a} = \left(\frac{\text{Scaling size}}{\text{Model scale}}\right)^{\mathrm{P}} \times \text{Model cost} \tag{2}$$

$$\mathrm{Cost_{2030}} = \left(\frac{\mathrm{CEPCI_{2030}}}{\mathrm{CEPCI_a}}\right) \times \mathrm{Cost_a} \tag{3}$$

The CAPEX of biological biogas upgrading facility and its associated technologies including electrolyzers, hydrogen storage, and biogas upgrading was taken from the recent study of Elyasi *et al.* (2021). FOPEX covers the running expenditure of employees, insurance, tax, maintenances, and depreciation. VOPEX is also attributed to the expenses of feedstock materials, power, chemicals, and water consumed.

The electricity used in the electrolyzer facility for hydrogen production is supplied from off-peak surplus wind electricity. According to Denmark's Energy Outlook (Danish Energy Agency 2019), Denmark will become an electricity exporter by 2030 due to the expansion of offshore wind turbines, which is accompanied by surplus wind electricity production (See more details in Elyasi *et al.* 2021). Accordingly, the price of surplus electricity was assumed to be zero for the economic assessment. However, the impact of various prices of surplus electricity on the economic performance of MP production was later investigated by the sensitivity analysis. In this context, the effect of increasing the surplus electricity price up to 50% of the conventional electricity price in Denmark was used for the sensitivity analysis. It should be pointed out that the Danish Energy Agency has projected that the electricity prices will gradually decrease, reaching 44.72 €/MWh in 2030 (Danish Energy Agency 2019).

The transportation costs were estimated based on the distance between the market and biorefinery and the amount of fuel consumed for such transports. The chemicals used in the cultivation medium are assumed to be transported from Copenhagen to the Avedøre plant with an average distance of 5 km. The biopulp is shipped from a biopulping facility standing at a distance of 15 km. According to Hagos & Ahlgren (2020), diesel will be still the major transportation fuel for heavy-duty vehicles in 2030. Accordingly, the price of diesel fuel in 2030 was predicted based on annual data published in three conditions: 1.89, 1.93, and 1.19 €/L for the base, pessimistic, andoptimistic conditions, respectively (Statista Data Platform) (see the supplementary file 'SI-4').

The cost of human labour is steadily rising and is projected to reach 26.28 €/h by 2030 (see supplementary file 'SI-5') (Statistics Denmark 2020). To include the cost of human labour, 8-hour working time per day was used, and the supervision cost was approximated at 4% of human labor expenditure (LaTurner *et al.* 2020). Depreciation was calculated using Equation (4) by considering the depreciated/salvage ratio of 20% of upfront capital expenditure within a lifelong of 20 years (Miyata 1980). In this respect, 6 and 1% ofdepreciated expenses were taken into account as insurance and property taxation, respectively (LaTurner *et al.* 2020).

$$\text{Depreciation expenses} = \frac{\text{CAPEX} - \text{Salvage value}}{\text{Lifelong}} \tag{4}$$

Salvage Value: 20% of CAPEX.

The price of MP was considered to be equal to its counterpart, i.e., soybean meal, in 2030 to make it competitive with conventional animal feed ingredients. Through the analysis of historical data between 2000 and 2020 (IndexMundi data portal), the sale prices of soybean meal for 2030 under various scenarios were projected; 502, 541, and 323 €/t for the base, optimistic, and pessimistic conditions,respectively. Oxygen is the by-product of hydrogen in the electrolyzer facility, which should be accounted for in the economic assessment. The price of electricity in the reference year was assumed to be 0.082 €/kg O_2 (Chen *et al.* 2018) with a 6% compound annual growth rate (CAGR) from 2020 to 2030 (Elyasi *et al.* 2021). Accordingly, themarket price for oxygen was calculated to be 0.17 €/kg O_2 in 2030. The economic performance of the MP production facility was evaluated using gross profit, internal rate of return (IRR), net present value (NPV), and return on investment (ROI) (more details in the supplementary file 'SI-6').

3. RESULT AND DISCUSSION

3.1. Environmental performance of developed biorefinery

The developed biorefinery showed promising results in two damage categories of ecosystem and resources scarcity where net negative impacts were attained, i.e., -1.07×10^{-2} species.yr/FU, and -1.68×10^{6} USD/FU, respectively. However, further improvements are still needed in the damage category of human health where a net impact of 2.37 DALY/FU was achieved. The results achieved herein demonstrated that a biorefinery platform developed on a municipal WWTP for the processing of WW and OFMS can be encouraging for closing the gaps in animal protein feed and supplying bioenergy. To identify the key hotspots in each damage category and potential possibilities for further improvements, the results of each damage category are discussed in detail separately in the following sub-sections.

3.1.1. Human health damage category

The contribution of various processes to the damage category of human health is shown in Figure 3. In this damage category, the unit process of anaerobic digestion had the highest contribution with a share of 41%. This impact attributes to methane leakage and the background emissions of heat and electricity production. Following the anaerobic digestion unit process, electrolysis and fermentation processes imposed the highest impacts on this damage category with a share of 27 and 15%, respectively. The impact from the electrolysis unit is caused by the consumption of electricity. Although wind electricity, which is known as an environmentally friendly and renewable energy source, is used for water electrolysis, it imposed a high environmental burden on the human health damage category. This attributes to the fact that the manufacture of

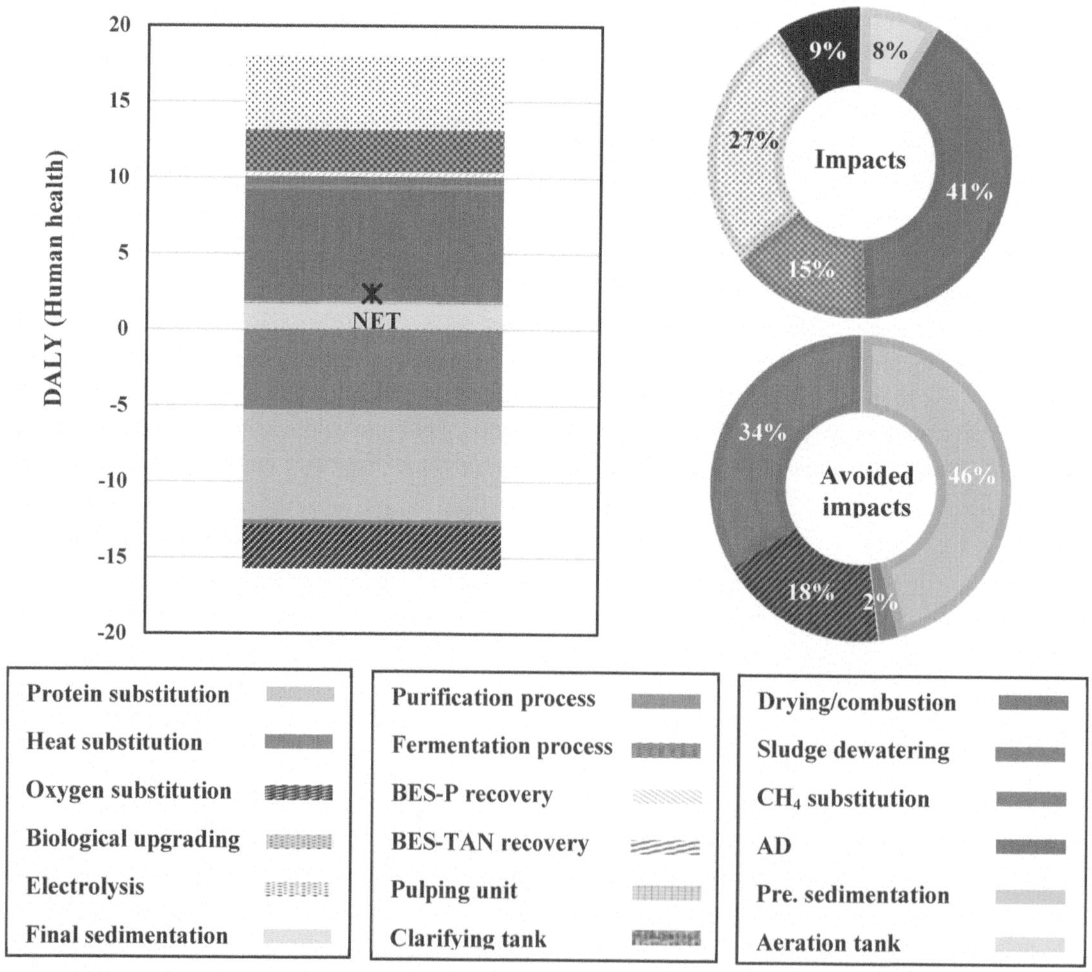

Figure 3 | Contribution of all unit processes to damage category of human health resulting from integrating WWTP (with sludge and biopulp co-digestion) and MP production.

wind turbines (both fixed and moving parts), as well as the maintenance operations, necessitate the use of materials such as concretes, steels, and glass-fibers (Ghenai 2012), which create such negative impacts within their life cycle from extraction to use phase. Delpierre (2019) already reported that wind electricity contributed most to the environmental impacts of electrolytic hydrogen production. In another study conducted by Elyasi *et al.* (2021), the same conclusion was also reached, demonstrating that wind electricity production contributes to the damage category of human health when surplus wind electricity is used for the biological upgrading of biogas.

The impacts from the fermentation process attributed to the electricity consumption (i.e., 1.4% of total imposed impacts) as well as the consumption of chemicals used for culture medium preparation. The background emissions from the production of such chemicals as well as their transportation are all shown under the fermentation process. As discussed earlier, the outflows of BESs do not contain any supplementary nutrients, needed for the growth of methanotrophic bacteria, except NH_4-N and P. Thereby, a balanced enriched nutrient culture medium is provided by the addition of various chemicals to prepare modified-dAMS. The detailed analysis demonstrated that P/potassium sources, used as the buffering agents, were the major contributors to the environmental impact of the fermentation process, with over 70% of total induced impacts. As biopulp is not rich in P (2.70 g /kg TS biopulp), the co-digestion of a P rich waste stream such as fruit-vegetable waste, livestock manure, and agricultural waste with sludge and biopulp and the subsequent extraction of struvite P can decrease demand for P sources and hence will improve the environmental profile of fermentation process. Moreover, optimizing the fermentation process and recycling the culture medium to ensure the complete consumption of nutrients can further improve the environmental performance in this damage category. Other processes such as aeration tank, also contributed to the human health damage category but to a letter extent, i.e., 1.54 DALY/FU. These impacts are rooted in on-site emissions from biological wastewater treatment.

As shown in Figure 3, the avoided impacts caused by the substitution of soybean meal led to the highest saving in the human health damage category. The substitution of soybean meal has a net impact of −7.22 DALY/FU, contributing to 46% of total net savings. Following that, the substitution of biomethane for natural gas and oxygen as a co-product of the electrolysis process, led to the highest savings in this damage category with a net impact of −5.30 and −2.87 DALY/FU, respectively. Although considerable amounts of environmental impacts in human health damage category could be avoided by the substituted products, it was not enough to reach a net negative impact in this damage category. Optimizing the TAN extraction process and enhancing the recovery efficiency of BES (above 61%) would be one of the possibilities to improve the environmental performance of the biorefinery. The higher TAN recovery, the higher MP production, and hence the higher substitution of conventional proteinaceous feed. Moreover, improving N-to-protein conversion rate and the optimization of MP fermentation will also bring about more savings in this damage category. Furthermore, it is highly suggested to investigate the possibilities of storage and use of the electricity which is generated during NH_4-N and P recovery in BES. The use of BES-driven electricity can contribute to the better environmental and economic performance of MP-based biorefinery platforms.

3.1.2. Ecosystem quality damage category

On the contrary to the damage category of human health, a net saving of −0.011 species.yr/FU was attained in ecosystem quality (Figure 4). Contribution analysis demonstrated that the anaerobic digestion and fermentation process had the greatest impacts, i.e., 0.047 and 0.016 species.yr/FU, contributing to 60 and 20% of total net impacts, respectively. The detailed contributional analysis demonstrated that heat consumption was the major contributor to the environmental impact of anaerobic digestion, i.e., about 80% of the total impact. As discussed in the previous sub-section, the impacts associated with the fermentation process originate from the consumption of chemicals and electricity demand. Faragò *et al.* (2021) reported that the utilization of electricity and chemicals for P recovery from sludge ash contributed to the impact category of water eutrophication and subsequently to the ecosystem quality damage category. Following AD and fermentation, the electrolysis process imposed the highest impact on this damage category with a share of 9% (i.e., 0.007 species.yr/FU).

3.1.3. Resource scarcity damage category

Similar to the ecosystem quality damage category, a net saving of -1.68×10^6 USD/FU was also attained in the resource scarcity damage category. In other words, the saving achieved by the avoided products under this damage category was 7.18 times as much as the impacts imposed by the unit processes. According to the results shown in Figure 5, electrolysis and fermentation unit processes dominated this damage category with a share of 32 and 28%, respectively. The impact from electrolysis

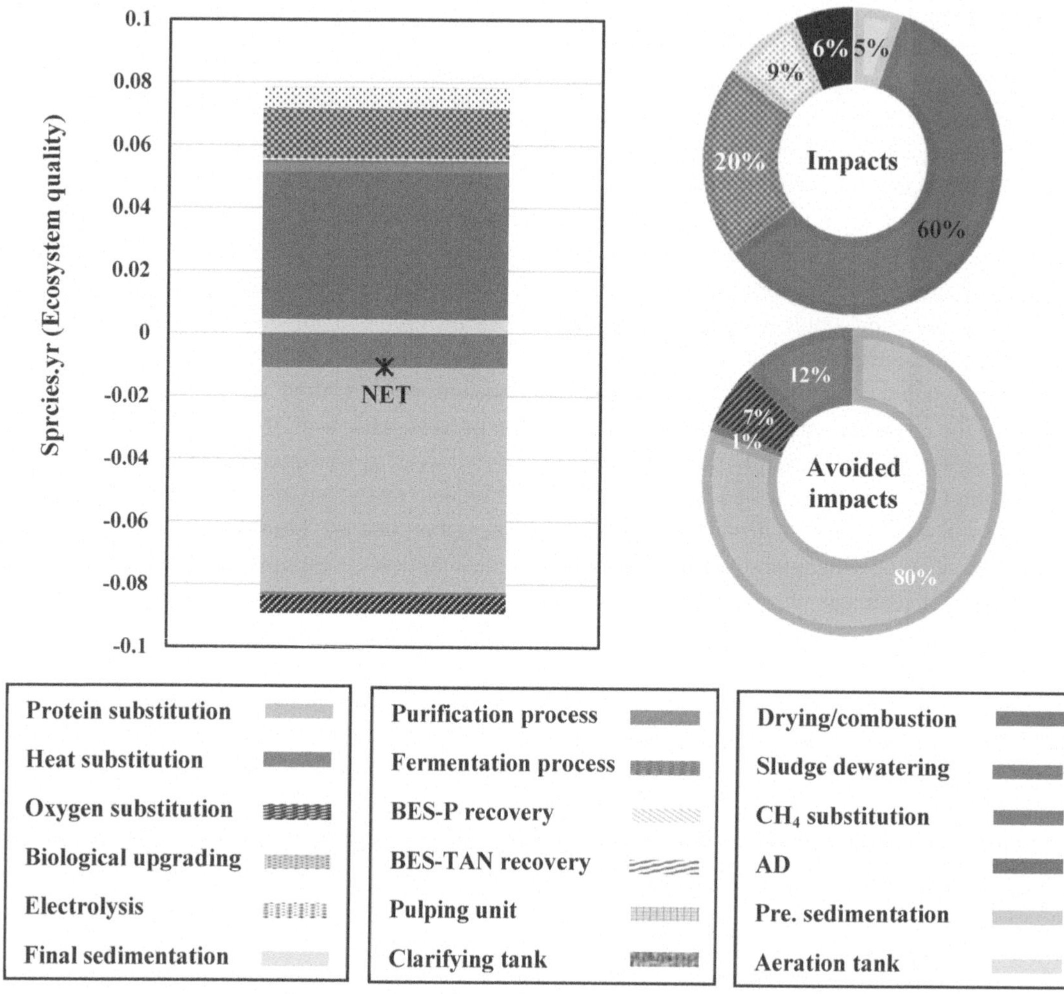

Figure 4 | Contribution of all unit processes to damage category of ecosystem quality resulting from integrating WWTP (with sludge and biopulp co-digestion) and MP production.

unit process attributes to the consumption of materials used for the manufacture of fixed and moving parts of offshore wind turbines. The impact from the fermentation process attributes to the background emissions induced by the production of chemicals and minerals needed for the preparation of modified-dAMS. As can be observed in Figure 5, the substitution of natural gas (the marginal source of biomethane production in 2030 in agreement with Elyasi *et al.* (2021) and Faragò *et al.* (2021)) caused the highest saving in the resource scarcity damage category, accounting for 93% of the total saving. The co-digestion of biopulp and sludge results in higher methane yield (Zha *et al.* 2020). Moreover, the biological biogas upgrading increases the total biomethane production of the facility. The whole biorefinery's heat and electricity demand are from biomass (wood chips and pellet) and surplus wind power, respectively, while a high portion of the biomethane (approximately 75%) can be injected into the gas grid and substitute the marginal biomethane sources. Accordingly, the substitution of natural gas had higher net savings in resource scarcity than soybean meal substitution.

3.1.4. Environmental cost of MP production

The environmental impact of 1 tonne pure protein from 2nd generation MP was calculated at 0.007 DALY in human health, 6.75×10^{-5} species.yr in ecosystem quality, −711 kg CO_2 in climate change, and −2,555 USD in resources damage categories. Thecommercialized methanotrophs-based MPs such as FeedKind® use natural gas and chemical nitrogen as their energy and nitrogen sources, respectively. According to Cumberlege *et al.* (2011), even with 100% use of biogas and renewables, FeedKind® MP, could not outperform the conventional protein sources such as soybean meal. The average global soybean meal and fishmeal production are considered herein to have an insight into the environmental profile of 2nd

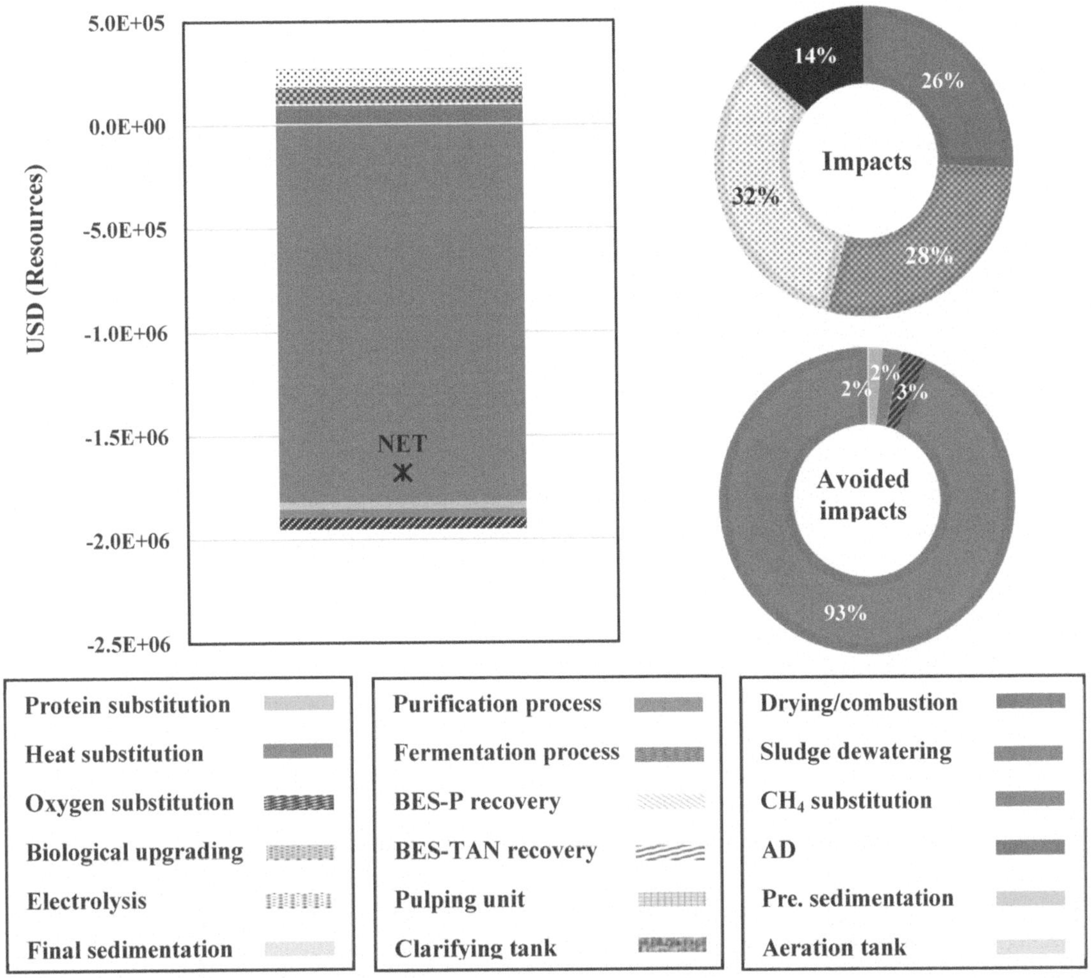

Figure 5 | Contribution of all unit processes to damage category of resources resulting from integrating WWTP (with sludge and biopulp co-digestion) and MP production.

generation MP compared to conventional protein sources (www.ecoinvent.org). The production of 1 tonne of protein from soybean meal and fishmeal has a net impact of 0.009 and 0.011 DALY in human health, 7.06×10^3 and 3.00×10^3 kg CO_2 in climate change, 9.84×10^{-5} and 1.51×10^{-5} species.yr in ecosystem quality, and 1.17×10^2 and 9.47×10^1 USD in resources damage categories, respectively. As can be seen, the production of MP from waste effluents is a promising approach to produce more environmentally friendly proteinaceous feed in the future.

3.1.5. Sensitivity analysis for LCA

Sensitivity analysis was performed to investigate to what extent the environmental profile of the WWTP-based biorefinery will change with respect to the assumptions made herein and the improvements which can be implemented. As discussed earlier, the use of chemicals in modified-dAMS solution as supplementary nutrients plays a key role in the overall environmental performance of the biorefinery. Moreover, improving the protein profile of MP and increasing its total protein content (by enhancing conversion rate from N to protein) can substitute more soybean meal as a competing product, hence, will help to improve the environmental performance. The results showed that (Table 4) the environmental profile of the biorefinery had the highest sensitivity to improving the conversion rate of N to protein. According to the lab experiments, 1 g of NH_4-N could be valorized into 4.36 g of protein. Theoretically, proteins contain 16% N (Shuuluka *et al.* 2013; Krul 2019). Hence, improving the protein profile of the produced biomass can increase the protein yield from a given amount of nitrogen. The enhancement of NH_4-N to protein conversion rate from 4.36 to 5.25 and optimization of the fermentation process and recycling of culture medium could to some extent decrease the impacts in the human health damage category. However, such

Table 4 | Sensitivity analysis over the reduction of dAMS solution by 5% (Sens. 5%), 10% (Sens. 10%), 20% (Sens. 20%); improve protein yield of NH_4-N (Sens. Pure protein); both a 20% reduction in dAMS and improve pure protein content (Sens. All)

	Human health		Ecosystem		Resources	
Parameter	Net saving/impact (DALY)	Improve (%)	Net saving/impact (Species.yr)	Improve (%)	Net saving/impact (USD)	Improve (%)
Base	2.366	–	-1.07×10^{-2}	–	-1.679×10^{6}	–
Sen.5%	2.204	6.85	-1.16×10^{-2}	7.64	-1.683×10^{6}	0.26
Sen.10%	2.072	12.44	-1.23×10^{-2}	15.00	-1.687×10^{6}	0.48
Sen.20%	1.803	23.78	-1.39×10^{-2}	29.77	-1.695×10^{6}	0.94
Sen. Pure protein	0.899	62.01	-2.53×10^{-2}	135.55	-1.686×10^{6}	0.41
Sens. All	0.336	85.78	-2.85×10^{-2}	165.32	-1.702×10^{5}	1.35

USD, United States dollar; DALY, Disability-adjusted life year.

a reduction in the environmental profile was not sufficient to achieve a net saving in this damage category. Considering both strategies, i.e., a 20% reduction in consumption of dAMS and enhancing pure protein content (Table 4; Sens. All), could decrease the net impact by 86% reaching 0.34 DALY/FU in human health.

The supply of plant's electricity demand from excess wind power and the upgrading of all biogas into biomethane for further injection to gas grids instead of combusting in CHP have higher environmental benefits (Faragò *et al.* 2021). However, the results achieved herein demonstrated that supplying the plant's heat demand from the grid would result in significant environmental burdens. As mentioned earlier the marginal heat sources are wood chips and wood pellets in 2030 whose combustion has a wide range of air emissions. Having considered to supply plant's heat demand (i.e., 19,784 MWh_{th}/year) through the combustion of part of biogas in a boiler instead of supplying from the grid, sensitivity analysis was performed over the source of heat production. By assuming the lower heat value (LHV) of biomethane (36 MJ/m^3 CH_4 or 10 kWh/m^3 CH_4 (Panpong *et al.* 2014; Firdaus *et al.* 2017)) and 95% projected efficiency of modern boilers in 2030 (Elyasi *et al.* 2021), approximately 44% of the biogas (i.e., 3,471 million m^3 per year biogas) shall be combusted in the boiler to meet plant's heat requirement (see boiler's emissions inventory in Table 5). It is worth bearing in mind that CO_2 and CH_4 emissions from biogas combustion were considered biogenic due to their non-fossil origin. The remaining biogas (i.e., 4,417 million m^3/year), after supplying the requirements of the MP facility, was upgraded then and injected into the natural gas grid to substitute natural gas. Supplying plant's heat demand from biogas combustion in the boiler could improve the environmental profile of the biorefinery by 79 and 73% in human health and ecosystem quality damage categories, respectively (Figure 6). On the contrary, lower saving was attained in the resource scarcity damage category. The detailed contributional analysis demonstrated (see the supplementary file 'SI-7') that the unit process of fermentation dominated in ecosystem quality and resources damage categories with a share of 45 and 46%, respectively.

3.2. Economic assessment

The following sub-sections will comprehensively discuss the economic performance of MP production in terms of expenditures, incomes, and subsequent economic indices.

3.2.1. Expenditures

As demonstrated in Table 6, equipment and more specifically fermenters had the highest contribution to the total CAPEX. The expenditure on fermenters is straightly appertained to providing sufficient hydraulic retention time (HRT) for MP fermentation and consequently the volume of the reactors (Skadborg 2018). Considering a 72-hours HRT to secure high MP yield and N to protein efficiency (Tsapekos *et al.* 2019) and an input flow rate of 1,384 m^3/day (i.e., extracted N-rich flow), a total fermenters' volume of 4,152 m^3 is needed. Among all the facilities used for biological biogas upgrading technology, the water electrolysis facility had the largest contribution to CAPEX. The results achieved herein are in line with those of Elyasi *et al.* (2021), who proved that, although learning technology and increasing the volume of the water electrolysis system will significantly reduce the investment cost in 2030 (Zauner *et al.* 2020), it will still have the largest contribution to the CAPEX of biological biogas upgrading technology.

Table 5 | Emission inventory for the boiler unit

Emission	Abbreviation	Emission to	Amount (kg/year)
Sulfur dioxide	SO_2	Air	1.87×10^3
Nitric oxides	NO_X	Air	2.10×10^3
Non-methane volatile organic compounds	NMVOC	Air	1.50×10^2
Methane	CH_4	Air	7.50×10^{1}[a]
Carbon monoxide	CO	Air	2.70×10^3
Carbon dioxide	CO_2	Air	6.31×10^{6}[a]
Nitrous oxide	N_2O	Air	7.50×10^0
Total suspended particles	TSP	Air	1.12×10^2
Particulate matter (aerodynamic diameters<=10 μm)	PM_{10}	Air	1.12×10^2
Particulate matter (aerodynamic diameters<=2.5 μm)	$PM_{2.5}$	Air	1.12×10^2
Black carbon	BC	Air	3.71×10^0
Chromium	Cr	Air	1.35×10^{-2}
Copper	Cu	Air	2.32×10^{-2}
Mercury	Hg	Air	9.00×10^{-3}
Zinc	Zn	Air	2.96×10^{-1}
Arsenic	As	Air	3.00×10^{-3}
Cadmium	Cd	Air	1.50×10^{-4}
Nickel	Ni	Air	1.72×10^{-2}
Lead	Pb	Air	3.75×10^{-4}
Selenium	Se	Air	1.57×10^{-2}

[a]Biogenic.

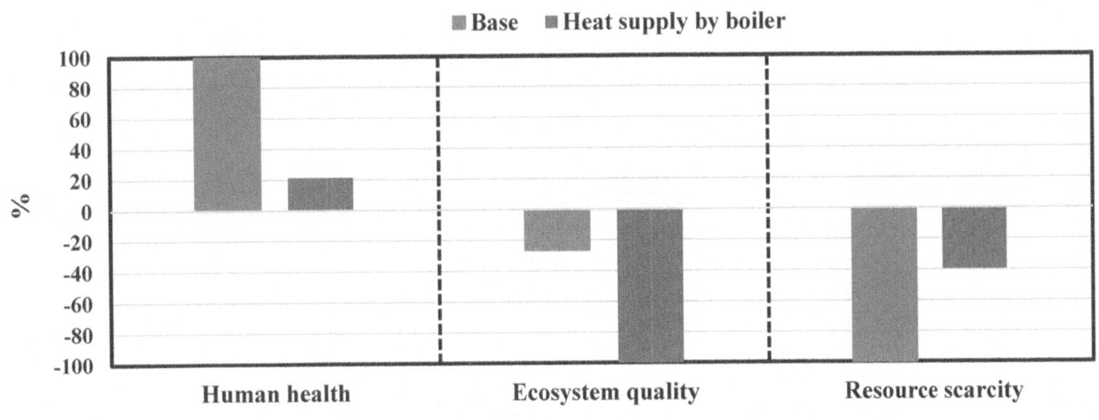

Figure 6 | Sensitivity analysis over the supply of heat demand by a boiler in ecosystem quality, human health, and resource scarcity damage categories.

VOPEX and FOPEX of the current process are profoundly calculated and shown in Table 7.In this regard, chemicals used in BES (for P precipitation and pH adjustment) and fermentation process (as supplementary nutrients for methanotrophs), had the highest contribution to OPEX with a share of 35%. The annual cost of chemicals adds up to 0.41 million €/year, which significantly affects the economic performance of MP production. On the contrary, OPEX of biological biogas upgrading made up a low proportion of total OPEX (i.e., 3.5%). This can be attributed to the fact that only 26.35% of the upgraded biogas is consumed within the MP facility (either in form of methane for MP production or in form of electricity), hence, the OPEX of biological biogas upgrading is allocated between the MP facility and wastewater treatment plant.

Table 6 | Total CAPEX of a microbial protein production facility and its complementary electrolyzer facility standing at the downstream of a WWTP treating wastewater and organic fraction of municipal solid waste

Item	Cost (M€)	Portion of total CAPEX (%)	Ref.	Description
Equipment expenses	6.4	27.96	–	Table SI-1
Installation part	2.1	9.32	Peters *et al.* (2003); Acién *et al.* (2012); Kwan *et al.* (2015)	33% of equipment capital
Process pipeline	1.3	5.59	Peters *et al.* (2003); Acién *et al.* (2012); Kwan *et al.* (2015)	20% of equipment capital
Controls and instrumentation	0.86	3.73	Peters *et al.* (2003); Acién *et al.* (2012); Kwan *et al.* (2015)	13% of equipment capital
Electrical systems	0.86	3.73	Peters *et al.* (2003); Acién *et al.* (2012); Kwan *et al.* (2015)	13% of equipment capital
Buildings	2.1	9.32	Peters *et al.* (2003); Acién *et al.* (2012); Kwan *et al.* (2015)	33% of equipment capital
Yard enhancements	0.43	1.86	Peters *et al.* (2003); Acién *et al.* (2012); Kwan *et al.* (2015)	7% of equipment capital
Assisting facilities	1.1	4.66	Peters *et al.* (2003); Acién *et al.* (2012); Kwan *et al.* (2015)	17% of equipment capital
Engineering and technical	1.3	5.59	Peters *et al.* (2003); Acién *et al.* (2012); Kwan *et al.* (2015)	20% of equipment capital
Construction expenses	1.7	7.46	Peters *et al.* (2003); Acién *et al.* (2012); Kwan *et al.* (2015)	27% of equipment capital
Legal expenses	0.43	1.86	Peters *et al.* (2003); Acién *et al.* (2012); Kwan *et al.* (2015)	7% of equipment capital
Contractor's payment	0.64	2.80	Peters *et al.* (2003); Acién *et al.* (2012); Kwan *et al.* (2015)	10% of equipment capital
Contingency	2.1	9.32	Peters *et al.* (2003); Acién *et al.* (2012); Kwan *et al.* (2015)	33% of equipment capital
Electrolysis system	0.73	3.18	Zauner *et al.* (2020)	500 €/kW_{el}
Hydrogen storage tank	0.06	0.25	Gorre *et al.* (2020)	33 €/m^3 H_2
Biological upgrading	0.43	1.88	Thema *et al.* (2019); Zauner *et al.* (2019)	300 €/kW_{el}
Install, design, and plan	0.34	1.49	Leonzio (2017)	28% of sum CAPEX
Total	**23.1**	**100**	–	–

M€, million euro; Construction expenses, construction of equipment and tools, safety, permits, etc.; Yard enhancements, site expansion and landscaped; Contingency, uncertainty in evaluation; Assisting facilities, non-process equipment, facilities, and packaging and distribution.

3.2.2. Economic indices

Having considered optimistic, base, and pessimistic scenarios, the economic indices are calculated and shown in Table 8. As can be seen, the gross profit for this process was estimated at 0.29 M€ and 0.22 M€ under optimistic and base scenarios, respectively. It is noteworthy that the economic indices under the pessimistic scenario were not reported as they were found to be unprofitable. Although the results of gross profit were slightly promising but the overall economic performance of MP production was not encouraging in terms of NPV and IRR. The high CAPEX associated with 2nd generation MP from waste streams has undermined its economic performance. For a project to be economically viable, IRR should be equivalent to the discount rate (i.e., 8%) and NPV should be positive. The detailed analysis demonstrated that CAPEX associated with fermenters (i.e., cost of fermenters) had the largest share. As discussed earlier, considering a 72-hours HRT to secure high MP yield and N to protein efficiency increased the total reactors' volume needed for the MP production plant, and subsequently raised the total CAPEX. The future advancement in the technology of fermenters and the development of U-Loop fermenters can play a significant role in decreasing the CAPEX of MP production in the future. Having compared with CSTRs used for the MP production in this study, U-Loop fermenters can perform better in terms of production rate while shortening the HRT

Table 7 | Total OPEX of MP production with reject water

Item	Amount	Unit	Cost (€/year)	Description	Refs.
Chemicals substance					
Potassium and phosphor[a]	417.2	ton	340,840*	Avg. unit price: 602 €/t	Taken from various sources such as chemical reports, peer-reviewed papers, and reputable supplier companies.
Magnesium source[a]	86.1	ton	29,712*	Avg. unit price: 210 €/t	
Calcium source	128.8	ton	32,661*	Avg. unit price: 252 €/t	
Iron source	5.53	ton	1,737*	Avg. unit price: 231 €/t	
Trace metals	0.39	ton	730*	Avg. unit price: 1,868 €/t	
Sum			**405,680**		
Portion of total OPEX (%)			**35.2**		
Employees costs					
Staff	4	Person	306,950	For a person: 26.28 €/hour	Estimated
Supervision and control	–	–	12,278	0.04 staff	LaTurner *et al.* (2020)
Sum			**319,228**		
Portion of total OPEX (%)			**27.7**		
Insurance, tax, and depreciation cost					
Depreciation	–	–	257,929	According to Equation (4)	Calculated
Property tax	–	–	2,579	0.01 of depreciation	LaTurner *et al.* (2020)
Insurance	–	–	1,548	0.06 of depreciation	LaTurner *et al.* (2020)
Sum			**262,056**		
Portion of total OPEX (%)			**22.7**		
Utility costs					
Electricity	851	MWh	0	0	Estimated
Biopulp	32,115	ton	111,756	Unit price: 17 €/t	Cimpan & Wenzel (2016)
Sum			**125,507***		
Portion of total OPEX (%)			**10.9**		
Biological biogas upgrading costs					
Electrolyzer service	–	–	14,689	2% of CAPEX	Zauner *et al.* (2019)
Hydrogen storage	–	–	577	1% of CAPEX	Zauner *et al.* (2019)
Biological upgrading	–	–	21,694	5% of CAPEX	Zauner *et al.* (2019)
Electricity	53,017	MWh	0	0	Estimated
Water	11,348	m^3	14,299	Unit price:1.26 €/m^3	Elyasi *et al.* (2021)
Sum			**39,886**		
Portion of total OPEX (%)			**3.5**		

*Including the cost of transportation.
[a]Used in BES and culture medium.

and decreasing the required reactor size. In this regard, the economic performance of MP under the assumption of replacing CSTRs with U-Loop fermenters has been investigated under sensitivity analysis.

3.2.3. Sensitivity analysis for economic assessment

Sensitivity analysis was performed by considering the potential economic changes due to the following modifications: (1) replacing conventional fermenters with U-loop reactors, (2) reducing the consumption of Modified-dAMS solution by 5, 10, and 20% through process optimization and proper recycling of culture medium, (3) improving pure protein yield on NH_4-N, and (4) considering of surplus electricity price. Table 9 summarises the expected changes in economic indices.

Table 8 | Revenue, expenditure, and economic profitability parameters under pessimistic, base, and optimistic scenarios

Item	Scenarios			Unit
	Optimistic	Base	pessimistic	
Revenue from products sales				
MP	0.88	0.81	0.52	M€/year
Oxygen	0.31	0.31	0.31	M€/year
Sum	1.18	1.12	0.83	M€/year
Expenditure of 1 ton of pure protein production				
CAPEX	35,293	35,293	35,293	€/t pure protein
FOPEX	946	946	946	€/t pure protein
VOPEX	817	829	830	€/t pure protein
Economic profitability indices				
Gross profit*	0.29	0.22	–	M€
ROI	5.14	4.87	–	%
NPV	−20.18	−20.88	–	M€
IRR	−11.03	−12.74	–	%

M€, million €.
*Gross profit does not take the property tax and depreciation into account; FOPEX, fixed operation cost; CAPEX, capital expenditure; VOPEX, variable operation cost; IRR, internal rate of return; ROI, return on investment; NPV, net present value.

Table 9 | Economic sensitivity analysis under an optimistic scenario

Parameters	Economic profitability indices							
	Gross profit* (M€)	Improve (%)	ROI (%)	Improve (%)	NPV (M€)	Improve (%)	IRR (%)	Improve (%)
Sen. Reactor	0.294	0.34	13.13	7.99	−6.14	69.59	−4.38	6.65
Sen. 5%	0.35	20.69	5.14	0	−19.57	3.02	−9.80	1.23
Sen. 10%	0.37	27.59	5.14	0	−19.39	3.91	−9.46	1.57
Sen. 20%	0.41	41.38	5.14	0	−19.01	5.80	−8.81	2.22
Sen. Pure protein	0.47	62.07	5.91	0.77	−18.43	8.67	−7.88	3.15
Sens. electricity price	−0.93	−420.68	5.14	0	–	–	–	–
Sens. All	0.59	101.33	15.10	9.96	−3.23	84.01	1.91	12.94

*Gross profit does not take the property tax and depreciation into account; M€, million €; ROI, return on investment; NPV, net present value; IRR; internal rate of return.

Having replaced CSTRs with U-loop fermenters, the required HRT is expected to decrease much less than 24 hours (Drejer *et al.* 2017; Petersen *et al.* 2019), so that the size of the reactors decreases without affecting the biomass production rate. Although, the amount of much less than 24 hours (up to 5 hours) has been reported by unpublished reports from companies working in this context, 24-hours HRT was opted for sensitivity analysis based on journal peer review. Accordingly, IRR, NPV, and ROI can be improved by 6.65, 69.59, and 7.99%, respectively, though changes in gross profit were negligible, i.e., 0.34%. Hence, the cost-efficiency of this MP production pathway was highly dependent on reactor volume and the HRT required to optimal production rate.

Although increasing the NH_4-N to protein conversion ratio up to 5.25 is needed to have a better environmental profile for 2nd generation MP and can raise gross profit by 62%, it was not sufficient to make MP production profitable. As shown in Table 9, by considering all possible changes together (Table 9; Sens. All), the economic performance of 2nd generation MP production can be significantly improved, although it will not be still economically profitable. Among all the economic indices calculated herein, ROI demonstrated the least sensitivity, i.e., 9.96%. This is because ROI is independent of the raw materials and utility cost (i.e., OPEX). Therefore, reducing the consumption of modified-dAMS and considering surplus electricity price did not affect the ROI. In respect to MP production for substituting soy-meal and fishmeal, LaTurner *et al.* (2020)reported that the market price of soy-meal undermines the economic feasibility of MP production. Therefore, granting subsidies and incentive schemes are needed to promote the production of 2nd generation MP.

4. CONCLUSION

In the current study, the environmental and economic feasibility of a biorefinery platform established on a WWTP was assessed. This platform has targeted to valorising N and carbon flows from wastewater treatment and anaerobic co-digestion of sludge and OFMSW into microbial protein and bioenergy. The environmental profile of such a biorefinery was found promising in ecosystem and resources scarcity damage categories where net negatives of -1.07×10^{-2} species.yr/FU, and -1.68×10^{6} USD/FU, respectively, were attained. Further system modifications and improvements are still needed to gain net savings in the damage category of human health. Under the baseline scenario a net impact of 2.37 DALY/FU was achieved in the human health damage category. Impacts from anaerobic digestion, electrolysis unit process, and background emissions from the production of supplementary chemicals that are used for the fermentation of MP had considerable impacts on theenvironmental profile. However, having compared 1 ton of pure protein from MP against conventional proteinaceous feed sources such as soybean and fishmeal, better perspectives are imagined for the future of MP. In terms of economic performance, although gross profit was slightly promising, but the overall economic performance of MP production was not encouraging. The high CAPEX has undermined the economic feasibility of MP production. More specifically, conventional and low-efficient fermenters increase investment costs. Sensitivity analysis indicated that improving pure protein yield on ammonium nitrogen and replacing conventional fermenters with U-loop reactors can to some extent increase the gross profit and decrease the investment costs, respectively, although still not enough to make the process economically profitable.

ACKNOWLEDGEMENT

We acknowledge support from the FUBAF project financed by the Danish EPA – MUDP (J.nr. Mst-11700508).

DATA AVAILABILITY STATEMENT

All relevant data are included in the paper or its Supplementary Information.

REFERENCES

Acién, F., Fernández, J., Magán, J. & Molina, E. 2012 Production cost of a real microalgae production plant and strategies to reduce it. *Biotechnology Advances* **30** (6), 1344–1353.

Acosta, N., Sakarika, M., Kerckhof, F.-M., Law, C. K. Y., De Vrieze, J. & Rabaey, K. 2020 Microbial protein production from methane via electrochemical biogas upgrading. *Chemical Engineering Journal* **391**, 123625.

Almatouq, A. & Babatunde, A. 2018 Identifying optimized conditions for concurrent electricity production and phosphorus recovery in a mediator-less dual chamber microbial fuel cell. *Applied Energy* **230**, 122–134.

Alvarado-Morales, M., Kuglarz, M., Tsapekos, P. & Angelidaki, I. 2021 Municipal biopulp as substrate for lactic acid production focusing on downstream processing. *Journal of Environmental Chemical Engineering* **9** (2), 105136.

Arredondo, M. R., Kuntke, P., Jeremiasse, A., Sleutels, T., Buisman, C. & Ter Heijne, A. 2015 Bioelectrochemical systems for nitrogen removal and recovery from wastewater. *Environmental Science: Water Research & Technology* **1** (1), 22–33.

Bao, Z., Sun, S. & Sun, D. 2016 Assessment of greenhouse gas emission from A/O and SBR wastewater treatment plants in Beijing, China. *International Biodeterioration & Biodegradation* **108**, 108–114.

Cavinato, C., Bolzonella, D., Pavan, P., Fatone, F. & Cecchi, F. 2013 Mesophilic and thermophilic anaerobic co-digestion of waste activated sludge and source sorted biowaste in pilot- and full-scale reactors. *Renewable Energy* **55**, 260–265.

Chemical Engineering 2020 *The Chemical Engineering Plant Cost Index*. Available from: https://www.chemengonline.com/pci-home/ (accessed 2020).

Chen, Q., Gu, Y., Tang, Z., Wei, W. & Sun, Y. 2018 Assessment of low-carbon iron and steel production with CO2 recycling and utilization technologies: a case study in China. *Applied Energy* **220**, 192–207.

Cimpan, C. & Wenzel, H. 2016 *Cost-efficient Management of Organic Household Waste: Collection, pre-Treatment and Digestion*. Centre for Life Cycle Engineering, SDU, Denmark.

Cumberlege, T., Blenkinsopp, T. & Clark, J. 2011 *Assessment of Environmental Impact of FeedKind Protein*. Carbon Trust, UK.

Dalgaard, R., Schmidt, J., Halberg, N., Christensen, P., Thrane, M. & Pengue, W. A. 2007 LCA of soybean meal. *The International Journal of Life Cycle Assessment* **13** (3), 240.

Danish Energy Agency 2019 *Denmark's Energy and Climate Outlook 2019 Copenhagen, Denmark*. Available from: http://www.ens.dk/outlook/ (accessed 2020).

Delgado, L., Catarino, A. S., Eder, P., Litten, D., Luo, Z. & Villanueva, A. 2009 *End-of-Waste Criteria*. JRC Scientific and Technical Reports.

Delpierre, M. 2019 *An Ex-Ante LCA Study on Wind-Based Hydrogen Production in the Netherlands*. Delft University of Technology, Leiden University.

Di Maria, F., Sordi, A., Cirulli, G. & Micale, C. 2015 Amount of energy recoverable from an existing sludge digester with the co-digestion with fruit and vegetable waste at reduced retention time. *Applied Energy* **150**, 9–14.

Drejer, A., Ritschel, T., Jørgensen, S. B. & Jørgensen, J. B. 2017 Economic optimizing control for single-cell protein production in a U-loop reactor. In: *Computer Aided Chemical Engineering*. Puigjaner, L., (ed.), Elsevier, DTU, Denmark, pp. 1759–1764.

Du, Y., Ge, Y., Ren, Y., Fan, X., Pan, K., Lin, L., Wu, X., Min, Y., Meyerson, L. A. & Heino, M. 2018 A global strategy to mitigate the environmental impact of China's ruminant consumption boom. *Nature Communications* **9** (1), 1–11.

Elyasi, S. N., He, L., Tsapekos, P., Rafiee, S., Khoshnevisan, B., Carbajales-Dale, M., Mohtasebi, S. S., Liu, H. & Angelidaki, I. 2021 Could biological biogas upgrading be a sustainable substitution for water scrubbing technology? A case study in Denmark. *Energy Conversion and Management* **245**, 114550.

Energy & Resources 2022 *Energy Producing Wastewater Treatment*. Available from: https://en.biofos.dk/produktion/energi-og-ressourcer (accessed 1 February 2022).

Faragò, M., Damgaard, A., Madsen, J. A., Andersen, J. K., Thornberg, D., Andersen, M. H. & Rygaard, M. 2021 From wastewater treatment to water resource recovery: environmental and economic impacts of full-scale implementation. *Water Research* **204**, 117554.

Finkbeiner, M., Inaba, A., Tan, R., Christiansen, K. & Klüppel, H.-J. 2006 The new international standards for life cycle assessment: ISO 14040 and ISO 14044. *The International Journal of Life Cycle Assessment* **11** (2), 80–85.

Firdaus, N., Prasetyo, B. T., Sofyan, Y. & Siregar, F. 2017 Part I of II: palm oil mill effluent (POME): biogas power plant. *Distributed Generation & Alternative Energy Journal* **32** (2), 73–79.

Ghenai, C. 2012 Life cycle analysis of wind turbine. In: Sustainable Development-Energy: Engineering and Technologies-Manufacturing and Environment Ghenai C (ed.), IntechOpen, Ocean and Mechanical Engineering Department, Florida Atlantic University USA.

Gorre, J., Ruoss, F., Karjunen, H., Schaffert, J. & Tynjälä, T. 2020 Cost benefits of optimizing hydrogen storage and methanation capacities for power-to-Gas plants in dynamic operation. *Applied Energy* **257**, 113967.

Gu, Y., Li, Y., Li, X., Luo, P., Wang, H., Wang, X., Wu, J. & Li, F. 2017 Energy self-sufficient wastewater treatment plants: feasibilities and challenges. *Energy Procedia* **105**, 3741–3751.

Guo, C. H., Stabnikov, V. & Ivanov, V. 2010 The removal of nitrogen and phosphorus from reject water of municipal wastewater treatment plant using ferric and nitrate bioreductions. *Bioresource Technology* **101** (11), 3992–3999.

Gustavsson, D., Syd, V. & Malmö, S. 2010 Biological sludge liquor treatment at municipal wastewater treatment plants–a review. *Vatten* **66** (3), 179–192.

Gylling, M. & Hermansen, J. 2018 *Kvantificering af forventede fremtidige proteinmarkeder og kortlægning af potentialer i forskellige nye proteinkilder*. University department of Food and Resource in Frederiksberg, Denmark.

Hagos, D. A. & Ahlgren, E. O. 2020 Exploring cost-effective transitions to fossil independent transportation in the future energy system of Denmark. *Applied Energy* **261**, 114389.

Hanum, F., Yuan, L. C., Kamahara, H., Aziz, H. A., Atsuta, Y., Yamada, T. & Daimon, H. 2019 Treatment of sewage sludge using anaerobic digestion in Malaysia: current state and challenges. *Frontiers in Energy Research* **7**, 19.

Herrero, M., Henderson, B., Havlík, P., Thornton, P. K., Conant, R. T., Smith, P., Wirsenius, S., Hristov, A. N., Gerber, P. & Gill, M. 2016 Greenhouse gas mitigation potentials in the livestock sector. *Nature Climate Change* **6** (5), 452–461.

Horan, N. J. 1994 *Nutrient Removal From Wastewaters*. CRC Press, Pennsylvania, US.

Huijbregts, M. A., Steinmann, Z. J., Elshout, P. M., Stam, G., Verones, F., Vieira, M., Zijp, M., Hollander, A. & van Zelm, R. 2017 Recipe2016: a harmonised life cycle impact assessment method at midpoint and endpoint level. *The International Journal of Life Cycle Assessment* **22** (2), 138–147.

Hwang, K.-L., Bang, C.-H. & Zoh, K.-D. 2016 Characteristics of methane and nitrous oxide emissions from the wastewater treatment plant. *Bioresource Technology* **214**, 881–884.

Index Mundi data 2020 *IndexMundi Data Portal, Agriculture, Denmark Proteinaceous Animal Feed Imports by Year*. Available from: https://www.indexmundi.com/agriculture/?country=dk&commodity (accessed 2020).

IndexMundi data portal 2020 *Soybean Meal Monthly Price*. Available from: https://www.indexmundi.com/commodities/?commodity=soybean-meal&months=240 (accessed 2020).

Ivy, J. 2004 *Summary of Electrolytic Hydrogen Production: Milestone Completion Report*. National Renewable Energy Lab., Golden, CO, US.

Jones, S. W., Karpol, A., Friedman, S., Maru, B. T. & Tracy, B. P. 2020 Recent advances in single cell protein use as a feed ingredient in aquaculture. *Current Opinion in Biotechnology* **61**, 189–197.

JRC 2010 *International Reference Life Cycle Data System (ILCD) Handbook – General Guide for Life Cycle Assessment -Provisions and Action Steps. EUR 24378 EN*. Publications Office of the EuropeanUnion, Luxembourg (Luxembourg).

Khoshnevisan, B., Tsapekos, P., Alvarado-Morales, M. & Angelidaki, I. 2018a Process performance and modelling of anaerobic digestion using source-sorted organic household waste. *Bioresource Technology* **247**, 486–495.

Khoshnevisan, B., Tsapekos, P., Alvarado-Morales, M., Rafiee, S., Tabatabaei, M. & Angelidaki, I. 2018b Life cycle assessment of different strategies for energy and nutrient recovery from source sorted organic fraction of household waste. *Journal of Cleaner Production* **180**, 360–374.

Khoshnevisan, B., Tsapekos, P., Zhang, Y., Valverde-Pérez, B. & Angelidaki, I. 2019 Urban biowaste valorization by coupling anaerobic digestion and single cell protein production. *Bioresource Technology* **290**, 121743.

Khoshnevisan, B., Dodds, M., Tsapekos, P., Torresi, E., Smets, B. F., Angelidaki, I., Zhang, Y. & Valverde-Pérez, B. 2020a Coupling electrochemical ammonia extraction and cultivation of methane oxidizing bacteria for production of microbial protein. *Journal of Environmental Management* **265**, 110560.

Khoshnevisan, B., Tabatabaei, M., Tsapekos, P., Rafiee, S., Aghbashlo, M., Lindeneg, S. & Angelidaki, I. 2020b Environmental life cycle assessment of different biorefinery platforms valorizing municipal solid waste to bioenergy, microbial protein, lactic and succinic acid. *Renewable and Sustainable Energy Reviews* **117**, 109493.

Khoshnevisan, B., Duan, N., Tsapekos, P., Awasthi, M. K., Liu, Z., Mohammadi, A., Angelidaki, I., Tsang, D. C., Zhang, Z. & Pan, J. 2021 A critical review on livestock manure biorefinery technologies: sustainability, challenges, and future perspectives. *Renewable and Sustainable Energy Reviews* **135**, 110033.

Khoshnevisan, B., He, L., Xu, M., Valverde-Pérez, B., Sillman, J., Mitraka, G.-C., Kougias, P. G., Zhang, Y., Yan, S. & Ji, L. 2022 From renewable energy to sustainable protein sources: advancement, challenges, and future roadmaps. *Renewable and Sustainable Energy Reviews* **157**, 112041.

KNOEMA 2016 *Enterprise Data Solutions, FAOTSJUL, Trade Statistics – Crops, Livestock Products & Live Animal*. Available from: https://knoema.com/FAOTSJUL2016/trade-statistics-crops-livestock-products-live-animal (accessed 2020).

Koch, K., Plabst, M., Schmidt, A., Helmreich, B. & Drewes, J. E. 2016 Co-digestion of food waste in a municipal wastewater treatment plant: comparison of batch tests and full-scale experiences. *Waste Management* **47**, 28–33.

Krul, E. S. 2019 Calculation of nitrogen-to-protein conversion factors: a review with a focus on soy protein. *Journal of the American Oil Chemists' Society* **96** (4), 339–364.

Kuckshinrichs, W., Ketelaer, T. & Koj, J. C. 2017 Economic analysis of improved alkaline water electrolysis. *Frontiers in Energy Research* **5**, 1.

Kuntke, P., Śmiech, K. M., Bruning, H., Zeeman, G., Saakes, M., Sleutels, T. H. J. A., Hamelers, H. V. M. & Buisman, C. J. N. 2012 Ammonium recovery and energy production from urine by a microbial fuel cell. *Water Research* **46** (8), 2627–2636.

Kuntke, P., Sleutels, T., Arredondo, M. R., Georg, S., Barbosa, S. G., Ter Heijne, A., Hamelers, H. V. & Buisman, C. 2018 (Bio) electrochemical ammonia recovery: progress and perspectives. *Applied Microbiology and Biotechnology* **102** (9), 3865–3878.

Kwan, T. H., Pleissner, D., Lau, K. Y., Venus, J., Pommeret, A. & Lin, C. S. K. 2015 Techno-economic analysis of a food waste valorization process via microalgae cultivation and co-production of plasticizer, lactic acid and animal feed from algal biomass and food waste. *Bioresource Technology* **198**, 292–299.

Kyung, D., Kim, M., Chang, J. & Lee, W. 2015 Estimation of greenhouse gas emissions from a hybrid wastewater treatment plant. *Journal of Cleaner Production* **95**, 117–123.

LaTurner, Z. W., Bennett, G. N., San, K.-Y. & Stadler, L. B. 2020 Single cell protein production from food waste using purple non-sulfur bacteria shows economically viable protein products have higher environmental impacts. *Journal of Cleaner Production* **276**, 123114.

Leonzio, G. 2017 Design and feasibility analysis of a power-to-gas plant in Germany. *Journal of Cleaner Production* **162**, 609–623.

Liu, Y., Cheng, X., Lun, X. & Sun, D. 2014 CH4 emission and conversion from A2O and SBR processes in full-scale wastewater treatment plants. *Journal of Environmental Sciences* **26** (1), 224–230.

Lynd, L. R., Wyman, C., Laser, M., Johnson, D. & Landucci, R. 2005 *Strategic Biorefinery Analysis: Analysis of Biorefineries*. National Renewable Energy Lab.(NREL), Golden, CO, United States.

Marami, H., He, L., Rafiee, S., Khoshnevisan, B., Tsapekos, P., Mobli, H., Elyasi, S. N., Liu, H. & Angelidaki, I. 2022 Bridging to circular bioeconomy through a novel biorefinery platform on a wastewater treatment plant. *Renewable and Sustainable Energy Reviews* **154**, 111895.

Masuda, S., Suzuki, S., Sano, I., Li, Y.-Y. & Nishimura, O. 2015 The seasonal variation of emission of greenhouse gases from a full-scale sewage treatment plant. *Chemosphere* **140**, 167–173.

Matassa, S., Batstone, D. J., Hülsen, T., Schnoor, J. & Verstraete, W. 2015a Can direct conversion of used nitrogen to new feed and protein help feed the world? *Environmental Science & Technology* **49** (9), 5247–5254.

Matassa, S., Boon, N. & Verstraete, W. 2015b Resource recovery from used water: the manufacturing abilities of hydrogen-oxidizing bacteria. *Water Research* **68**, 467–478.

Matassa, S., Boon, N., Pikaar, I. & Verstraete, W. 2016a Microbial protein: future sustainable food supply route with low environmental footprint. *Microbial Biotechnology* **9** (5), 568–575.

Matassa, S., Verstraete, W., Pikaar, I. & Boon, N. 2016b Autotrophic nitrogen assimilation and carbon capture for microbial protein production by a novel enrichment of hydrogen-oxidizing bacteria. *Water Research* **101**, 137–146.

Matassa, S., Papirio, S., Pikaar, I., Hülsen, T., Leijenhorst, E., Esposito, G., Pirozzi, F. & Verstraete, W. 2020 Upcycling of biowaste carbon and nutrients in line with consumer confidence: the 'full gas' route to single cell protein. *Green Chemistry* **22** (15), 4912–4929.

Mehmeti, A., Angelis-Dimakis, A., Arampatzis, G., McPhail, S. J. & Ulgiati, S. 2018 Life cycle assessment and water footprint of hydrogen production methods: from conventional to emerging technologies. *Environments* **5** (2), 24.

Miyata, E. S. 1980 *Determining Fixed and Operating Costs of Logging Equipment*. Department of Agriculture, Forest Service, North Central Forest Experiment, Department of Agriculture, US.

Morales Pereira, N. 2014 *Novel Technologies for WWTP Optimization in Footprint, Nutrients Valorization, and Energy Consumption*.

Naroznova, I., Møller, J. & Scheutz, C. 2016 Global warming potential of material fractions occurring in source-separated organic household waste treated by anaerobic digestion or incineration under different framework conditions. *Waste Management* **58**, 397–407.

Nghiem, L. D., Koch, K., Bolzonella, D. & Drewes, J. E. 2017 Full scale co-digestion of wastewater sludge and food waste: bottlenecks and possibilities. *Renewable and Sustainable Energy Reviews* **72**, 354–362.

Nguyen, T. K. L., Ngo, H. H., Guo, W., Chang, S. W., Nguyen, D. D., Nghiem, L. D., Liu, Y., Ni, B. & Hai, F. I. 2019 Insight into greenhouse gases emissions from the two popular treatment technologies in municipal wastewater treatment processes. *Science of the Total Environment* **671**, 1302–1313.

OEC 2018 *Soybean Meal in Denmark, Observatory of Economic Complexity (OEC)*. Available from: https://oec.world/en/profile/bilateral-product/soybean-meal/reporter/dnk (Accessed 2020).

Panpong, K., Srisuwan, G., Sompong, O. & Kongjan, P. 2014 Anaerobic co-digestion of canned seafood wastewater with glycerol waste for enhanced biogas production. *Energy Procedia* **52**, 328–336.

Park, N. D., Thring, R. W., Garton, R. P., Rutherford, M. P. & Helle, S. S. 2011 Increased biogas production in a wastewater treatment plant by anaerobic co-digestion of fruit and vegetable waste and sewer sludge–a full scale study. *Water Science and Technology* **64** (9), 1851–1856.

Peters, M. S., Timmerhaus, K. D. & West, R. E. 2003 *Plant Design and Economics for Chemical Engineers*. McGraw-Hill, New York.

Petersen, L. A., Lieven, C., Nandy, S. K., Villadsen, J., Jørgensen, S. B., Christensen, I. & Gernaey, K. V. 2019 Dynamic investigation and modeling of the nitrogen cometabolism in Methylococcus capsulatus (Bath). *Biotechnology and Bioengineering* **116** (11), 2884–2895.

Pöschl, M., Ward, S. & Owende, P. 2010 Evaluation of energy efficiency of various biogas production and utilization pathways. *Applied Energy* **87** (11), 3305–3321.

Rodríguez Arredondo, M., Kuntke, P., Jeremiasse, A. W., Sleutels, T. H. J. A., Buisman, C. J. N. & ter Heijne, A. 2015 Bioelectrochemical systems for nitrogen removal and recovery from wastewater. *Environmental Science: Water Research & Technology* **1** (1), 22–33.

Schmidt, J. H. & Weidema, B. P. 2008 Shift in the marginal supply of vegetable oil. *The International Journal of Life Cycle Assessment* **13** (3), 235.

Shuuluka, D., Bolton, J. J. & Anderson, R. J. 2013 Protein content, amino acid composition and nitrogen-to-protein conversion factors of Ulva rigida and Ulva capensis from natural populations and Ulva lactuca from an aquaculture system, in South Africa. *Journal of Applied Phycology* **25** (2), 677–685.

Sillman, J., Uusitalo, V., Ruuskanen, V., Ojala, L., Kahiluoto, H., Soukka, R. & Ahola, J. 2020 A life cycle environmental sustainability analysis of microbial protein production via power-to-food approaches. *The International Journal of Life Cycle Assessment* **25** (11), 2190–2203.

Skadborg, M. M. 2018 *Tailoring the Macromolecular Composition of Methanotrophic Bacteria Cultivated in Bubble Free Continuous Flow Reactors*. Bachelor Project, Technical University of Denmark Dtu | Kongens Lyngby, Department of Environmental Engineering, Denmark.

Statista Data Platform Average prices of diesel fuel in Denmark from 2000 to 2021. Available from: https://www.statista.com/statistics/603698/diesel-fuel-prices-denrmark/ (accessed 2020).

Statistics Denmark 2020 *The Statistics on the Labor Salary in Denmark*. Available from: https://www.statbank.dk/statbank5a/selectvarval/saveselections.asp (accessed 2020).

Strong, P. J., Xie, S. & Clarke, W. P. 2015 Methane as a resource: can the methanotrophs add value? *Environmental Science & Technology* **49** (7), 4001–4018.

Sun, S., Bao, Z. & Sun, D. 2015 Study on emission characteristics and reduction strategy of nitrous oxide during wastewater treatment by different processes. *Environmental Science and Pollution Research* **22** (6), 4222–4229.

Thema, M., Bauer, F. & Sterner, M. 2019 Power-to-Gas: Electrolysis and methanation status review. *Renewable and Sustainable Energy Reviews* **112**, 775–787.

Tian, X., Richardson, R. E., Tester, J. W., Lozano, J. L. & You, F. 2020 Retrofitting municipal wastewater treatment facilities toward a greener and circular economy by virtue of resource recovery: techno-economic analysis and life cycle assessment. *ACS Sustainable Chemistry & Engineering* **8** (36), 13823–13837.

Tsapekos, P., Khoshnevisan, B., Zhu, X., Zha, X. & Angelidaki, I. 2019 Methane oxidising bacteria to upcycle effluent streams from anaerobic digestion of municipal biowaste. *Journal of Environmental Management* **251**, 109590.

Tsapekos, P., Zhu, X., Pallis, E. & Angelidaki, I. 2020 Proteinaceous methanotrophs for feed additive using biowaste as carbon and nutrients source. *Bioresource Technology* **313**, 123646.

Tsapekos, P., Khoshnevisan, B., Alvarado-Morales, M., Zhu, X., Pan, J., Tian, H. & Angelidaki, I. 2021 Upcycling the anaerobic digestion streams in a bioeconomy approach: a review. *Renewable and Sustainable Energy Reviews* **151**, 111635.

Tsapekosa, P., Khoshnevisana, B., Valverde-Péreza, B., Zhanga, Y. & Angelidakia, I. 2019 Single cell protein production using biogas and effluent produced from anaerobic digestion of urban bio-waste. In *16th Anaerobic Digestion World Congress*.

Uwizeye, A., de Boer, I. J., Opio, C. I., Schulte, R. P., Falcucci, A., Tempio, G., Teillard, F., Casu, F., Rulli, M. & Galloway, J. N. 2020 Nitrogen emissions along global livestock supply chains. *Nature Food* **1** (7), 437–446.

Verbeeck, K., De Vrieze, J., Pikaar, I., Verstraete, W. & Rabaey, K. 2020 Assessing the potential for up-cycling recovered resources from anaerobic digestion through microbial protein production. *Microbial Biotechnology* **14** (3), 897–910.

Verbeeck, K., De Vrieze, J., Pikaar, I., Verstraete, W. & Rabaey, K. 2021 Assessing the potential for up-cycling recovered resources from anaerobic digestion through microbial protein production. *Microbial Biotechnology* **14** (3), 897–910.

Vinnerås, B., Palmquist, H., Balmér, P. & Jönsson, H. 2006 The characteristics of household wastewater and biodegradable solid waste – a proposal for new Swedish design values. *Urban Water Journal* **3** (1), 3–11.

WATER VISION DENMARK 2021 *Economic Benefits of Energy Efficiency in Danish Wastewater Treatment.* Available from: https://www.vandvision.dk/wp-content/uploads/2021/09/Memo-on-the-Economic-Benefits-of-Energy-Efficiency-in-Danish-Wastewater-Treatment-002.pdf (accessed 1 February 2022).

Wehner, M., Lichtmannegger, T., Robra, S., Lopes, A. D. C. P., Ebner, C. & Bockreis, A. 2021 The economic efficiency of the co-digestion at WWTPs: a full-scale study. *Waste Management* **133**, 110–118.

Weidema, B. P., Frees, N. & Nielsen, A.-M. 1999 Marginal production technologies for life cycle inventories. *The International Journal of Life Cycle Assessment* **4** (1), 48–56.

Wong, J. W., Kaur, G., Mehariya, S., Karthikeyan, O. P. & Chen, G. 2018 Food waste treatment by anaerobic co-digestion with saline sludge and its implications for energy recovery in Hong Kong. *Bioresource Technology* **268**, 824–828.

Xavier, L. D., Cammarota, M. C., Yokoyama, L. & Volschan, I. 2014 Study of the recovery of phosphorus from struvite precipitation in supernatant line from anaerobic digesters of sludge. *Water Science and Technology: Water Supply* **14** (5), 751–757.

Yang, Z., Tsapekos, P., Zhang, Y., Zhang, Y., Angelidaki, I. & Wang, W. 2021 Bio-electrochemically extracted nitrogen from residual resources for microbial protein production. *Bioresource Technology* **337**, 125353.

Zauner, A., Böhm, H., Rosenfeld, D. C. & Tichler, R. 2019 *Innovative Large-Scale Energy Storage Technologies and Power-to-Gas Concepts After Optimization. D7. 7. Analysis on Future Technology Options and on Techno-Economic Optimization.* Report Ares, 4315303.

Zauner, A., Böhm, H., Rosenfeld, D. & Tichler, R. 2020 *D7. 7–Analysis on Future Technology Options and on Techno-Economic Optimization*. EU Horizon.

Zha, X., Tsapekos, P., Alvarado-Morales, M., Lu, X. & Angelidaki, I. 2020 Potassium inhibition during sludge and biopulp co-digestion; experimental and model-based approaches. *Waste Management* **113**, 304–311.

Zha, X., Tsapekos, P., Zhu, X., Khoshnevisan, B., Lu, X. & Angelidaki, I. 2021 Bioconversion of wastewater to single cell protein by methanotrophic bacteria. *Bioresource Technology* **320**, 124351.

Zhang, Y. & Li, H. 2019 Energy recovery from wastewater treatment plants through sludge anaerobic digestion: effect of low-organic-content sludge. *Environmental Science and Pollution Research* **26** (30), 30544–30553.

Zorpas, A. A. 2016 Sustainable waste management through end-of-waste criteria development. *Environmental Science and Pollution Research* **23** (8), 7376–7389.

Zou, S. & He, Z. 2018 Efficiently 'pumping out' value-added resources from wastewater by bioelectrochemical systems: a review from energy perspectives. *Water Research* **131**, 62–73.

First received 30 November 2021; accepted in revised form 7 March 2022. Available online 18 March 2022

doi: 10.2166/wst.2022.052

Evaluating the potential impact of energy-efficient ammonia control on the carbon footprint of a full-scale wastewater treatment plant

Riccardo Boiocchi and Giorgio Bertanza *

DICATAM – Department of Civil, Environmental, Architectural Engineering and Mathematics, Universita degli Studi di Brescia, via Branze 43, 25123, Brescia, Italy

*Corresponding author. E-mail: giorgio.bertanza@unibs.it

RB, 0000-0003-2462-1031; GB, 0000-0002-2965-023X

ABSTRACT

An assessment was performed for elucidating the possible impact of different aeration strategies on the carbon footprint of a full-scale wastewater treatment plant. Using a calibrated model, the impact of different aeration strategies was simulated. The ammonia controller tested showed its ability in ensuring effluent ammonia concentrations compliant with regulation along with significant savings on aeration energy, compared to fixed oxygen set point (DO_{sp}) control strategies. At the same time, nitrous oxide emissions increased due to accumulation of nitrification intermediates. Nevertheless, when coupled with the carbon dioxide emissions due to electrical energy consumption for aeration, the overall carbon footprint was only marginally affected. Using the local average CO_2 emission factor, ammonia control slightly reduced the carbon footprint with respect to the scenario where DO_{sp} was fixed at 2 $mg \cdot L^{-1}$. Conversely, no significant change could be detected when compared against the scenarios where the DO_{sp} was fixed. Overall, the actual impact of ammonia control on the carbon footprint compared to other aeration strategies was found to be strictly connected to the sources of energy employed, where the larger amount of low CO_2-emitting energy is, the higher the relative increase in the carbon footprint will be.

Key words: carbon footprint, energy saving, fuzzy, modelling, nitrogen cycle, nitrous oxide

HIGHLIGHTS

- Oxygen and ammonia control are compared using a calibrated full-scale wastewater model.
- Ammonia control is found to enable savings on aeration energy (7–23%) while increasing N_2O emissions.
- Overall WWTP carbon footprint is affected by assumed CO_2 emission factors for electrical energy.
- Assuming the average local CO_2 emission factor, ammonia and oxygen control show a similar carbon footprint.

1. INTRODUCTION

In wastewater treatment, effluent ammonia nitrogen is an important parameter that needs to be compliant with prescribed regulatory limits. Over time, the strong influent wastewater fluctuations with respect to flow rate and concentrations have increased the need for automatic control applications aimed at stabilizing process performance (Olsson 2007). In the aerated tanks of wastewater treatment plants (WWPTs), oxygen controllers are now commonly applied (Khatri *et al.* 2020). These controllers automatically change the oxygen or air supply in order to minimize the difference between a fixed dissolved oxygen set point (DO_{sp}) and the measured DO concentration. By keeping the oxygen concentration to a particular level, the operator attempts to ensure a good nitrification rate at all times, which is needed to keep effluent ammonia compliant with regulations. However, the choice of the optimal DO_{sp} is not always straightforward and frequent manual adjustments may need to take place to ensure good effluent quality under unfavourable influent and environmental conditions. As a matter of fact, under nitrogen loading higher than usual, the chosen DO_{sp} may be lower than the optimal one, leading to effluent ammonia nitrogen higher than the limit. On the other hand, under low nitrogen loading conditions the chosen DO_{sp} may be higher than the optimal one, thus leading to a waste of energy linked to air supply. To automate the change of DO_{sp}, feedback ammonium controllers were later developed (Medinilla *et al.* 2020). Ammonia control is based on automatically changing the DO_{sp} in the function of the discrepancy between a set point for effluent ammonium concentration and its measurement. While ensuring effluent ammonia concentrations compliant with regulations, this control strategy can actually save on energy for the air supply system. In turn, the emissions of carbon dioxide (CO_2) due to the production of energy are expected to decrease. However, the carbon footprint of WWTPs is given not only by the CO_2 emissions related to power consumption but also by the emissions of methane and nitrous oxide (N_2O). The latter is a powerful greenhouse gas with a global warming potential about 300 times larger than that of CO_2 (IPCC 2007). In the last couple of decades, the problem of N_2O emissions from WWTPs has increasingly gained attention (Law *et al.* 2012; Daelman *et al.* 2015; Mannina *et al.* 2016; Boiocchi *et al.* 2017a). There are various pathways according to which N_2O is found to be generated in a WWTP (Kampschreur *et al.* 2009). Low oxygen concentrations in aerated tanks are known to trigger autotrophic denitrification, where ammonia-oxidizing bacteria (AOB) use nitrite nitrogen instead of oxygen as an electron acceptor for ammonia oxidation and thus produce N_2O as an end product of nitrite enzymatic reduction (Peng *et al.* 2014). Based on this, an ammonia control strategy reducing the oxygen concentration to what is strictly needed to have effluent ammonia concentration compliant with regulation may potentially trigger N_2O emissions. In addition, an incomplete heterotrophic denitrification due to excessive oxygen and/or low biodegradable carbon in the anoxic compartment can lead to N_2O accumulation and thus increase N_2O emissions (Law *et al.* 2012). Furthermore, another AOB-mediated N_2O-producing pathway – known as incomplete hydroxylamine (NH_2OH) oxidation – was found to be favoured by very high oxygen concentrations (Domingo-Félez *et al.* 2017). Therefore, it is not clear from a theoretical perspective how the application of an ammonia control strategy will affect the N_2O emissions from WWTPs and, considering the possible reduction in the CO_2 emissions due to expected aeration energy savings, how the overall carbon footprint will be. With the increasing demand for carbon neutral WWTP operation, it is important to understand the overall behaviour of the carbon footprint in WWTPs under different aeration regulation patterns. A control strategy may be able to reduce N_2O emissions on one hand, while increasing the aeration energy consumption and the CO_2 emitted thereby on the other. In some of these cases, the advantage in terms of N_2O emission reduction may be negatively compensated by increased CO_2 emissions linked to aeration energy (Santín *et al.* 2017). It can also be that in some scenarios reducing the N_2O emissions will not lead to an increased energy demand. For these reasons, the present work aims at elucidating the combined effect that different aeration control strategies may have on the carbon footprint of WWTPs. This is performed according to a model-based integrated approach where other basic performance parameters such as the effluent quality and the energy consumption are considered as well. This is because, when studying the carbon footprint of WWTPs, the main target of ensuring good effluent quality within reasonable energy costs is still of great importance.

2. MATERIALS AND METHODS

To achieve the goal presented in the introduction, the work was structured as follows: first, a model mathematical structure for the WWT system described in Section 2.1 was defined as presented in Section 2.2; secondly, the model was calibrated as described in Section 2.3; and, thirdly, simulations with different aeration setups were run using the calibrated model and the results obtained thereby were compared against one another as defined in Section 2.4.

2.1. System description

In this work, a model was developed and calibrated in order to describe the main biological and physical processes occurring in one of the six activated sludge lines of a WWTP located in Northern Italy with an overall treatment capacity of 330,000 p.e. As shown in Figure 1, the modelled plant layout consists of a biological compartment with an overall volume of 7250 m^3 followed by two secondary clarifiers in parallel, each having a horizontal area of 934 m^2 and a side water depth of approximately 4 meters. The biological compartment is made of four non-aerated compartments in series followed by a sequence of eight aerated compartments. These last are equipped with an air supply system consisting of a single air blower providing air to fine-bubble membrane diffusers.

The WTTP must comply with local regulations in force (Regional Council of Veneto 2009).

2.2. Definition of the model mathematical structure

The biological processes occurring in the biological compartment were mathematically described by the Activated Sludge Model for Nitrogen by Hiatt & Grady (2008) updated with the AOB-mediated processes by Pocquet *et al.* (2016). While the former incorporates the N_2O production by heterotrophic biomass, the latter includes the production of N_2O according to both AOB denitrification and incomplete NH_2OH oxidation pathways. Thus, all the main N_2O production pathways were

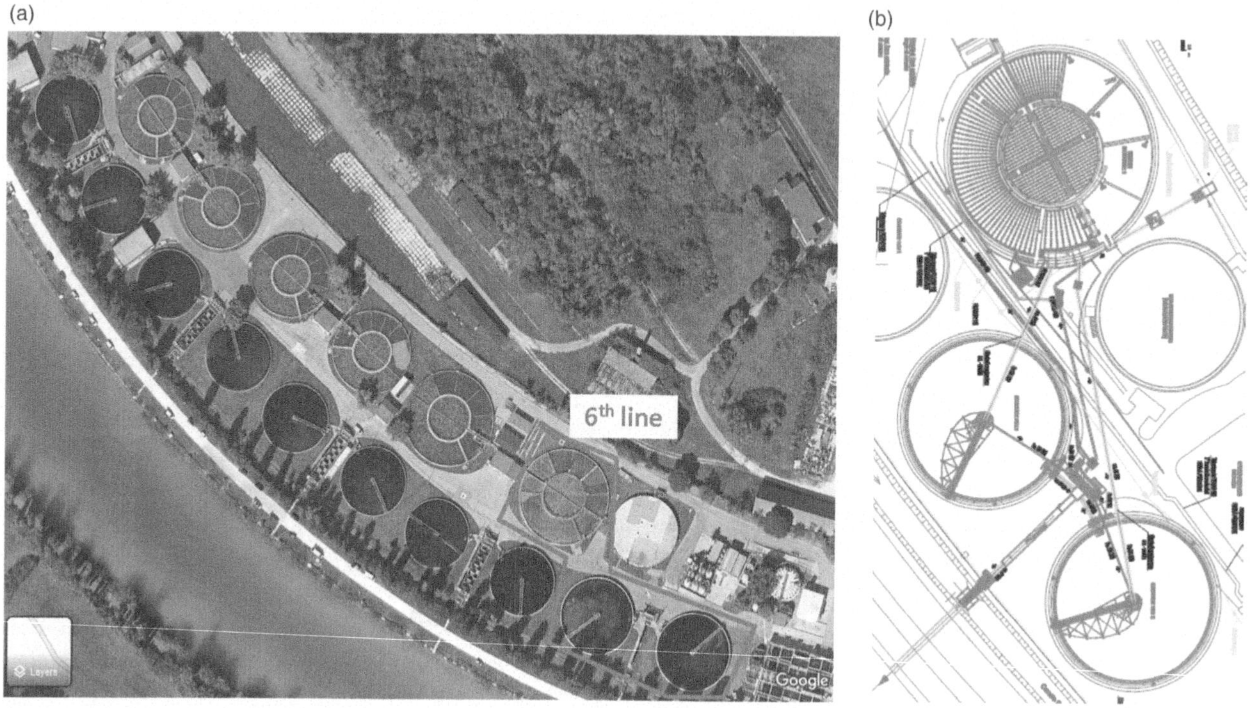

Figure 1 | (a) Layout of the overall WWTP (from Google Maps), and (b) detail of the sixth line modelled.

considered. Abiotic N_2O-producing processes were not included in virtue of their low rates under the typical pH values established during the typical nitrogen removal processes (Su *et al.* 2019). The Petersen matrix of the resulting biological model used in this work is presented in Supplementary Material (A). In addition, the liquid-gas mass transfer of N_2O was included as a function of the oxygen mass transfer coefficient and the ratio between oxygen and N_2O molecular diffusivities in water according to Foley *et al.* (2010). As carried out by Flores-Alsina *et al.* (2014), the mass transfer of nitric oxide and dinitrogen was modelled with the same approach of N_2O.

In an effort to realistically reproduce its actual operation, the air supply system was modelled using algebraic equations defining first the standard oxygen transfer rate (SOTR) (see Equation (1)) and, secondly, the oxygen mass transfer coefficient (see Equation (2)):

$$SOTR = SOTE \cdot \rho_{O_2} \cdot f_{O_2,air} \cdot Q_{air,tot} \tag{1}$$

$$k_L a = \frac{SOTR \cdot \alpha \cdot \theta^{T-20}}{(\delta \cdot C_{sat}) \cdot V} \tag{2}$$

In Equations (1) and (2), SOTE is the Standard Oxygen Transfer Efficiency, α is the alpha-factor, ρ_{O_2} is the density of oxygen, $f_{O_2,air}$ is the mass percentage of oxygen in the aspirated air, $Q_{air,tot}$ is the total volumetric air flow rate diffused in standard conditions, θ is the temperature correction factor for the oxygen mass transfer, δ is the correction factor for the oxygen saturation concentration (C_{sat}), and V is the liquid wastewater volume.

Additionally, with regards to Equation (1), the characteristic relationship between the SOTE and the air flow rate from each diffuser was implemented based on the diffuser data sheet. Furthermore, real minimum and maximum boundaries for the total air flow rate carried to the biological system were retrieved from the operating curve of the air blower and thus considered during simulation. Finally, although more work should be done to characterize the relationship between the α-factor and the mixed liquor qualitative characteristics, a constant α-factor of 0.7 was chosen for all the aerated tanks for the sake of simplicity. In the absence of specific data, introducing complex formulas for the estimation of α-factor would increase the model uncertainty. The choice of using the value of 0.7 was taken based on the works by Jiang *et al.* (2017) and Ahmed *et al.* (2021). Ahmed *et al.* (2021) found a relationship between the α-factor and the reactor concentration of soluble COD. Using this relationship with the soluble COD concentration predicted in our system, the α-factor results are equal or higher than 0.7. The same work showed much lower values only when the soluble COD concentration is much higher (up to 350 mg $COD{\cdot}L^{-1}$). This scenario of very high COD concentrations typically occurs at the beginning of the cycle in sequencing-batch reactors while the system modelled in this work is continuous. Similar considerations can be made by using the correlation between COD concentration and α-factor proposed by Jiang *et al.* (2017). For better choice in the future, more effort should be made to characterize the variability of α-factor in the function of reactor concentrations of soluble COD, dissolved and suspended solids.

Thanks to these modelling efforts, the actual operation and limitation of the air supply system were taken into account in this work. This allows a more realistic quantification of the electrical energy spent and, in turn, of the carbon footprint linked to aeration.

Finally, the secondary clarifiers were modelled as flat bottom unreactive tanks with constant area along the height according to the one-dimensional settling model by Takacs *et al.* (1991). Thus, N_2O production in secondary clarifiers was neglected. This was in virtue of much more limited mixing conditions inside clarifiers compared to those in the biological tanks, which means that, even if N_2O were produced, there would be very low emissions of it due to a low stripping capability. Furthermore, the same poor mixing conditions limit the contact between substrate and biomass, thus leading to rather low kinetic rates for the anoxic N_2O producing processes. In addition, high denitrification rates within secondary clarifiers should lead to problems related to rising sludge and consequent poor settling efficiencies due to high nitrogen gas production (Henze *et al.* 1993). However, the TSS data used in our work do not suggest any sludge settling issues. In the literature, evidence about the contribution of N_2O emissions from secondary clarifiers compared to the contributions from the biological tanks seem limited and contradictory. As a matter of fact, only one published work found significant N_2O emission contribution from secondary clarifiers (Mikola *et al.* 2014) whereas Caniani *et al.* (2019) found the opposite.

2.3. Model calibration

Two independent data sets were employed for model calibration: one collected during the plant operation under fixed DO_{sp} control (Data Set 1) and the other collected during the plant operation with the ammonia fuzzy-logic controller described in Bertanza *et al.* (2021a) (Data Set 2). It is important to point out that the DO_{sp} could not always be tracked due to physical limitations of the air supply system where the lower saturation limit of the air flow rate was reached at low influent TKN loads and consequently the oxygen concentration became higher than the set point. The use of both data sets obtained under different oxygenation regulations allowed calibrating the biomass response to a highly heterogeneous oxygen profile, which in turn could augment the model descriptive reliability. Both the first and the second data sets comprise the daily average influent concentrations of ammonium, TKN, COD, BOD, the daily average temperature with varying frequencies within a week, and the influent flow rate with a 15-minute frequency. For Data Set 1, mean and standard deviation values for ammonium, TKN, COD, BOD, temperature and flow rate are as follows: 11.5 ± 4.9 g $N{\cdot}m^{-3}$, 18.1 ± 6.9 g $N{\cdot}m^{-3}$, 202 ± 66 g $COD{\cdot}m^{-3}$, 111.1 ± 36.2 g $BOD_5{\cdot}m^{-3}$, 15.8 ± 3.6 °C, $36{,}946 \pm 8900$ $m^3{\cdot}d^{-1}$, respectively, while for Data Set 2 the same variables are as follows: 12.1 ± 5.2 g $N{\cdot}m^{-3}$, 19.8 ± 7.9 g $N{\cdot}m^{-3}$, 177.7 ± 36.7 g $COD{\cdot}m^{-3}$, 88.9 ± 37 g $BOD_5{\cdot}m^{-3}$, 18.6 ± 4.7 °C, $39{,}772 \pm 9694.8$ $m^3{\cdot}d^{-1}$, respectively. Unmeasured daily average influent data were estimated by linear interpolation between the two closest known values. Then, the mapping of the influent data thereby obtained into the state variables of the model described in Section 2.1 was performed as illustrated in Supplementary Material (B). After the mapping, in an effort to include more realism, daily dynamic profiles for the influent concentrations of pollutants and influent temperature were created based on the measured dynamic profile of the flow rate. A more detailed description of this procedure is provided in Supplementary Material (C).

Besides the influent characterization, operating conditions such as the instantaneous oxygen concentrations in the sixth aerobic compartment and the daily average TSS concentrations in the last aerobic compartment and in the stream recycled from the secondary clarifier were provided. To reproduce the same operating conditions during calibration, an ideal proportional-integral (PI) controller having dynamic set points for the DO in the sixth aerobic tank equal to the instantaneous measured ones was implemented, while another PI controller was employed to regulate the wastage flow rate at the aim of establishing the TSS concentrations in the last aerobic tank equal to measured ones. The internal recycle flow rate was deduced based on reported working days of dedicated pumps. Regarding the effluent data, both Data Set 1 and Data Set 2 provided the effluent daily average concentrations of ammonia, nitrite and nitrate nitrogen while only Data Set 1 additionally provided the effluent daily average TSS concentrations. No data regarding N_2O emissions was available.

Starting from their default values reported in literature (Hiatt & Grady 2008; Guo & Vanrolleghem 2014; Pocquet *et al.* 2016), the model parameters related to the solid settling velocity and the aerobic activity of AOB and nitrite-oxidising bacteria (NOB), as well as the anoxic reduction factor for the heterotrophic yield and the nitrate half-saturation coefficient for heterotrophs, were iteratively changed until a sufficiently good match between model predictions and the corresponding effluent data was achieved. Root mean squared error (RMSE) was computed in order to quantify the discrepancy between data and model outputs, as carried out by Vangsgaard *et al.* (2013). The parameters describing the N_2O production by AOB were kept the same as those by Pocquet *et al.* (2016) while those describing the N_2O production by HB were kept the same as those by Guo & Vanrolleghem (2014). Given the relatively low effluent concentrations of ammonium and nitrite measured, the inhibition by free ammonia and free nitrous acid on the autotrophic biomass activity originally modelled by Hiatt & Grady (2008) were suppressed during calibration.

Based on its high computational speed, the software chosen here was GPS-X 8.1 (Hydromantis).

2.4. Model simulations and scenario comparison

Using the influent characterization of Data Set 1 along with the related time series of external and internal recycle flow rates, different simulations were run with the following aeration supply setups: fixed DO_{sp} PI control with set points at 0.5, 1 and 2 $mg \cdot L^{-1}$; the ammonia fuzzy-logic control described in Bertanza *et al.* (2021a); fixed air flow rates set at the averages of those established by the controllers during the previous simulation scenarios. The wastage flow rate was kept the same as the ideal one identified by the TSS controller described in Section 2.3. For each scenario, the following quantities were evaluated: the effluent concentrations of ammonium, nitrite and nitrate nitrogen; the N_2O emissions; the electrical energy consumed by air supply system; and the carbon footprint. Specifically, the instantaneous electrical energy consumed for air supply (P_{aer}) was calculated according to Metcalf & Eddy (1991), as mathematically expressed by Equation (3):

$$P_{aer} = \frac{Q_{air}^{act} \cdot R \cdot T_{asp} \cdot \left[\left(\frac{p_{out}}{p_{in}} \right)^{0.283} - 1 \right]}{29.7 \cdot n \cdot \eta_{comp} \cdot 0.84} \cdot N_h \tag{3}$$

In Equation (3), Q_{air}^{act} is the actual mass air flow rate provided in standard conditions ($kg \cdot s^{-1}$), R is the molar gas constant equal to 8.314 ($J \cdot K^{-1} \cdot mol^{-1}$), T_{asp} is the air temperature in the air aspiration pipe (K), p_{out} is the absolute pressure in the air delivery pipe (atm), p_{in} is the absolute pressure in the air aspiration pipe (atm), n is the air constant equal to 0.283, η_{comp} is the efficiency of the compressor system assumed equal to 69%, and N_h is number of working hours per day ($h \cdot d^{-1}$).

For the calculation of the carbon dioxide (CO_2) emitted due to the production of electrical energy linked to the aeration supply system, different emission factors ($EF_{CO_2}^{elec}$) were used given the diversity of values in function of the country-specific energy-mix. Different energy mixes occur due to the employment of energy sources such as fossil, renewable (e.g. wind, photovoltaic) and nuclear energy having specific CO_2 emissions. According to the data by the European Environment Agency (2017), while Sweden had an emission factor of only 9.27 g $CO_2 \cdot kWh^{-1}$, Estonia in the same year had an emission factor of 922.4 g $CO_2 \cdot kWh^{-1}$ which is two orders of magnitude larger. To take into account this variability, the following values were chosen for $EF_{CO_2}^{elec}$ in this study:

- the value of 258.8 g $CO_2 \cdot kWh^{-1}$ characterizing the average emission factor for Italy where the plant described in Section 2.1 is located;
- the values of 9.27 g $CO_2 \cdot kWh^{-1}$ and 922.4 g $CO_2 \cdot kWh^{-1}$ representing two realistic extremes with regards to the emission factors for electrical energy.

In order to merge the carbon footprint contributions due to the production of electrical energy for air supply and due to the emission of N_2O, Equation (4) was used:

$$GHG_{tot} = P_{aer} \cdot EF_{CO_2}^{elec} + N_2O \cdot CO_{2,eq}^{N_2O} \tag{4}$$

In Equation (4), GHG_{tot} is the overall daily load of CO_2 equivalent emitted for the production of electricity for air supply and due to N_2O emissions from the plant mainstream, N_2O is the overall daily nitrous oxide load emitted from the mainstream activated sludge compartment as daily mass load, and $CO_{2,eq}^{N_2O}$ is the CO_2 equivalent for N_2O. Although the IPCC report recommends 265 as the CO_2 equivalent for N_2O (IPCC 2014), a value of 300 g $CO_2 \cdot g^{-1}$ N_2O was chosen for $CO_{2,eq}^{N_2O}$, which is very close to the one suggested by the statistical office of the European Union (i.e. 298 g $CO_2 \cdot g^{-1}$ N_2O). More insights on how this choice can change the overall carbon footprint predictions could be achieved in the future.

It can be observed that the CO_2 contributions linked to the electrical energy consumed by systems other than the one for air supply, such as the various pumping devices (e.g. the sludge recycle, the internal recycle and the sludge wastage pumps), as well as the energy consumed in the pre-treatment compartment, were not included in Equation (4). This is because, as previously mentioned, these contributions were kept the same from one scenario to the other and only the air supply was regulated differently. Hence, the energy consumed by electrical devices other than the air supply system does not change from one scenario to the other. Besides this, the effect of different aeration setups on the carbon dioxide emissions in the side-stream treatment compartments is assumed negligible here. As a matter of fact, although more studies could be performed in the future for verification, the production of wastage sludge and the corresponding methane losses from the anaerobic digester as well as the energy used for sludge dewatering were assumed to be impacted only marginally by the different aeration setups in the mainstream of the same WWTP.

With regards to the software employed, GPS-X could not be used to simulate the fuzzy-logic controller. Hence, MATLAB R2020b (The MathWorks, Natick, MA) was used for all the simulations described in this section.

3. RESULTS AND DISCUSSION

In this section, using the model whose calibration results are presented in Supplementary Material (D), a comparison among the different simulation scenarios described in Section 2.4 in terms of effluent quality, nitrous oxide emissions, energy consumption and carbon footprint is presented. The results obtained are analysed in terms of: effluent nitrogen species such as ammonium, nitrite and nitrate (Section 3.1), N_2O emissions (Section 3.2), power consumption by the air supply system (Section 3.3), and total carbon footprint (Section 3.4). As mentioned in Section 2.4, besides the scenarios where the air flow rate is automatically changed to achieve either a prefixed DO_{sp} (in the sixth aerobic tank) or a fixed value for effluent ammonium concentration, four other scenarios were run with air flow rate constant over time. The values of these flow rates were found by calculating the averages of the same variables obtained from the previously described scenarios as follows: 82,896, 92,033, and 100,879 $Nm^3 \cdot d^{-1}$, respectively from the scenarios with DO_{sp} fixed at 0.5, 1 and 2 $mg \cdot L^{-1}$, and 77,841 $Nm^3 \cdot d^{-1}$ from the ammonia control scenario.

3.1. Effluent nitrogen species

Table 1 shows the annual average effluent concentrations for ammonium, nitrite, and nitrate. In addition, the dynamic profiles for the effluent concentrations of the same N species during a high TKN loading rate period between the 230th and the 240th days) are shown in Figure 2. During that 10-day period, the average influent TKN load is 974 kg $N \cdot d^{-1}$, namely 50% larger than the overall average TKN load of 663 kg $N \cdot d^{-1}$.

Among the scenarios where the air flow rate is changed over time, effluent ammonia nitrogen concentrations are the highest when oxygen concentration is controlled at a set point of 0.5 $mg \cdot L^{-1}$. However, as the set point is increased, effluent ammonia concentrations are drastically reduced. When ammonia control is applied in place of the control strategies with constant DO_{sp} at 1 or 2 $mg \cdot L^{-1}$, the average effluent ammonia concentration increases. Nevertheless, based on the dynamic behaviour of effluent ammonia nitrogen, it was observed that the limit of 5 mg $NH_4^+ \cdot L^{-1}$ imposed by the local regulation in force is never exceeded. This indicates that, when ammonia control is adopted, oxygen is supplied in an amount strictly needed for ammonia not to overcome the local regulatory limits. On the other hand, when the control strategy with fixed

Table 1 | Effluent annual average concentrations of ammonium, nitrite, and nitrate in case of: control with fixed DO_{sp} at 0.5, 1 and 2 $mg \cdot L^{-1}$, ammonia control, constant air flow rates equal to 77,841, 82,896, 92,033, and 100,879 $Nm^3 \cdot d^{-1}$, respectively

	Units of measurement	DO_{sp} = 0.5 $mg \cdot L^{-1}$	DO_{sp} = 1 $mg \cdot L^{-1}$	DO_{sp} = 2 $mg \cdot L^{-1}$	NH_4^+ control	Q_{air} = 77,841 $Nm^3 \cdot d^{-1}$	Q_{air} = 82,896 $Nm^3 \cdot d^{-1}$	Q_{air} = 92,033 $Nm^3 \cdot d^{-1}$	Q_{air} = 100,879 $Nm^3 \cdot d^{-1}$
Ammonium	$mg N \cdot L^{-1}$	0.53	0.07	0.02	1.03	1.44	1.03	0.55	0.3
Nitrite	$mg N \cdot L^{-1}$	0.05	0.022	0.016	0.088	0.105	0.09	0.064	0.048
Nitrate	$mg N \cdot L^{-1}$	5.74	6.28	6.42	4.95	4.57	4.95	5.49	5.85

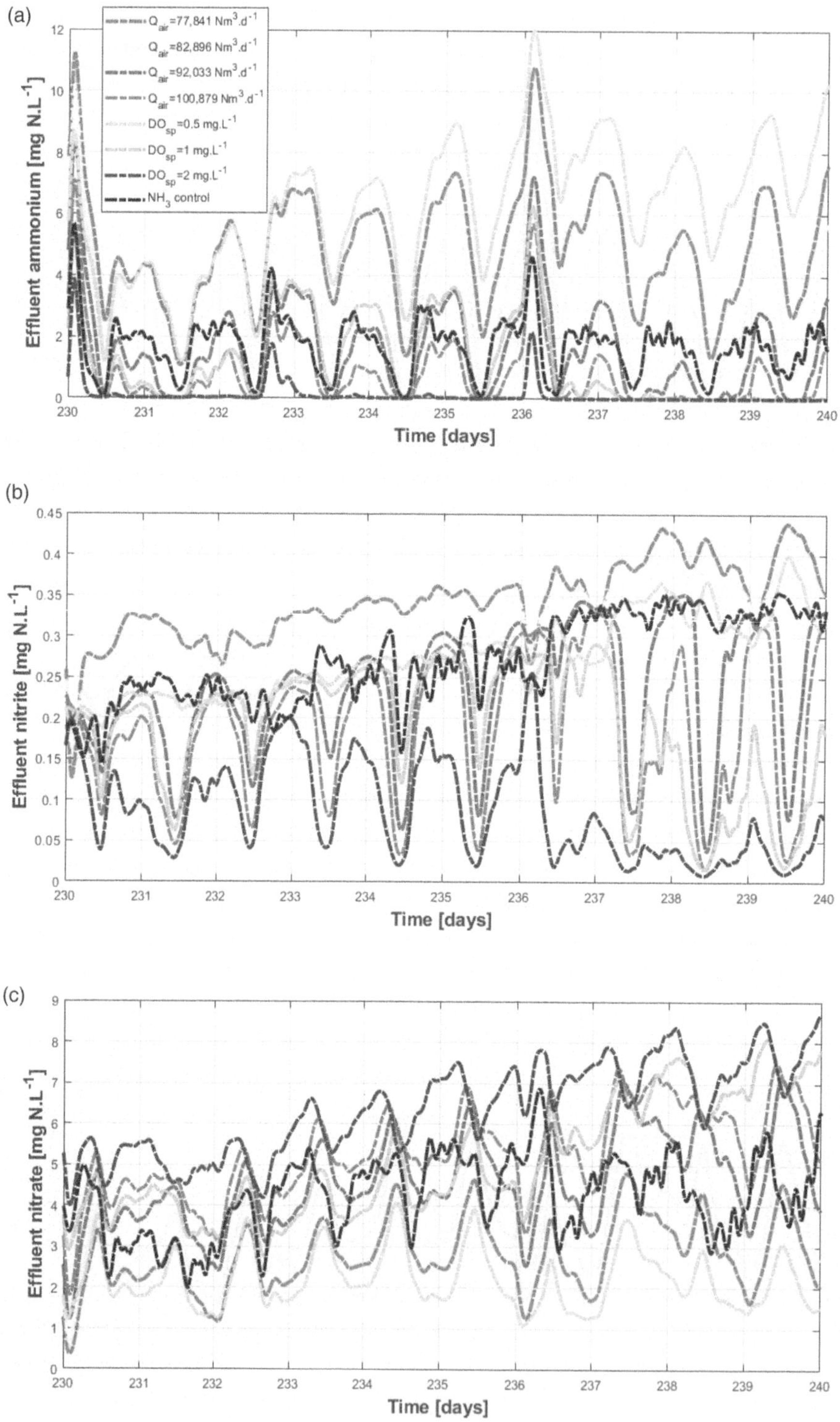

Figure 2 | Dynamic instantaneous effluent concentrations of: (a) ammonium, (b) nitrite, and (c) nitrate, obtained from the different simulation scenarios described in Section 2.4 (for interpretation of the references to colour in this figure legend, the reader is referred to the web version of this article).

DO_{sp} of 2 mg·L^{-1} is applied, oxygen is supplied excessively to the amount strictly needed, thus leading to effluent ammonia nitrogen concentrations abundantly lower than the limit. With regards to those cases where an air flow rate constant in time is applied, effluent ammonia concentrations are much higher than any other cases where the air flow rate is not constant, with the exception of the case when the air flow rate is set to the average one achieved with fixed DO_{sp} equal to 2 $mg{\cdot}L^{-1}$ (see Figure 2(a)).

Besides ammonia, it can be observed that, among the control strategies changing the oxygen supply over time, the highest nitrite concentrations in the effluent occur when the ammonia controller and the controller with DO_{sp} equal to 0.5 $mg{\cdot}L^{-1}$ are applied (see also Figure 2(b)). It can further be noted that effluent nitrite decreases with the increase in the DO_{sp}. This can be explained by the fact that the larger the oxygen concentration is, the higher the NOB activity is, the lower the effluent nitrite concentration is expected. Based on these considerations, the results suggest that, although nitrite resulted below the national regulation limit of 0.6 $mg\ N{\cdot}L^{-1}$ (Legislative Decree 3rd April n.152 2006), effluent nitrite concentrations must be kept under close surveillance.

Finally, the general behaviour of effluent nitrate can be observed to be negatively correlated with the effluent ammonia: the lower the effluent ammonia is, the higher the effluent nitrate is expected (see Figure 2(c)). This is because, when the influent characterization is the same, when more ammonia is oxidized, more nitrate is produced, provided nitrite is oxidized at a high rate.

3.2. Nitrous oxide emissions

Table 2 summarizes the annual average N_2O emissions from the entire activated sludge unit and the emission factors (i.e. N_2O emitted per unit of influent as well as per unit of removed TKN). Compared to the N_2O emission factors per unit of influent TKN load from full-scale WWTPs reported in the literature review by Daelman *et al.* (2015), these simulated values can be considered to fall within the literature typical ranges. As can be further observed, among the scenarios where a constant air flow rate is applied, N_2O emissions and the emission factors decrease at increasing influent air flow rates. Similarly, N_2O emissions decrease at increasing DO set points for those scenarios where the DO set point is kept constant. Furthermore, when the fuzzy-logic ammonia controller is implemented, N_2O emissions are significantly higher compared to any of those cases with fixed DO_{sp} control.

In order to better understand the underlying environmental conditions determining the various N_2O emission predictions, one-by-one correlations between the total N_2O emitted and the average aerobic concentrations of nitrification intermediates (such as nitrite and hydroxylamine) and of oxygen were evaluated, as presented in Figure 3. The reason for considering average values for the correlation study was in order to avoid the confounding effect of time delay between the instantaneous accumulation of a nitrogen species such as hydroxylamine or nitrite and the consequent instantaneous N_2O emission (Bertanza *et al.* 2021b). In fact, by considering the average over long periods, the time delay effect on the correlation gets largely reduced.

As can be seen in Figure 3, the N_2O emissions appear to be strongly positively correlated with the average concentrations of both the nitrification intermediates, and negatively correlated with the average oxygen concentration. By comparing the obtained regression coefficients (R^2) with one another as done by Daelman *et al.* (2015), the correlation of N_2O with the

Table 2 | Average annual N_2O emitted from the entire activated sludge unit and the emission factors per unit of influent TKN and per unit of removed TKN in case of: control with fixed DO_{sp} at 0.5, 1 and 2 $mg{\cdot}L^{-1}$, ammonia control, constant air flow rates equal to 77,841, 82,896, 92,033, and 100,879 $Nm^3{\cdot}d^{-1}$

	Units of measurement	DO_{sp} = 0.5 $mg{\cdot}L^{-1}$	DO_{sp} = 1 $mg{\cdot}L^{-1}$	DO_{sp} = 2 $mg{\cdot}L^{-1}$	NH_4^+ control	Q_{air} = 77,841 $Nm^3{\cdot}d^{-1}$	Q_{air} = 82,896 $Nm^3{\cdot}d^{-1}$	Q_{air} = 92,033 $Nm^3{\cdot}d^{-1}$	Q_{air} = 100,879 $Nm^3{\cdot}d^{-1}$
Overall N_2O emitted	$g\ N{\cdot}d^{-1}$	286.3	222	204.2	385.5	436.5	405.8	352	308.2
Emission factor per unit of influent TKN	% $g\ N_2O\text{-}N{\cdot}g^{-1}\ TKN_{in}$	0.042	0.033	0.03	0.057	0.064	0.06	0.052	0.045
Emission factor per unit of removed TKN	% $g\ N_2O\text{-}N{\cdot}g^{-1}\ TKN_{rem}$	0.046	0.034	0.032	0.061	0.073	0.067	0.056	0.048

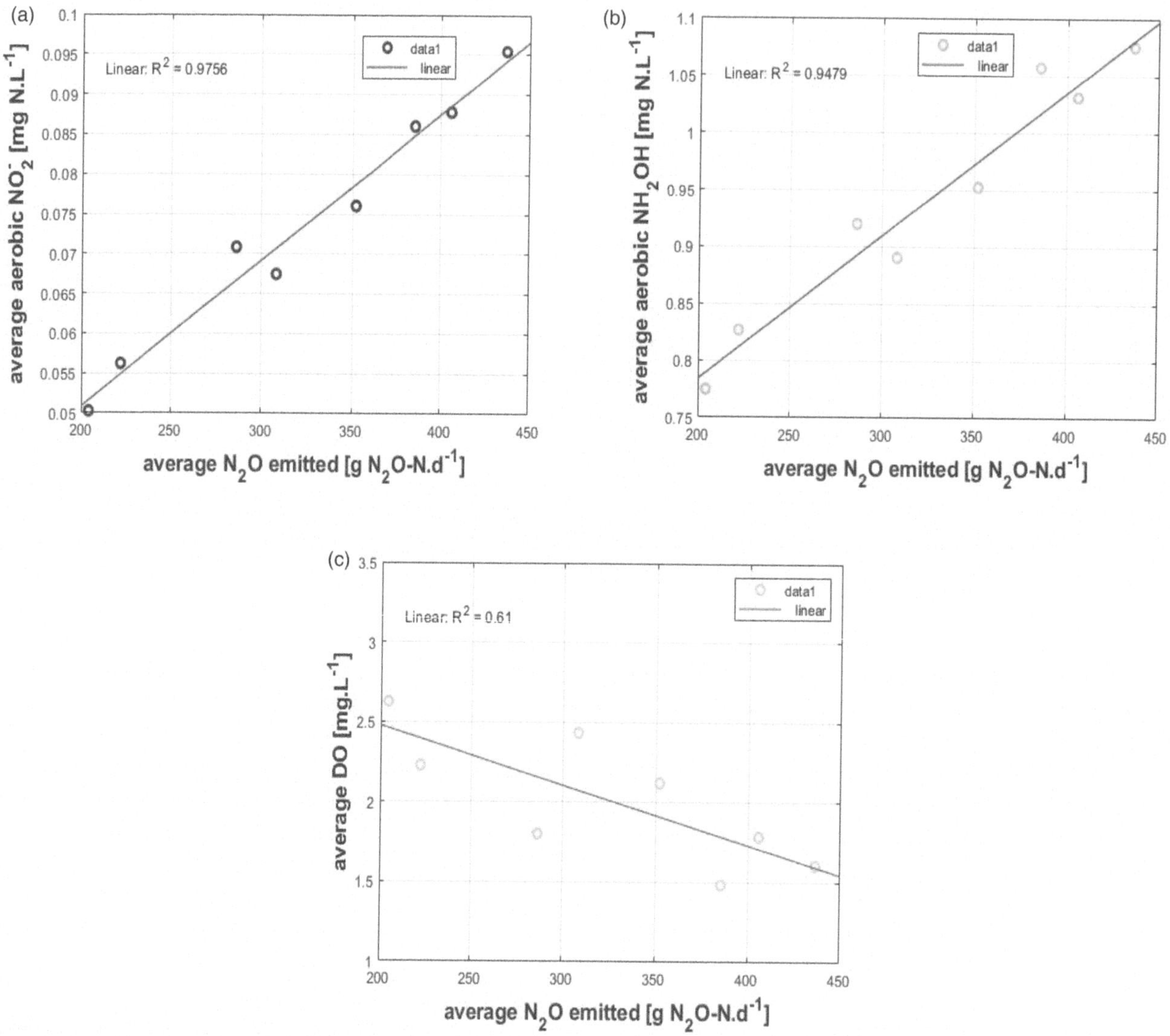

Figure 3 | Correlations between the average N_2O emissions and: (a) the average NO_2^- concentration, (b) the average NH_2OH concentration, and (c) the average DO concentration in the aerobic zone.

oxygen concentration can be considered weaker than the one with the nitrification intermediates. Nevertheless, these results suggest that suboptimal oxygen concentrations can be the leading cause for high N_2O emissions by promoting the accumulation of nitrification intermediates. This is confirmed by a large number of studies found in the literature (Tallec *et al.* 2008; Stein 2011; Peng *et al.* 2014, 2015; He *et al.* 2017; Kuokkanen *et al.* 2021). It must be acknowledged that this is only a preliminary analysis and that more investigations about the N_2O dynamics predicted by the model used in this work need to be conducted.

3.3. Aeration energy requirements

The estimated annual average electrical energy consumptions by the air supply system are reported in Table 3 while Figure 4 shows the dynamic profiles of the air flow rate supplied according to the different control strategies along with the influent TKN load under high and low influent TKN load conditions (Figure 4(a) and 4(b), respectively).

As can be seen from Table 3, for the scenarios where a fixed DO_{sp} is imposed, aeration energy consumption increases at increased set points. Similarly, for the scenarios with constant air supply, energy consumptions increase at increased air flow rate supplied. Importantly, the ammonia control is found to lead to a significant saving on aeration energy compared to the cases with fixed DO_{sp} control. Specifically, the average savings over one year of operation were estimated to range between 7

Table 3 | Daily energy consumption for air supply, calculated as average on an yearly basis, in case of: control with fixed DO_{sp} at 0.5, 1, and 2 $mg{\cdot}L^{-1}$, ammonia control, constant air flow rates equal to 77,841, 82,896, 92,033, and 100,879 $Nm^3{\cdot}d^{-1}$

	Units of measurement	DO_{sp} = 0.5 $mg{\cdot}L^{-1}$	DO_{sp} = 1 $mg{\cdot}L^{-1}$	DO_{sp} = 2 $mg{\cdot}L^{-1}$	NH_4^+ control	Q_{air} = 77,841 $Nm^3{\cdot}d^{-1}$	Q_{air} = 82,896 $Nm^3{\cdot}d^{-1}$	Q_{air} = 92,033 $Nm^3{\cdot}d^{-1}$	Q_{air} = 100,879 $Nm^3{\cdot}d^{-1}$
Energy consumed by the air supply system	$kWh{\cdot}d^{-1}$	1737.8	1932.5	2119.2	1630	1624.1	1729.5	1920.2	2104.7

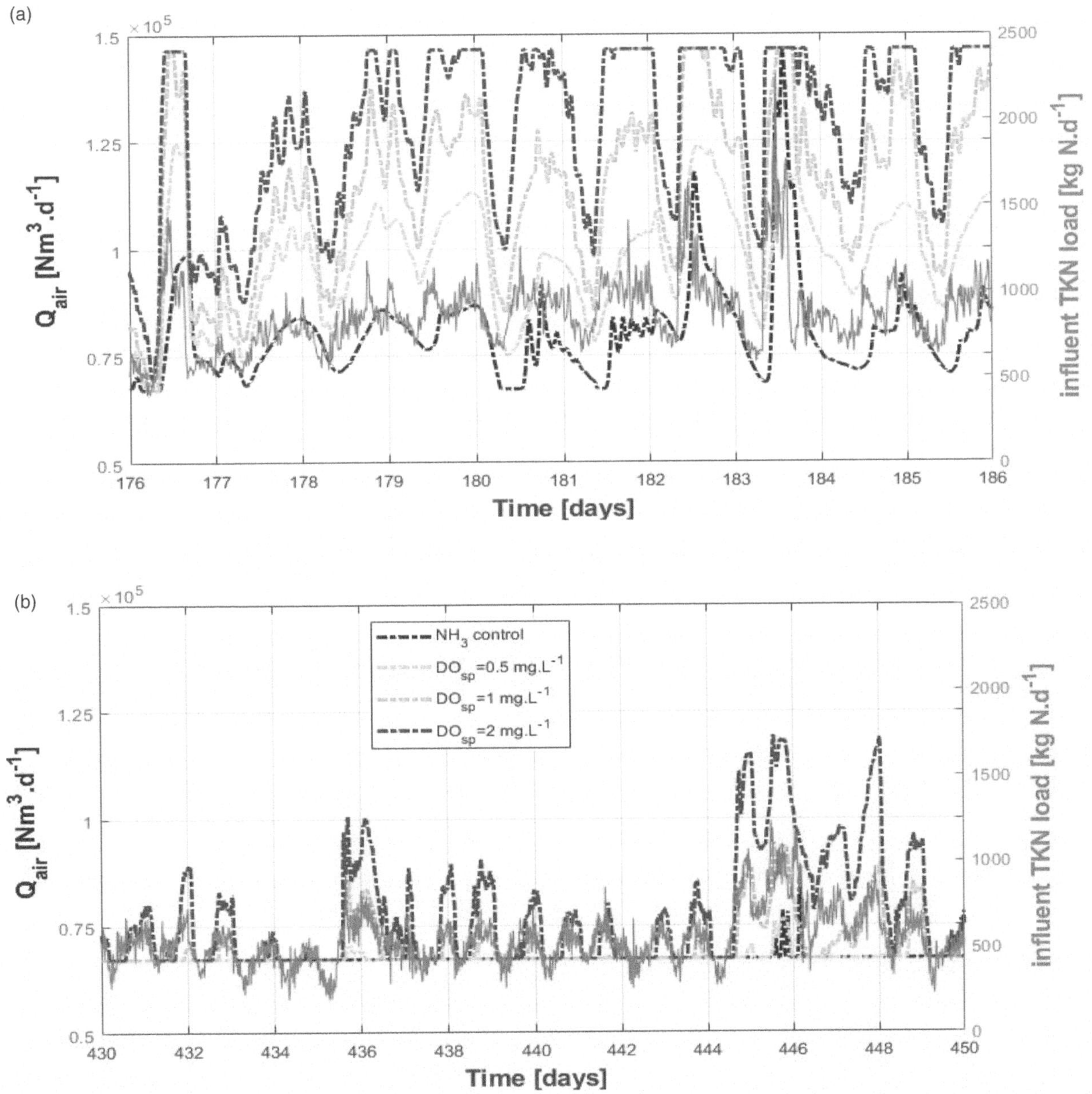

Figure 4 | Dynamic profiles of the total air flow rate supplied when control with constant DO_{sp} at 0.5, 1 and 2 $mg{\cdot}L^{-1}$ (respectively in cyan, green, and blue colours) is applied and when ammonia control is applied (in black) and of the influent TKN load (in orange). Figure 4(a) is a scenario of high TKN load while Figure 4(b) is a scenario of low TKN load (for interpretation of the references to colour in this figure legend, the reader is referred to the web version of this article).

and 23% depending on whether the comparison is made against the control strategy with fixed DO_{sp} equal to 0.5 mg·L^{-1} or against the control strategy with fixed DO_{sp} equal to 2 mg·L^{-1}, respectively. As can be seen in Figure 4, the higher the influent TKN load is, the larger the difference among the various air flow rates become. Conversely, when the influent TKN load is low, the controllers reach the lower boundary specific of the blower and the air flow rate values become closer to one another (see Figure 4(b)). It follows that the advantages in terms of energy savings through ammonia control occur mainly at high influent TKN loads. Finally, it is important to acknowledge that these results could change if a different α-factor were assumed during the simulations. More work and data are needed to identify the true relationship between α-factor and environmental conditions such as reactor concentrations of solids and soluble COD altogether.

3.4. Carbon footprint

The total carbon footprints calculated as the sum of the contribution by the N_2O emissions and the CO_2 emitted for the production of electrical energy for aeration by assuming different CO_2 emission factors for electrical energy production and their relative contributions are displayed in Figure 5. As can be seen in Figure 5(b), obtained by assuming emission factors equal to 0.2588 kg CO_2·kWh^{-1} (the average emission factor for Italy related to the year 2017), the contribution by the aeration energy is significantly larger than the one by N_2O emissions. This happens regardless of the aeration regulation setup applied. Given the preponderant contribution by aeration energy, the carbon footprint increases when the fixed DO_{sp} increases (see Figure 5(a)). Similarly, when the constant air flow rate is increased, the overall carbon footprint increases. The impact of applying ammonia control instead of oxygen control on the carbon footprint is found to be limited. Specifically, only when compared to the scenario with fixed DO_{sp} equal to 2 mg·L^{-1}, ammonia control shows its potential to reduce the carbon footprint. This is because the advantage in terms of reduction of CO_2 emissions linked to aeration overcomes the disadvantage linked to increased N_2O emissions. However, the same advantages and disadvantages seem to balance each other out when the ammonia control scenario is compared against the control scenarios with fixed DO_{sp} at 0.5 and 1 mg·L^{-1}. It was found that assuming an emission factor only slightly lower than the average Italian one (i.e. 0.256 kg CO_2·kWh^{-1}) could equate the carbon footprint led by the implementation of ammonia control to the one led by the implementation of oxygen control with set point at 1 mg·L^{-1}. Nevertheless, this situation drastically changes for a different energy mix. For instance, if the emission factor assumed were the same as the one of Sweden (i.e. 0.00927 kg CO_2·kWh^{-1}), N_2O emissions would become the largest influence on the carbon footprint (see Figure 5(c)). Specifically, the contribution by N_2O emissions on the total carbon footprint changes from 20% in the Italian case to 80–90% in the Swedish case. Due to the increased N_2O emission contribution, the ammonia controller results in an increase of the overall carbon footprint if implemented in place of a fixed DO_{sp} controller (see Figure 5(a)). A converse effect would result if the emission factor assumed were the same as the one of Estonia (i.e. 0.922 kg CO_2·kWh^{-1}) due to the clear preponderant contribution by the aeration energy (i.e. 90%) as shown in Figure 5(d).

These results suggest that the evaluation of the overall carbon footprint is very case specific depending – among several factors – on the source of energy which in turn affects the overall value for the CO_2 emission factors for electrical energy. CO_2 emissions linked to aeration energy could be drastically reduced by employing more carbon neutral energy sources while plant operation should be optimized to minimize N_2O emissions.

Besides the impact of CO_2 emission factor, it is also important to note that the reported variability of those model parameters used to describe the rates of N_2O producing processes (Pocquet *et al.* 2016) can represent another relevant source of uncertainty for the estimation of the overall carbon footprint. Additional simulations can be run with other model parameter values perturbed within their uncertainty range to see how the predicted carbon footprint results can change.

4. NOVELTIES AND LIMITATIONS OF THE WORK

Although there is already a significant amount of modelling work regarding full-scale N_2O emissions and carbon footprint from wastewater treatment plants (Corominas *et al.* 2012; Guo *et al.* 2012; Flores-Alsina *et al.* 2014), substantial steps towards a better N_2O model prediction have been taken in this work. Specifically, the model used here was calibrated with respect to the main nitrogen species involved in the production of N_2O such as ammonium, nitrite and nitrate using long-term data obtained during diverse full-scale WWTP operation modes. This represents the basis from which N_2O production and emission can be reliably predicted. As a matter of fact, while ammonium oxidation represents the starting process for N_2O production, nitrite accumulation and the ratio between nitrate produced and ammonium consumed in the aerobic zone of

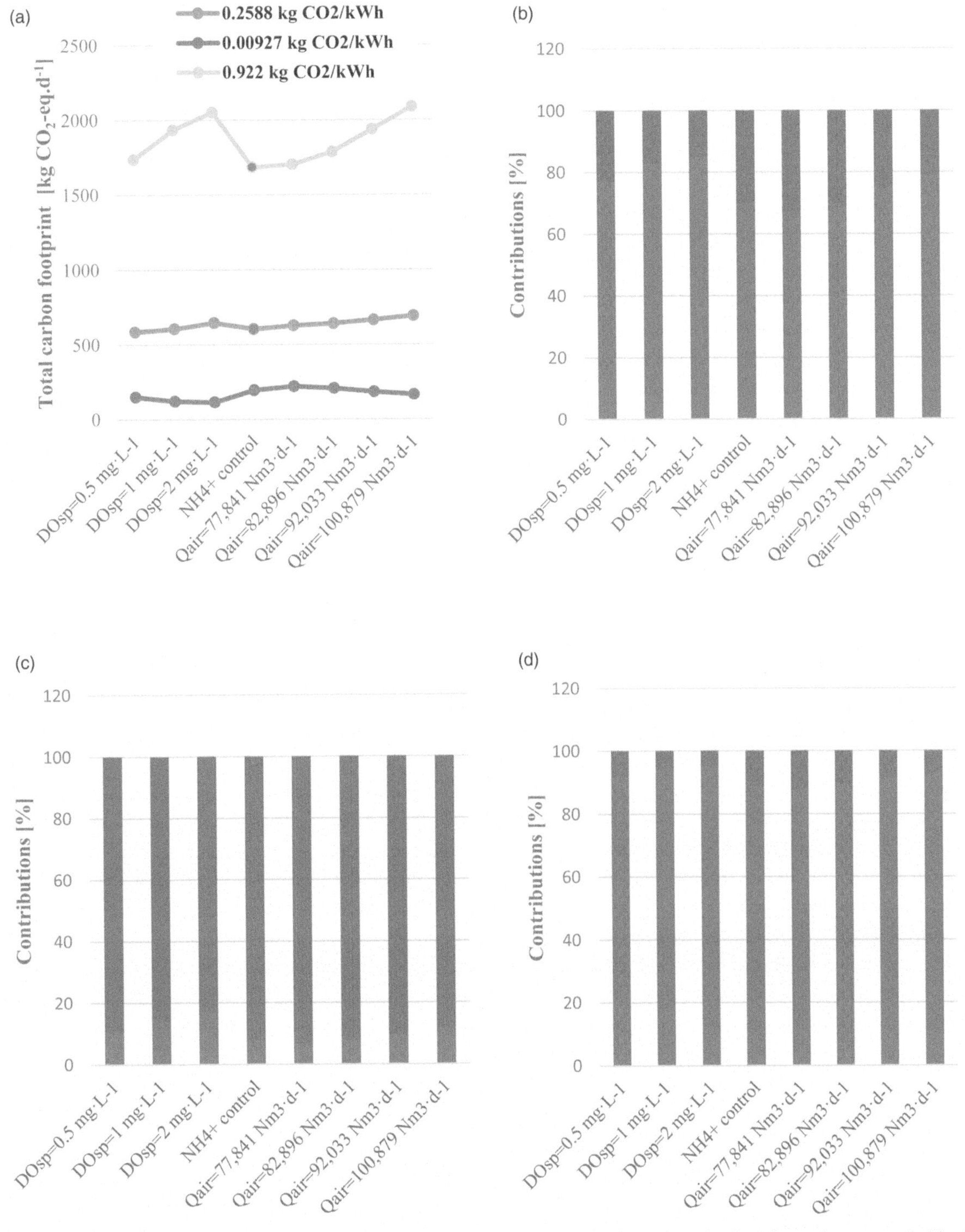

Figure 5 | Percent contributions by the N_2O emissions (in red) and by the CO_2 emitted for the production of aeration energy (in blue) obtained assuming the following CO_2 emission factors for electrical energy: (a) total carbon footprint expressed in emissions CO_2 equivalent according to the three CO_2 emission factors (red dot, the case of ammonia control), (b) 0.2588 kg $CO_2 \cdot kWh^{-1}$, (c) 0.00927 kg $CO_2 \cdot kWh^{-1}$, and (d) 0.922 kg $CO_2 \cdot kWh^{-1}$ (for interpretation of the references to colour in this figure legend, the reader is referred to the web version of this article).

a WWTP were found to play a key role (Daelman *et al.* 2015; Boiocchi *et al.* 2017a, 2017b). Conversely, works about full-scale N_2O modelling applications such as the one by Guo *et al.* (2012), Flores-Alsina *et al.* (2014) and Corominas *et al.* (2012) use models not calibrated against any full-scale data, not even with respect to full-scale ammonium, nitrite and nitrate concentrations. Snip *et al.* (2014) discussed the importance of calibrating full-scale wastewater models with data from full-scale WWTPs rather than from lab-scale reactors and the relevance of modelling the nitritation process correctly for reliable N_2O predictions. Furthermore, while incorporating the N_2O production by heterotrophic and autotrophic denitrification (apart from Corominas *et al.* (2012) who considered only heterotrophic denitrification), they neglect the incomplete NH_2OH oxidation pathway. In this regard, it is important to note that incomplete NH_2OH oxidation pathway may still be triggered under some operational pattern in wastewater treatment plants and may even become preponderant over other N_2O producing processes (Tumendelger *et al.* 2014, 2019).

Limitations of the results are essentially connected to the reliability of the model used in terms of description of the biological processes determining the effluent quality and the N_2O production and emission. Supplementary Material (D) deals with the challenges that can lead to poor model descriptive reliability. To overcome this weakness, detailed monitoring campaigns in full-scale WWTPs need to be properly scheduled (Vasilaki *et al.* 2019). It is important to point out that this is an intrinsic problem related to the calibration of full-scale models as the monitoring is usually carried out with the main aim of verifying compliance with regulation in force (Regional Council of Veneto 2009). Conversely, given its high demand in terms of costs and staff to be involved, long-term detailed monitoring for modelling purposes is opted more sparingly and restricted to few treatment units and short time periods (Liwarska-Bizukojc *et al.* 2011; Boiocchi *et al.* 2020). Furthermore, emerging issues such as N_2O emissions are not always the top priority. The other weakness is related to the fact that the results may be context specific for the design, location and influent characterization of the WWTP considered. For instance, in future plants under pressure for smaller construction than the WWTP modelled in this study in light of embodied carbon reduction (Hammond & Jones 2011), ammonia oxidation may occur at lower rates than the one modelled here and this could in turn change aeration energy requirements, N_2O emissions and the carbon footprint.

5. CONCLUSIONS

The present work has dealt with the potential advantages and disadvantages of implementing an ammonia controller in place of an oxygen controller. The model-based results achieved in this work suggest that, while enabling the respect of effluent regulations in terms of ammonium concentrations, the ammonia controller can lead to significant energy savings. However, with ammonia control, nitrite can build up in the system and the overall nitrous oxide emissions were found to increase. Based on correlations, the build-up of nitrification intermediates in the aerobic zone of the system along with suboptimal oxygen levels were suggested as the leading cause for increased N_2O production. At the same time, when coupling the contribution by N_2O with the one by the aeration system, no conclusive statement about the overall carbon footprint could be made. Specifically, the variation in carbon footprint from one scenario to the other was found to be strongly affected by the source of energy, which in turn affects the average CO_2 emission factor for electrical energy. Regardless, when the CO_2 emission factor was assumed as the average one for Italy, the increased carbon footprint due to N_2O emissions when ammonia control was implemented in place of oxygen control with a set point of 1 $mg{\cdot}L^{-1}$ was compensated by the reduction in the CO_2 emissions linked to aeration energy. The results from this study suggest the need of adopting integrated evaluations for the determination of greenhouse gas emissions upon implementation of both conventional and unconventional control strategies.

ACKNOWLEDGEMENTS

This work was partially funded by the Italian Ministry for University and Research, within the program 'Smart Cities and Communities and Social Innovation', Project SWaRM-Net – Smart Water Resource Management – Networks (SCN_00198). The authors warmly acknowledge the management staff of the studied WWTP, as well as the management company (which expressed intention to remain anonymous), for technical and financial support.

DATA AVAILABILITY STATEMENT

All relevant data are included in the paper or its Supplementary Information.

REFERENCES

Ahmed, A. S., Khalil, A., Ito, Y., van Loosdrecht, M. C. M., Santoro, D., Rosso, D. & Nakhla, G. 2021 Dynamic impact of cellulose and readily biodegradable substrate on oxygen transfer efficiency in sequencing batch reactors. *Water Res.* **190**, 116724. https://doi.org/10.1016/j.watres.2020.116724.

Bertanza, G., Baroni, P., Garzetti, S. & Martinelli, F. 2021a Reducing energy demand by the combined application of advanced control strategies in a full scale WWTP. *Water Sci. Technol.* **83**, 1813–1823. https://doi.org/10.2166/wst.2021.109.

Bertanza, G., Boiocchi, R. & Pedrazzani, R. 2021b Improving the quality of wastewater treatment plant monitoring by adopting proper sampling strategies and data processing criteria. *Sci. Total Environ.* **806**, 150724. https://doi.org/10.1016/j.scitotenv.2021.150724.

Boiocchi, R., Gernaey, K. V. & Sin, G. 2017a Understanding N2O formation mechanisms through sensitivity analyses using a plant-wide benchmark simulation model. *Chem. Eng. J.* **317**, 935–951. https://doi.org/10.1016/j.cej.2017.02.091.

Boiocchi, R., Gernaey, K. V. & Sin, G. 2017b A novel fuzzy-logic control strategy minimizing N2O emissions. *Water Res.* **123**, 479–494. https://doi.org/10.1016/j.watres.2017.06.074.

Boiocchi, R., Behera, C. R., Sherratt, A., Degroot, C. T., Gernaey, K. V., Sin, G. & Santoro, D. 2020 Dynamic model validation and advanced polymer control for rotating belt filtration as primary treatment of domestic wastewaters. *Chem. Eng. Sci.* **217**, 115510. https://doi.org/10.1016/j.ces.2020.115510.

Caniani, D., Caivano, M., Pascale, R., Bianco, G., Mancini, I. M., Masi, S., Mazzone, G., Firouzian, M. & Rosso, D. 2019 CO_2 and N_2O from water resource recovery facilities: evaluation of emissions from biological treatment, settling, disinfection, and receiving water body. *Sci. Total Environ.* **648**, 1130–1140. https://doi.org/10.1016/j.scitotenv.2018.08.150.

Corominas, L., Flores-Alsina, X., Snip, L. & Vanrolleghem, P. A. 2012 Comparison of different modeling approaches to better evaluate greenhouse gas emissions from whole wastewater treatment plants. *Biotechnol. Bioeng.* **109**, 2854–2863. https://doi.org/10.1002/bit.24544.

Daelman, M. R. J., van Voorthuizen, E. M., van Dongen, U. G. J. M., Volcke, E. I. P. & van Loosdrecht, M. C. M. 2015 Seasonal and diurnal variability of N_2O emissions from a full-scale municipal wastewater treatment plant. *Sci. Total Environ.* **536**, 1–11. https://doi.org/10.1016/j.scitotenv.2015.06.122.

Domingo-Félez, C., Calderó-Pascual, M., Sin, G., Plósz, B. G. & Smets, B. F. 2017 Calibration of the comprehensive NDHA-N_2O dynamics model for nitrifier-enriched biomass using targeted respirometric assays. *Water Res.* **126**, 29–39. https://doi.org/10.1016/j.watres.2017.09.013.

European Environment Agency 2017 *CO_2 Intensity of Electricity Generation.*

Flores-Alsina, X., Arnell, M., Amerlinck, Y., Corominas, L., Gernaey, K. V., Guo, L., Lindblom, E., Nopens, I., Porro, J., Shaw, A., Snip, L., Vanrolleghem, P. A. & Jeppsson, U. 2014 Balancing effluent quality, economic cost and greenhouse gas emissions during the evaluation of (plant-wide) control/operational strategies in WWTPs. *Sci. Total Environ.* **466–467**, 616–624. https://doi.org/10.1016/j.scitotenv.2013.07.046.

Foley, J., de Haas, D., Yuan, Z. & Lant, P. 2010 Nitrous oxide generation in full-scale biological nutrient removal wastewater treatment plants. *Water Res.* **44**, 831–844. https://doi.org/10.1016/j.watres.2009.10.033.

Guo, L. & Vanrolleghem, P. A. 2014 Calibration and validation of an activated sludge model for greenhouse gases no. 1 (ASMG1): prediction of temperature-dependent N_2O emission dynamics. *Bioprocess Biosyst. Eng.* **37**, 151–163. https://doi.org/10.1007/s00449-013-0978-3.

Guo, L., Porro, J., Sharma, K. R., Amerlinck, Y., Benedetti, L., Nopens, I., Shaw, A., Van Hulle, S. W. H., Yuan, Z. & Vanrolleghem, P. A. 2012 Towards a benchmarking tool for minimizing wastewater utility greenhouse gas footprints. *Water Sci. Technol.* **66**, 2483–2495. https://doi.org/10.2166/wst.2012.495.

Hammond, G. & Jones, C. 2011 Embodied Carbon: The Inventory of Carbon and Energy (ICE). A BSRIA Guid. 136.

He, Z., Yuan, L., Wei, Y. & Nan, Y. 2017 N_2O emission and hydroxylamine oxidase (HAO) activity in a nitrogen removal process based on activated sludge with three COD/NH_4^+ ratios. *Water Environ. Res.* **89**, 387–395. https://doi.org/10.2175/106143017x14839994523668.

Henze, M., Dupont, R., Grau, P. & de la Sota, A. 1993 Rising sludge in secondary settlers due to denitrification. *Water Res.* **27**, 231–236. https://doi.org/10.1016/0043-1354(93)90080-2.

Hiatt, W. C. & Grady, C. P. L. 2008 An updated process model for carbon oxidation, nitrification, and denitrification. *Water Environ. Res.* **80**, 2145–2156.

IPCC 2007 Climate Change 2007: Synthesis Report. Contribution of Working Groups I, II and III to the Fourth Assessment Report of the Intergovernmental Panel on Climate Change.

IPCC 2014 Climate Change 2014: Synthesis Report. Contribution of Working Groups I, II and III to the Fifth Assessment Report of the Intergovernmental Panel on Climate Change.

Jiang, L. M., Garrido-Baserba, M., Nolasco, D., Al-Omari, A., DeClippeleir, H., Murthy, S. & Rosso, D. 2017 Modelling oxygen transfer using dynamic alpha factors. *Water Res.* **124**, 139–148. https://doi.org/10.1016/j.watres.2017.07.032.

Kampschreur, M. J., Temmink, H., Kleerebezem, R., Jetten, M. S. M. & van Loosdrecht, M. C. M. 2009 Nitrous oxide emission during wastewater treatment. *Water Res.* **43**, 4093–4103. https://doi.org/10.1016/j.watres.2009.03.001.

Khatri, N., Khatri, K. K. & Sharma, A. 2020 Enhanced energy saving in wastewater treatment plant using dissolved oxygen control and hydrocyclone. *Environ. Technol. Innov.* **18**, 100678. https://doi.org/10.1016/j.eti.2020.100678.

Kuokkanen, A., Blomberg, K., Mikola, A. & Heinonen, M. 2021 Unwanted mainstream nitritation–denitritation causing massive N_2O emissions in a continuous activated sludge process. *Water Sci. Technol.* **83**, 2207–2217. https://doi.org/10.2166/wst.2021.127.

Law, Y., Ye, L., Pan, Y. & Yuan, Z. 2012 Nitrous oxide emissions from wastewater treatment processes. *Philos. Trans. R. Soc. B* **367**, 1265–1277. https://doi.org/10.1098/rstb.2011.0317.

Legislative Decree 3rd April n.152, 2006 'Norme in materia ambientale' [Regulations on environmental issues] Official Journal of Italy n. 88, 14th April 2006, Ordinary Supplement n. 96.

Liwarska-Bizukojc, E., Olejnik, D., Biernacki, R. & Ledakowicz, S. 2011 Calibration of a complex activated sludge model for the full-scale wastewater treatment plant. *Bioprocess Biosyst. Eng.* **34**, 659–670. https://doi.org/10.1007/s00449-011-0515-1.

Mannina, G., Ekama, G., Caniani, D., Cosenza, A., Esposito, G., Gori, R., Garrido-Baserba, M., Rosso, D. & Olsson, G. 2016 Greenhouse gases from wastewater treatment – a review of modelling tools. *Sci. Total Environ.* **551–552**, 254–270. https://doi.org/10.1016/j.scitotenv.2016.01.163.

Medinilla, V. R., Sprague, T., Marseilles, J., Burke, J., Deshmukh, S., Delagah, S. & Sharbatmaleki, M. 2020 Impact of ammonia-based aeration control (ABAC) on energy consumption. *Appl. Sci.* **10**, 5227. https://doi.org/10.3390/APP10155227.

Metcalf and Eddy 1991 *Wastewater Engineering: Treatment, Disposal and Reuse*, 3rd ed. McGraw-Hill, New York, USA.

Mikola, A., Heinonen, M., Kosonen, H., Leppañen, M., Rantanen, P. & Vahala, R. 2014 N_2O emissions from secondary clarifiers and their contribution to the total emissions of the WWTP. *Water Sci. Technol.* **70**, 720–728. https://doi.org/10.2166/wst.2014.281.

Olsson, G. 2007 Automation development in water and wastewater systems. *Environ. Eng. Res.* **12**, 197–200.

Peng, L., Ni, B.-J., Erler, D., Ye, L. & Yuan, Z. 2014 The effect of dissolved oxygen on N_2O production by ammonia-oxidizing bacteria in an enriched nitrifying sludge. *Water Res.* **66**, 12–21. https://doi.org/10.1016/j.watres.2014.08.009.

Peng, L., Ni, B.-J., Erler, D., Ye, L. & Yuan, Z. 2015 The combined effect of dissolved oxygen and nitrite on N_2O production by ammonia oxidizing bacteria in an enriched nitrifying sludge. *Water Res.* **73**, 29–36. https://doi.org/10.1016/j.watres.2014.08.009.

Pocquet, M., Wu, Z., Queinnec, I. & Spérandio, M. 2016 A two pathway model for N_2O emissions by ammonium oxidizing bacteria supported by the NO/N_2O variation. *Water Res.* **88**, 948–959. https://doi.org/10.1016/j.watres.2015.11.029.

Regional Council of Veneto, 2009 Water Protection Plan – Technical Standards for Implementation.

Santín, I., Barbu, M., Pedret, C. & Vilanova, R. 2017 Control strategies for nitrous oxide emissions reduction on wastewater treatment plants operation. *Water Res.* **125**, 466–477. https://doi.org/10.1016/j.watres.2017.08.056.

Snip, L. J. P., Boiocchi, R., Jeppsson, U. & Gernaey, K. V. 2014 Challenges encountered when expanding activated sludge models: a case study based on N_2O production. *Water Sci. Technol.* **70**, 1251–1260. https://doi.org/10.2166/wst.2014.347.

Stein, L. Y. 2011 Surveying N_2O-producing pathways in bacteria. *Methods Enzymol.* **486**, 131–152. https://doi.org/10.1016/B978-0-12-381294-0.00006-7.

Su, Q., Domingo-Félez, C., Jensen, M. M. & Smets, B. F. 2019 Abiotic nitrous oxide (N_2O) production is strongly pH dependent, but contributes little to overall N_2O emissions in biological nitrogen removal systems. *Environ. Sci. Technol.* **53**, 3508–3516. https://doi.org/10.1021/acs.est.8b06193.

Takacs, I., Patry, G. G. & Nolasco, D. 1991 A dynamic model of the clarification-thickening process. *Water Res.* **25**, 1263–1271.

Tallec, G., Garnier, J., Billen, G. & Gousailles, M. 2008 Nitrous oxide emissions from denitrifying activated sludge of urban wastewater treatment plants, under anoxia and low oxygenation. *Bioresour. Technol.* **99**, 2200–2209. https://doi.org/10.1016/j.biortech.2007.05.025.

Tumendelger, A., Toyoda, S. & Yoshida, N. 2014 Isotopic analysis of N_2O produced in a conventional wastewater treatment system operated under different aeration conditions. *Rapid Commun. Mass Spectrom.* **28**, 1883–1892.

Tumendelger, A., Alshboul, Z. & Lorke, A. 2019 Methane and nitrous oxide emission from different treatment units of municipal wastewater treatment plants in Southwest Germany. *PLoS One* **14**, 1–17. https://doi.org/10.1371/journal.pone.0209763.

Vangsgaard, A. K., Mutlu, A. G., Gernaey, K. V., Smets, B. F. & Sin, G. 2013 Calibration and validation of a model describing complete autotrophic nitrogen removal in a granular SBR system. *J. Chem. Technol. Biotechnol.*, 2007–2015. https://doi.org/10.1002/jctb.4060.

Vasilaki, V., Massara, T. M., Stanchev, P., Fatone, F. & Katsou, E. 2019 A decade of nitrous oxide (N_2O) monitoring in full-scale wastewater treatment processes: a critical review. *Water Res.* **161**, 392–412. https://doi.org/10.1016/j.watres.2019.04.022.

First received 11 December 2021; accepted in revised form 31 January 2022. Available online 14 February 2022

doi: 10.2166/wst.2022.049

Life cycle assessment of paper mill wastewater: a case study in Viet Nam

Hiep Nghia Bui[a], Yi-Ching Chen[a], Anh Thi Pham[b], Si Ling Ng[c], Kun-Yi Andrew Lin[d], Ngan Quang Viet Nguyen[e] and Ha Manh Bui[f,*]

[a] Department of Environmental Engineering, Dayeh University, Changhua 51591, Taiwan
[b] Institute for Environmental and Transport Studies, Ho Chi Minh City University of Transport, Ho Chi Minh City 700000, Vietnam
[c] School of Chemical Sciences, Universiti Sains Malaysia, Pulau Pinang 11800, Malaysia
[d] Department of Environmental Engineering & Innovation and Development Center of Sustainable Agriculture, National Chung Hsing University, Taichung 40227, Taiwan
[e] University of Social Sciences and Humanities, Vietnam National University of Ho Chi Minh City, Ho Chi Minh City 700000, Vietnam
[f] Department of Environmental Sciences, Saigon University, Ho Chi Minh City 700000, Vietnam
*Corresponding author. E-mail: manhhakg@yahoo.com.vn

HMB, 0000-0002-0647-3111

ABSTRACT

Although in a critical position in the economy, the paper industry releases a lot of wastewater that requires adequate treatment for sustainable development. This study presents an application of Life cycle assessment (LCA) with the ReCiPe tool on the wastewater treatment plant (WWTP) of a paper factory in Vietnam to evaluate the environmental effect of the individual techniques in WWTP, especially the internal circulation (IC) reactor, a pioneer and practical anaerobic technology. Both Midpoint and Endpoint categories results demonstrated that chemical use and electricity consumption mainly contributed to the environmental impact in the WWTP. The Dissolved air flotation (DAF) and Moving bed biofilm reactor (MBBR) are classified as effective techniques to reduce the impacts on the environment. Moreover, the comparison of LCA between IC and up-flow anaerobic sludge bed (UASB) shows that IC is the better practically green technique for the environment.

Key words: internal circulation reactor, life cycle assessment, ReCiPe, recycled paper wastewater, UASB

HIGHLIGHTS

- Life cycle assessment was applied in a paper mill wastewater treatment plant.
- Midpoint and Endpoint indicators were employed to evaluate environmental impact.
- Two scenarios of electricity generating from biogas between Internal circulation and up-flow anaerobic sludge bed were investigated.
- Chemicals use and energy consumption were determined to be the most factors to damage indicators.

GRAPHICAL ABSTRACT

1. INTRODUCTION

The pulp and paper mills are among the most critical industries in Vietnam. Approximately 300 pulp and paper mill factories have been established in Vietnam since 1920. The paper industry contributes around 1.5% of Vietnam's gross domestic product with over 1.97 billion $ in 2020 and is expected to reach 2.89 billion $ in 2026 (ReportLinker 2021). However, it discharges large volumes of waste into the environment and is considered the third-largest industrial waste source in the world (Liang *et al.* 2021). Afzal *et al.* (2008) indicated that each ton of produced paper releases approximately 60 m^3 of wastewater. The massive pulp wastewater contains a large number of organic and inorganic compounds expressed in high color, chemical oxygen demand (COD), biological oxygen demand (BOD), suspended solids (SS), xenobiotic compounds, heavy metals, and so on, which may cause long-term adverse effects on living creatures; therefore, wastewater treatment is needed to prevent pollution (Tiku *et al.* 2007; Farooqi & Basheer 2017). Most pulp and paper mill treatment plants integrate an innovative biological treatment with physical or chemical to eliminate the wastewater's contaminants (Liang *et al.* 2021). While physical or chemical units; that is, dissolved air flotation (DAF), sedimentation, oxidation tank, and so on remove most color, SS, heavy metals or decompose xenobiotic compounds in the wastewater, biological treatment is considered a key unit to eradiate organic compounds in the pulp and paper mill wastewater (Farooqi & Basheer 2017). The conventional biological method in a paper mill wastewater treatment plant is an up-flow anaerobic sludge bed (UASB) reactor, which has a higher organic load rate (OLR) than aerobic techniques (Larsson *et al.* 2015; Bakraoui *et al.* 2020). Larsson *et al.* (2015) had applied UASB to treat alkaline bleaching wastewater from a kraft mill. They stated that approximately 65% TOC was eliminated after

120 days of operation with OLR ranging from 0.9 to 2.2 g TOC/day and a stable treatment process was obtained at hydraulic retention time (HRT) of 7–14 h. While Bakraoui *et al.* (2020) engaged UASB to mitigate recycled paper mill wastewater in Morocco, this system eliminated around 80.6% COD with OLR ranging from 1.7 to 2.2 g VS/ day at HRT 15.14 h and operated for 130 days without any overload problems. Even UASB could be applied for treating the organic content of wastewater from paper mill plants; however, the exceptionally high organic content (over 20,000 mg/L of COD) in this wastewater had limited UASB systems (Tiku *et al.* 2007). The high organic content in the influent required high OLR during the operational phase or required more area for bigger UASB to eliminate the wastewater; therefore, some improvement should be conducted in the UASB system to successfully treat pollutants from paper mill wastewater (Ruan *et al.* 2017; Ha & Ha 2019). Recently, one of the successors to the UASB, the internal circulation (IC) reactor, has been developed to treat various high organic content influents such as refineries (Wang *et al.* 2020), breweries (Chen *et al.* 2021), dairy (Charalambous *et al.* 2020), molasses (Li & Li 2019) and paper mill (Ruan *et al.* 2017). With two sets of separator modules and internal circulation as compared with one set of a separator of UASB, the granular sludges in the IC reactor could efficiently mix with and expeditiously dilute influent; therefore, the IC reactor could retain high biomass or operate at high OLR with low HRT that significantly improve the organic treatment performance compared to UASB (Liu *et al.* 2020). Moreover, the IC reactor entails lower investment costs and requires less ground space; hence, the IC reactor has gradually begun replacing the UASB in the wastewater treatment field (Ha & Ha 2019; Wang *et al.* 2020). Even there have been studies of IC reactors in some wastewaters (Charalambous *et al.* 2020; Wang *et al.* 2020; Chen *et al.* 2021), the application of IC reactors on real paper mill wastewater treatment plants is still limited (Ruan *et al.* 2017; Ha & Ha 2019).

Beyond the importance of technical feasibility, the environmental impacts of the treatment method should be evaluated to determine the practical application of these techniques. Recently, several studies have focused on life cycle assessment (LCA), an approach to qualify the environmental impacts of the wastewater treatment process in both design and operation phases; therefore, the application of LCA could reduce the adverse effects on the environment (Corominas *et al.* 2020). Although some authors have given studies on the application of LCA in industrial wastewater treatment plants (Naushad 2018; Tabesh *et al.* 2019; Corominas *et al.* 2020), there is only a few authors attempted on specific wastewaters such as petroleum (Vlasopoulos *et al.* 2006), agro-industrial (Chen *et al.* 2020b) and olive mill (Chatzisymeon *et al.* 2013). Up to now, there have been no previous LCA studies on paper mill wastewater treatment plants.

This work aims to evaluate the environmental impacts of a paper mill wastewater treatment plant in Vietnam using the ReCiPe, a harmonized life cycle impact assessment method at Midpoint and endpoint level (Huijbregts *et al.* 2017). The scenarios of IC and UASB in anaerobic stages were also carried out to elucidate the environmental role of the IC reactor in the treatment plant.

2. METHODOLOGY

The LCA was conducted according to the ISO14040-44 guidelines with four steps (Klöpffer 2012). The assumptions and data are detailed in the following sections, starting with the goal and scope definition, the inventory analysis, the impact assessment and interpretation for treatment routes. The round outline of the LCA procedure in this study is displayed in Figure 1.

2.1. Goal and scope definition

This LCA study aims to assess the environmental benefits and drawbacks of a paper mill wastewater treatment plant (Ba Ria – Vung Tau Province, Vietnam). All effects of each treatment stage at the wastewater treatment plant (WWTP) and secondary emissions were valorized through chemical consumption and transportation. The comparison scenario between IC reactor and reference UASB at the anaerobic treatment unit was also investigated.

2.1.1. Functional unit

The daily average influent flow at the WWTP is 7,500 m^3/d; therefore, the functional unit can also be converted to 1 m^3 of wastewater treated by a scaling factor of 1/7,500 (Rebello *et al.* 2021).

2.1.2. Descriptive of influent characteristic

There are included two producing procedures in this mill paper factory; that is, testliner paper and tissue paper (Fig. S1). The full capacity of the paper mill could reach 2,73,000 tons of paper a year. Moreover, approximately 60% of 'white water' in the pulp washing influent is reused daily to save chemicals and water (Hubbe 2007). Therefore, the influent's characteristic was varied compared to traditional paper mill factories and it was not suitable for applying to biological treatment directly

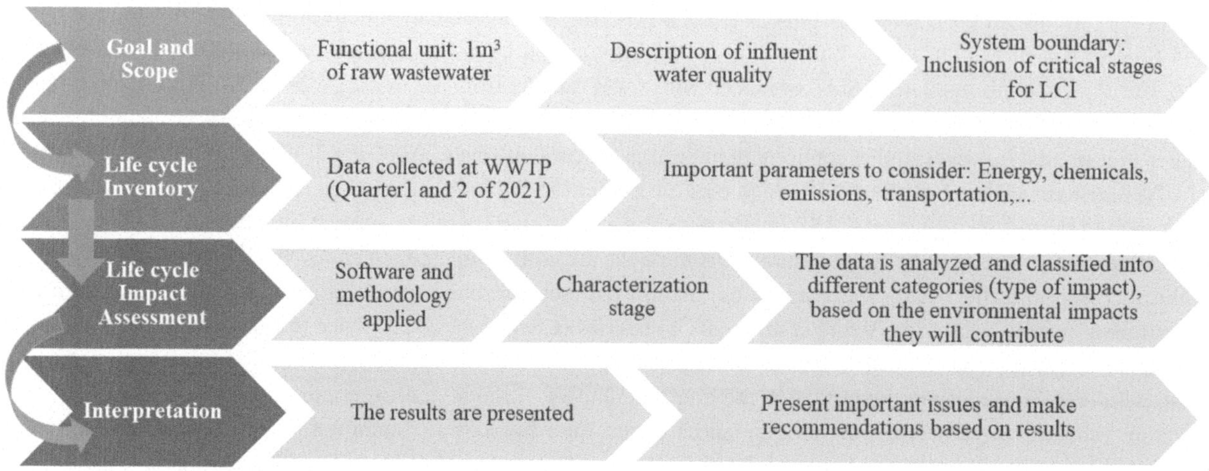

Figure 1 | The framework of the LCA performed in this study.

(BOD_5/COD ratio<0.3) (Dubey *et al.* 2012). To eliminate the contaminants from this mill paper, the WWTP was built in 2018 based on the estimated flow rate of 17,000 m^3/day and total COD (tCOD) of 15,000 mg/L to meet column A of Vietnam National technical regulation on the effluent of pulp and paper mills (QCVN 12:2015/BTNMT). After the acclimatization period, the soluble COD (sCOD) treatment capacity was always stable above 75% when the organic load fluctuated from 5 to 20 kg/m^3/day. When the organic load is lower than 5 kg/m^3/day, sCOD treatment efficiency will decrease gradually. The output sCOD of the IC system is the input of the moving bed biofilm reactor (MBBR) system. Usually, the MBBR inlet sCOD should be less than 1,000 mg/L for optimum performance (Hansen *et al.* 1999). Therefore, IC systems are designed with loads of 15–18 kg/m^3/day, with lower loads not taking full advantage of the system's capabilities. However, due to the covid-19 pandemic, the paper production is low and the organic load only reached an average of 10.5 kg/m^3/ day (approximately flow rate of 7,500 m^3/day) with tCOD value of 7,000 mg/L since the beginning of 2021. Some other average influent components of the wastewater were tabulated in Table 1.

The wastewater treatment processes shown in Figure 2 include physical, chemical and biological treatments through primary and secondary treatment stages. While equalization is used to remove SS partially in the influent, the primary dissolved air flotation (DAF) tank have applied chemicals (mixed of Polyaluminium Chloride -PAC and Polyacrylamide -PAM) to eliminate most SS and insoluble COD. The key treatment function in this WWTP is the internal circulating anaerobic (IC) reactor, which degrades most soluble COD (the regular operation of HRT fluctuates about 4–5 hours with COD: N: P=800: 5: 1 pollutant ratio (Li & Li 2019). This retention time is 2–4 times lower than that of a traditional UASB tank (Chen *et al.* 2020a; Liu *et al.* 2020). According to operational data from the WWTP, this IC reactor solely eliminates approximately 77% of sCOD from the influent before it lets in another biological unit, aerobic activated sludge (MBBR), which could remove 82% of the remaining sCOD. The wastewater then goes to the final DAF to separate the activated sludge; the sludge-free

Table 1 | Influent and effluent quality of the paper mill WWTP

Parameters	Value (mg/L)	
	Influent	Effluent[a]
pH[a]	5.9–7.2	6–9
tCOD	6,500–9,800	–
sCOD	3,000–4,520	67
BOD_5	900–1,340	19
TSS	2,600–2,800	60
TN	20.3–23.7	2.7
TP	5.6–6.4	1.8

[a]Vietnam National technical regulation on the effluent of pulp and paper mills, column A (QCVN12:2015/BTNMT), pH unit is not specific.

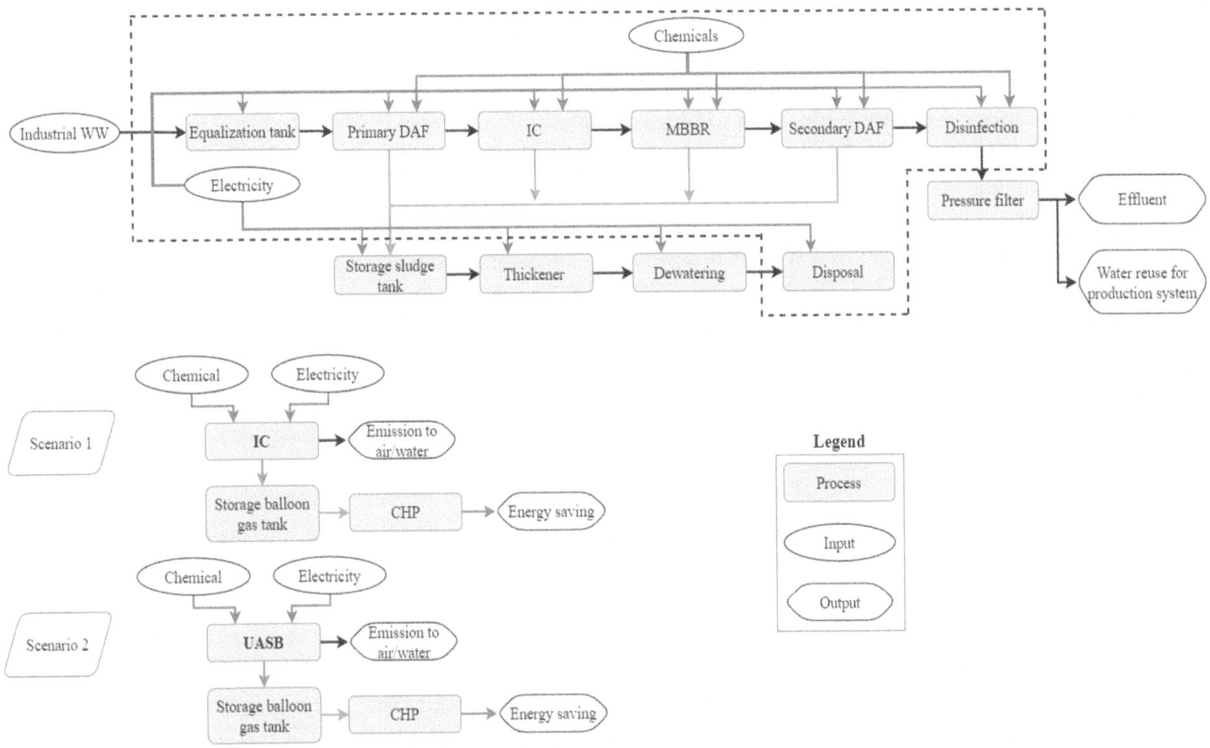

Figure 2 | The schematic flow diagram of WWTP and its system LCA boundaries (upper) with two comparison scenarios IC and UASB at the anaerobic treatment system (lower). The colored lines represent the input chemicals (red), sludge flow (orange), energy-saving (green), electricity flow (blue), and wastewater and emissions (black), gas flow (pink). Please refer to the online version of this paper to see this figure in colour: http://dx.doi.org/10.2166/wst.2022.049.

wastewater is disinfected and finally filtrated by a pressure filter before reusing or discharging to the environment. The treatment performance of each mentioned stage is listed in Table S1.

2.1.3. System boundaries

In the WWTP, there are three processes; that is, construction, operation, and demolition, which could embody high energy and material consumption and fugitive emissions. In this study, construction and demolitiion impacts were not included in the boundaries since there was insufficient data. As such, this was a gate-to-gate study beginning with the entry of wastewater into the primary dissolved air flotation (DAF) and ending with the effluent after pressure filter (Figure 2). In this case, the assumption is that the process inputs are stable, the pollutants are different according to COD removal efficiency, and the WWTP's electricity is supplied from the government grid. The sludge from the physical and the biological units is collected, dewatered, and dried before transferring to the local sludge incineration company. The incineration process

achieves a complete stabilization of the sludge and delivers thermal energy and a solid residue in ash; subsequently, the sludge is involved in LCA calculation at the Midpoint. Moreover, a UASB operating scenario instead of IC in the biological unit is calculated and compared through the environmental impacts and the conversion of biogas generated into electricity to highlight the IC performance.

2.2. Life cycle inventory

Following the goal and scope step, a life cycle inventory analysis was performed regarding all the energy, chemicals, and materials flows into or out across the system boundaries. In this study, the database on several processes such as power generation, chemical production, transportation, etc. was taken from Ecoinvent 3 (Raghuvanshi *et al.* 2017), while remaining wastewater treatment stages, i.e., UASB, IC, MBBR, etc. which could not find from the tool were retrieved from EMEP standard (Adams *et al.* 2013) or provided from WWTP source data and lab analysis.

The energy was inventoried through the form of electricity, which involved total purchased electricity from the Vietnamese national grid, power consumption of main equipment and generation of biogas (mostly CH_4) to heat and energy (CHP) from anaerobic digestion (IC or UASB) every day. The biogas emissions are determined using a flowrate measurement controlled by SCADA software and stored at a balloon gas tank. Approximately 12,000 m^3 biogas emissions come from the WWTP every day. According to research by Carnevale & Lombardi (2015), the power generation efficiency of CHP is about 32% and generates about 2.1 kWh of electricity for every cubic meter of biogas. In addition, they also assume that 1% CH_4 will be lost from the engine and emission air NOx, SO_2, PM and CO of 1.89, 0.00638, 0.0191 and 0.829 g per cubic meter of biogas, respectively.

Chemicals used as coagulants, flocculants, nutrients or disinfectants in the WWTP were determined and the secondary impacts from these chemicals were also assessed using the Ecoinvent database (Raghuvanshi *et al.* 2017). All used chemicals are produced by the same plant, located 42 km from the WWTP. The dewatered sludge (approximately 3 tons/day) was collected and burned at a local sludge incineration company far away (30 km) from the WWTP. The transportation of chemicals and the sludge was carried out by lorry with a capacity of 16 tons with Euro 4 emission standards, while the transportation background data were also analyzed using this Ecoinvent database. The inventory database of this study since the beginning of 2021 is summarized in Table 2 (Akbarjon *et al.* 2019). The Covid-19 pandemic has interfered with the regular operation of the WWTP; therefore, it could not operate at full capacity, which could lead to a slight difference in wastewater or dewatered sludge characteristics.

2.3. LCIA

This Life Cycle Impact Assessment (LCIA) step identifies and describes the environmental relevance and significance of the treatment system through the input and output data from the previous LCI process. Before analyzing, the collected data were classified into different categories (impact categories) according to their environmental contribution (Pennington *et al.* 2004; Ahmed 2010) by several widely standardized methods; that is, CML-IA, TRACI, ReCiPe, IMPACT 2002+, Eco-Indicator99, etc. (Hauschild *et al.* 2018). Unlike the CML-IA methods, TRACI focuses on assessing the environmental impacts at the middle stage through different individual influences (Midpoint). Then typical techniques such as IMPACT 2002+, Eco-Indicator99, evaluate environmental impacts through groups of end effects (endpoints) attributed or calculated from individual impacts (Hauschild *et al.* 2018). Regarding the choice of LCA method, CML-IA is the widely used LCA model, one of the oldest and most popular methods and it is still

Table 2 | Summary of LCI of WWTP per functional unit

	Parameters	Unit	Primary DAF	IC	MBBR	Final DAF	Disinfection
Input	sCOD	g	3,760	3,500	800	143	67
	BOD	g	1,120	1,064	218	19	19
	SS	g	2,700	200	200	200	60
	TN	g	0.220	0.188	5.9	2.7	2.7
	TP	g	6	5.5	2.3	1.8	1.8
	Energy consumption	Wh	210	178	990	150	59.2
	Chemical						
	Polymer	g	5.830	–	–	11.9	–
	PAC	g	260	–	–	111	–
	Urea	g	–	51.4	0.317	–	–
	DAF	g	–	19	–	–	–
	NaOH	g	–	179	–	–	–
	Javel	g	–		–	–	7.62
	Land use	m^2	0.120	0.038	0.200	0.068	0.040
Output	**Emission to water**						
	sCOD	g	3,500	800	143	67	67
	BOD	g	1,064	218	19	19	19
	SS	g	200	200	200	60	60
	TN	g	0.188	5.9	2.7	2.7	2.7
	TP	g	5.5	2.3	1.8	1.8	1.8
	Emission to air						
	CO_2	g	260	485	426	76	–
	N_2O	g	–	0.8	0.0413	–	–
	CH_4	g	–	10.4	–	–	–
	Produced biogas	m^3					
	CO_2 in Biogas	kg	–	0.485	–	–	–
	CH_4 in Biogas	m^3	–	1.39	–	–	–

'–': nonapplicable.

one of the preferred approaches. However, if research is directed towards applying the endpoint approach, this method is not suitable, especially when applied to WWTPs. Also, note that the ReCiPe method comes in second position (Fig. S2). The method dominantly reflects the environmental impact of WWTP both at the middle (Midpoint) and the end (Endpoint), based on CML-IA and Eco-Indicator99 (de Oliveira Schwaickhardt *et al.* 2017; Lutterbeck *et al.* 2017). Therefore, the ReCiPe method with midpoint version 2016 (H) was implemented to evaluate the environmental impact of this WWTP. This method categorized LCI data into two main groups: the potential impacts (Midpoint) and the potential damage (Endpoint). The Midpoint category involved 18 environmental impacts such as Global warming (GW), Stratospheric ozone depletion (SOD), Ionizing radiation (IR), Ozone formation – Human Health (OHH), Fine particulate matter formation (FPMF), Ozone formation-Terrestrial ecosystems (OTE), Terrestrial acidification (TA), Freshwater eutrophication (FE), Marine eutrophication (ME), Terrestrial ecotoxicity (TET), Freshwater ecotoxicity (FET), Marine ecotoxicity (MET), Human carcinogenic toxicity (HCT), Human non-carcinogenic toxicity (HNT), Land use (LU), Mineral resource scarcity (MRS), Fossil resource scarcity (FRS), Water consumption (WC). At the Endpoint, the final three impact categories are Human health (HH), Ecosystems (E), Resources (R). Figure 3 illustrates the categories and pathways of action covered in the ReCiPe methodology in this study. Midpoint indicators focus on individual environmental issues, such as climate change, eutrophication. Endpoint indicators show environmental impact across the three higher aggregate levels. Converting the Midpoint to the Endpoint simplifies the interpretation of the LCIA.

2.4. Interpretation

This is the final phase of an LCA study in which the results are presented. Thereby, the main types of impacts selected to replace or minimize are presented. Interpretation involves consideration of all stages of the LCA. From there, the quality of the data and the consistency of the hypotheses can be checked. Two of the three principal steps of the interpretation according to the ISO 14043 standard are identifying critical issues based on the inventory and LCA impact assessment stages, testing and conclusions, recommendations, and reporting (Ahmed 2010).

3. RESULTS AND DISCUSSION

3.1. Mid assessment

The environmental impacts at 18 different categories for the WWTP processes at individual components are described in Figure 4. For most types of effects, the primary DAF stage is the predominant factor. The primary DAF was used in

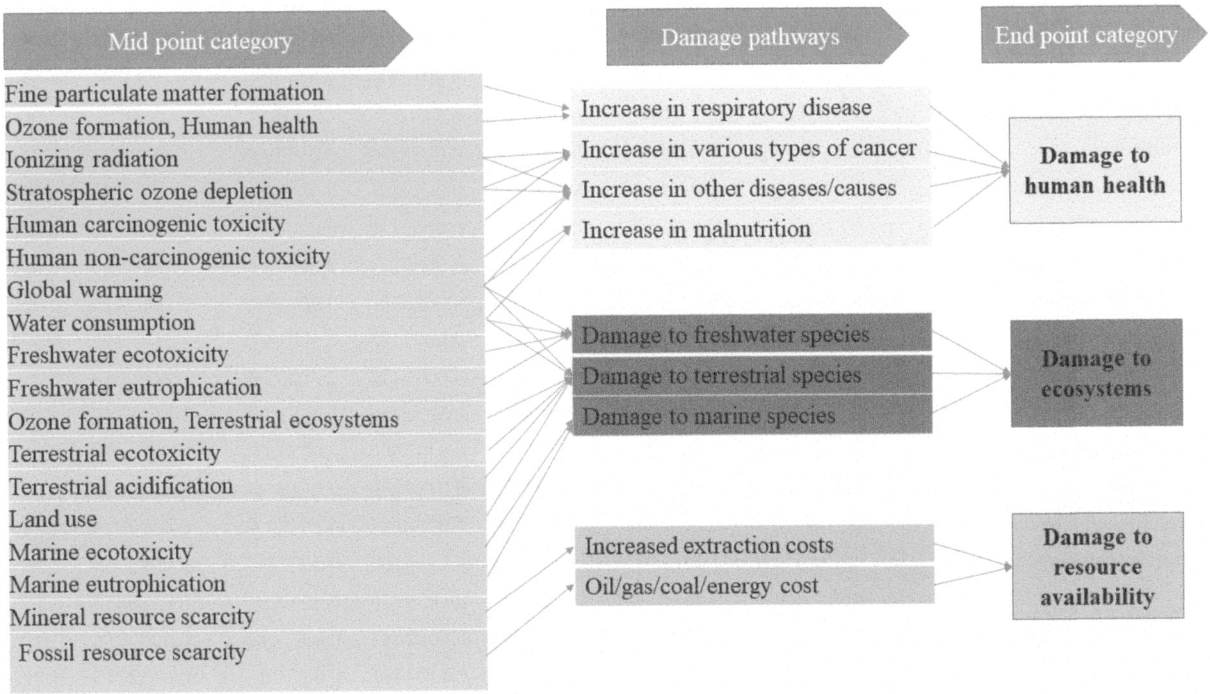

Figure 3 | Representation of the relationships between impact categories (Midpoint) and the areas of protection (Endpoint) in ReCiPe 2016.

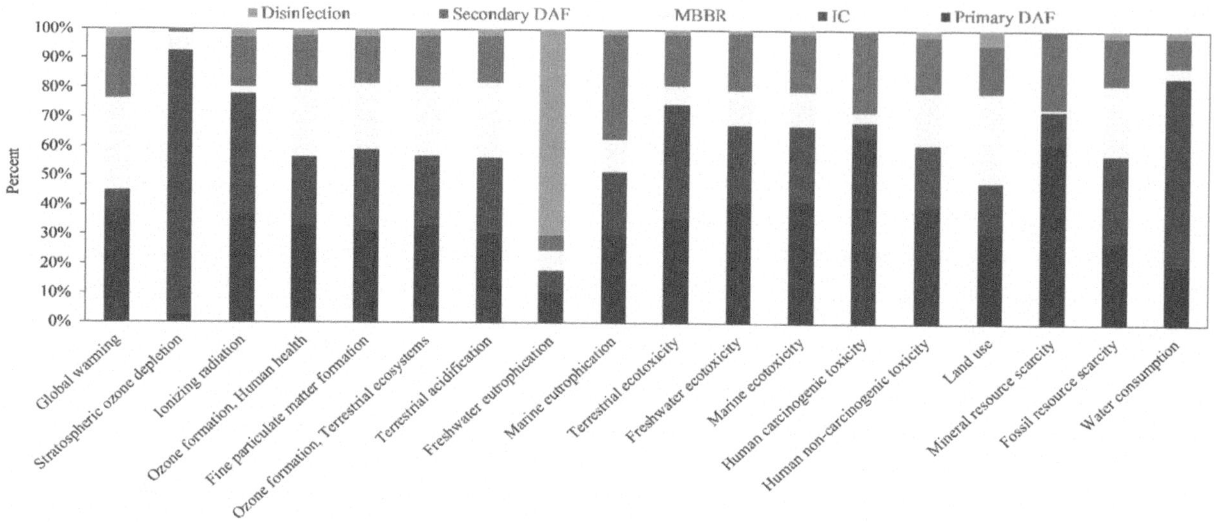

Figure 4 | The contribution of each activity to the environmental impact indicators midpoint.

WWTP of the paper industry to eliminate the high SS content in the influent; however, the process consumes a lot of electricity and chemicals. Therefore, the primary DAF should be the most critical indicator for environmental evaluation (Sharaai *et al.* 2009). Related to eutrophication impact, the sterilization phase contributes 70.36% for all the stages, proving the critical roles of the WWTP process in disinfecting and reusing the wastewater. However, the most identified stratospheric ozone depletion and water consumptions are 89.79% and 63.62% for the IC process. The findings could be explained that a high amount of nutrients is required and high sludge generation during the IC phase could increase the water consumption for nutrient production and stratospheric ozone depletion from handling and decomposing the generated sludge (Yang *et al.* 2020). The environmental impacts of each treatment stage can be shown in Table 3 and Fig. S3.

Table 3 | Midpoint environmental impacts in various phases of the WWTP

Impact category	**Unit**	**Primary DAF**	**IC**	**MBBR**	**Secondary DAF**	**Disinfection**	**Total**
Global warming	kg CO_{2eq}	0.5895635	0.100	0.4910533	0.3116983	0.0515795	1.5438946
Stratospheric ozone depletion	kg $CFC11_{eq}$	2.7×10^{-7}	9.2×10^{-6}	6.1×10^{-7}	1.3×10^{-7}	3.1×10^{-8}	0.0000102
Ionizing radiation	kBq $Co\text{-}60_{eq}$	0.0293007	0.0327965	0.0018390	0.0133913	0.0024745	0.0798020
Ozone formation, Human health	kg NO_{xeq}	0.0016812	0.0011838	0.0012227	0.0008491	0.0001242	0.0050610
Fine particulate matter formation	kg $PM2.5_{eq}$	0.0012794	0.0011135	0.0009146	0.0006446	0.0001048	0.0040569
Ozone formation, Terrestrial ecosystems	kg NO_{xeq}	0.0016987	0.0012001	0.0012274	0.0008590	0.0001250	0.0051102
Terrestrial acidification	kg SO_{2eq}	0.0030487	0.0025607	0.0025394	0.0015942	0.0002415	0.0099845
Freshwater eutrophication	kg P_{eq}	0.0002732	0.0001878	0.0001735	0.0001333	0.0018225	0.0025902
Marine eutrophication	kg N_{eq}	0.0000304	0.0000215	0.0000113	0.0000358	0.0000018	0.0001007
Terrestrial ecotoxicity	kg 1,4-DCB	2.2081680	2.3902854	0.3956286	1.0681006	0.1048587	6.1670413
Freshwater ecotoxicity	kg 1,4-DCB	0.0500275	0.0316026	0.0145596	0.0235024	0.0016854	0.1213776
Marine ecotoxicity	kg 1,4-DCB	0.0670157	0.0414685	0.0192239	0.0314345	0.0023010	0.1614436
Human carcinogenic toxicity	kg 1,4-DCB	0.2464588	0.0185621	0.0130226	0.1072000	0.0018824	0.3871259
Human non-carcinogenic toxicity	kg 1,4-DCB	1.1801783	0.6206391	0.5359196	0.5597949	0.0664968	2.9630287
Land use	m^2a $crop_{eq}$	0.1846798	0.1012295	0.1828276	0.0959685	0.0318859	0.5965913
Mineral resource scarcity	kg Cu_{eq}	0.0181347	0.0033865	0.0002644	0.0078780	0.0000878	0.0297513
Fossil resource scarcity	kg oil_{eq}	0.1528594	0.1580457	0.1304397	0.0877319	0.0132276	0.5423043
Water consumption	m^3	0.0056423	0.0177047	0.0009925	0.0028350	0.0006562	0.0278308

3.1.1. Primary DAF stage

According to Figure 4, primary DAF accounts for a relatively large proportion of factors affecting the environment in the WWTP. More than 80% of the primary DAF effect comes from the used chemicals (PAC and PAM), while electricity only accounts for approximately 15%. The primary DAF itself affects land use with only 47.4% contribution (Figure 5). The significant impact of the DAF process may come from the emissions in the production of chemicals and energy (Sharaai *et al.* 2009).

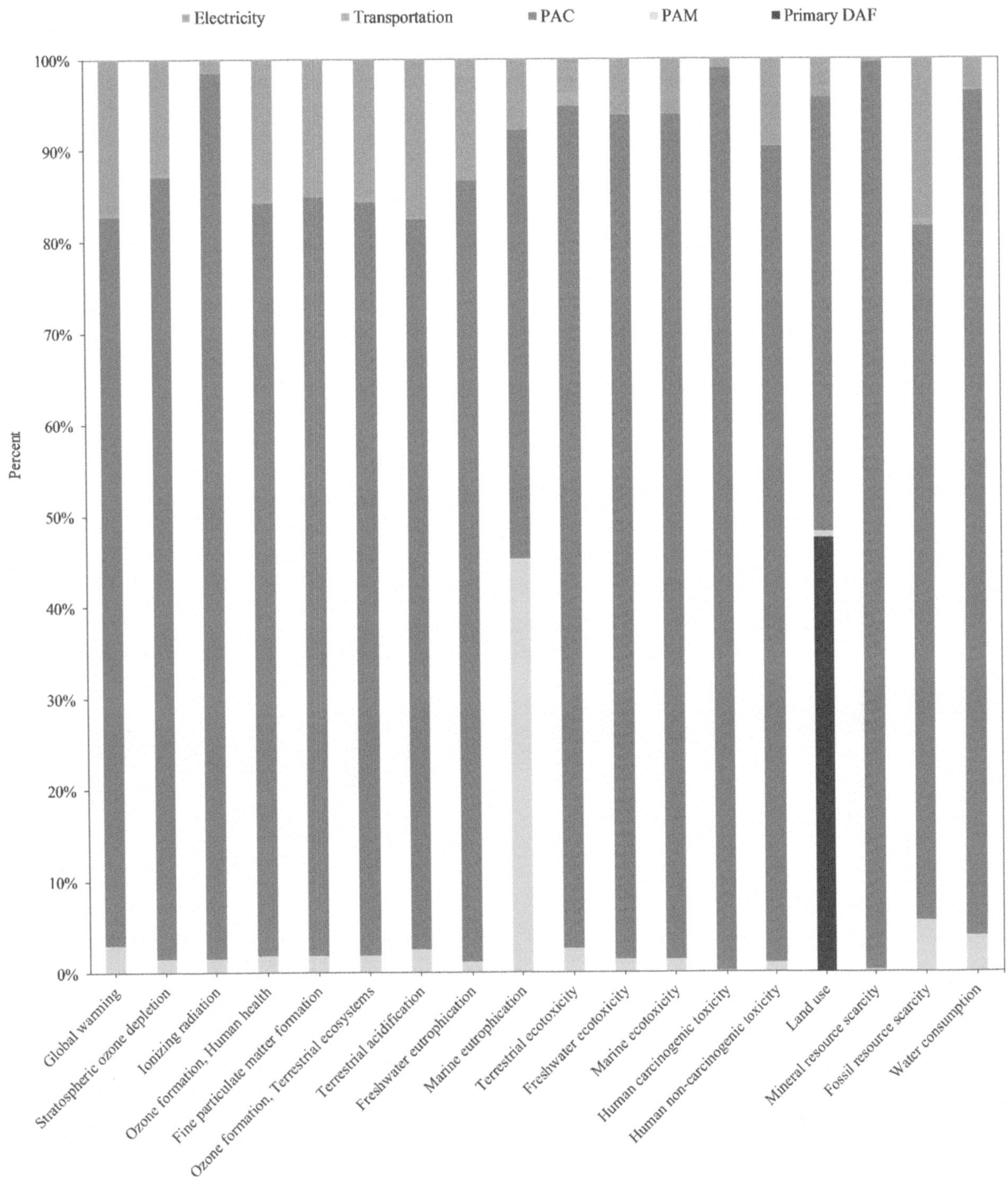

Figure 5 | Most significant processes for environmental impacts during the primary DAF process.

Although in the previous studies, it was presented that DAF will create many impacts on the environment, actually, the DAF technique was comprehended compared with the ultrafiltration (UF) membrane technique.

3.1.2. IC stage

The effect of the IC process on the environment in WWTP can be depicted in Figure 6. The most influential factors to the environment could be determined as the used chemicals (NaOH, Diammonium Phosphate and Urea) for the anaerobic process. The start-up of the IC reactor requires a lot of nutrients (Diammonium Phosphate and Urea) to cultivate anaerobic bacteria that form granular sludges and release organic acid that has to be controlled by dosing NaOH (Charalambous *et al.* 2020; Wang *et al.* 2020; Chen *et al.* 2021). This chemical consumption accounts for approximately 60% of the environmental impact of the IC process, while the IC stage only significantly contributes to land use, global warming, and stratospheric ozone depletion. High greenhouse gas emissions could be explained by these contributions (approximately 27%) during anaerobic treatment and the land for building IC systems or capturing gas. The significant effect of the greenhouse gas from the anaerobic process on the environment has been mentioned elsewhere (Lopes *et al.* 2018; Gallego-Schmid & Tarpani 2019).

3.1.3. MBBR stage

MBBR is a biological treatment that uses plastic carriers covered in biofilm to decompose waste with aeration conditions. In this process, the bacteria consume oxygen, while a large amount of N_2O is released from denitrification (Sabba *et al.* 2018). Revell *et al.* (2012) stated that N_2O is the essential ozone-depleting gas. This is in line with our findings in Figure 7; MBBRs contribute approximately 74.04% to stratospheric ozone depletion, are dominant (79.86%) in land use and contribute in a minor way to global warming (0.21%). Surprisingly, the highest impact on the environment at the MBBR stage is the power consumption for generating and distributing oxygen in the systems. High electricity consumption in the MBBR process was also obtained by Teow *et al.* (2021) in a palm oil mill WWTP.

3.1.4. Secondary DAF stage

In the WWTP, the secondary DAF plays a role in removing decayed or excess sludge generated from previous biological treatment units. This process used the same equipment and chemicals as primary DAF. It is not surprising that the chemicals (PAC and PAM) are the greatest environmental impact followed by electricity; in contrast, transportation is considered neglectable environmental impact (Figure 8). The secondary DAF only significantly causes land use (51.8%) due to the high required area

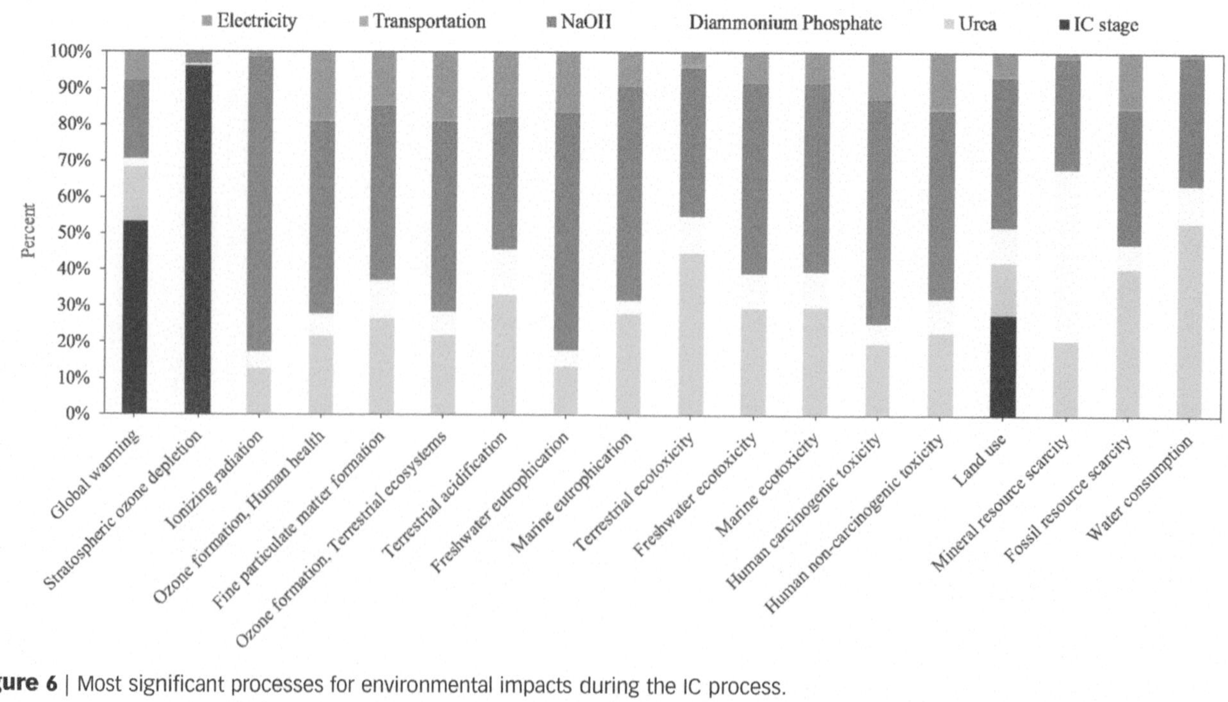

Figure 6 | Most significant processes for environmental impacts during the IC process.

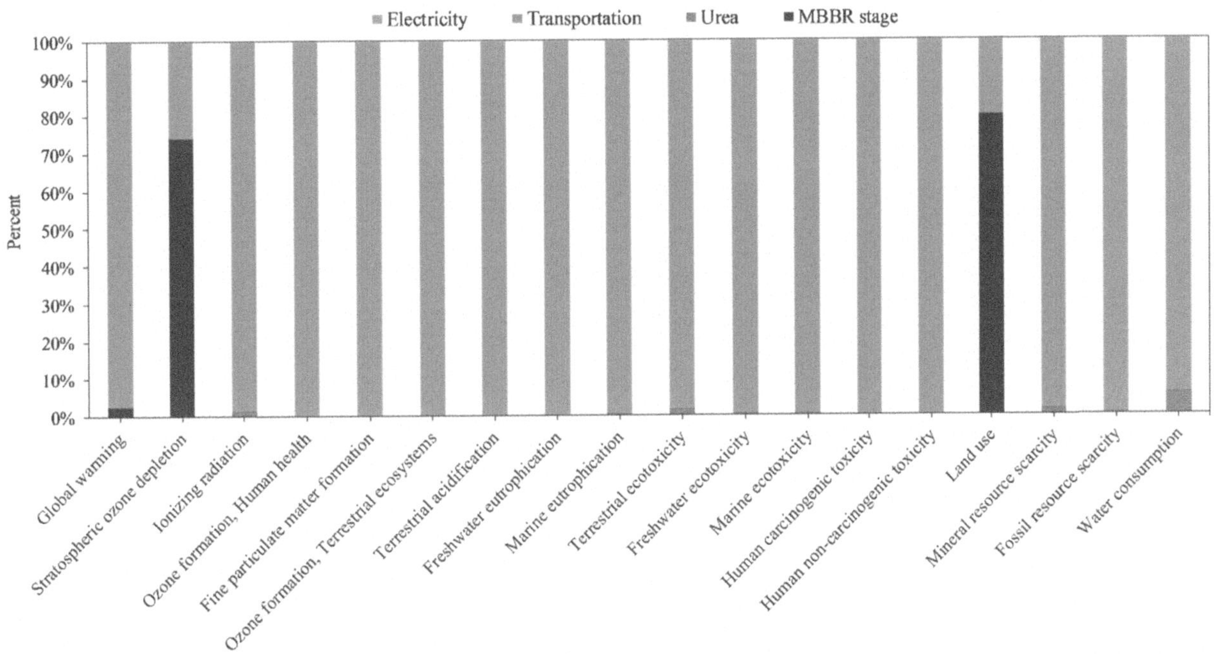

Figure 7 | Most significant processes for environmental impacts during the MBBR process.

for operating (Sharaai *et al.* 2009). However, compared with primary DAF, the percentage contribution of chemicals in the DAF process is slightly lower and the energy consumption is a little higher. The results may come from the lower pollutant concentration, especially SS in the effluent of secondary DAF and, therefore, a reduction in chemical usage.

3.1.5. Disinfection stage

The disinfection in this WWTP is required for effluent wastewater to reach as defined by the WWTP safely. This process applied the most widely used disinfectant, sodium hypochlorite (NaOCl), as a disinfectant to eliminate viruses, bacteria, fungi or mycobacterium. The concentration of 0.008 mg/L of NaOCl was directly injected into the disinfection tank and

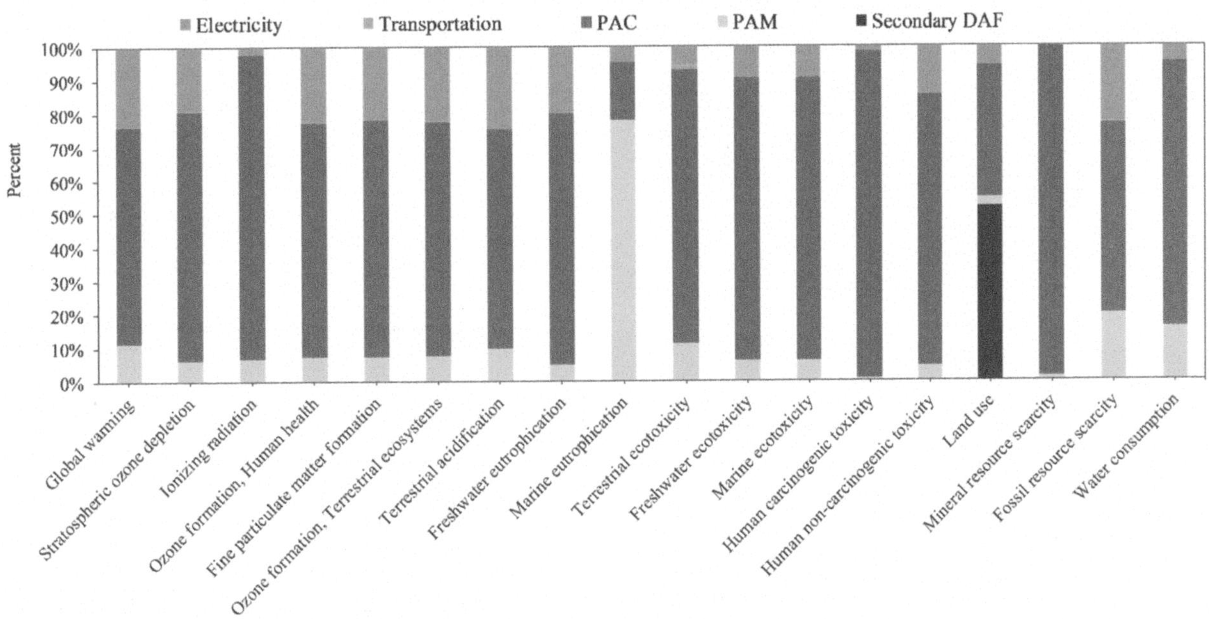

Figure 8 | Most significant processes for environmental impacts during the secondary DAF process.

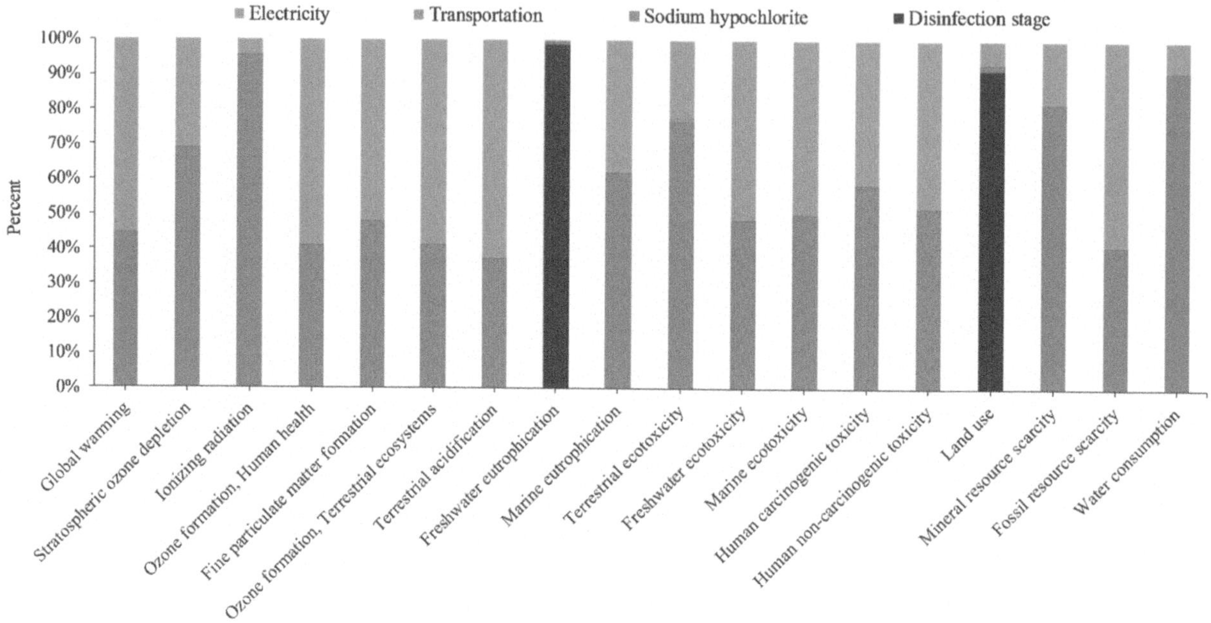

Figure 9 | Most significant processes for environmental impacts during the disinfection process.

mixed by continuous pumping. Major percentage contributors to the environment are NaOCl, electricity production followed by the transportation and disinfection stage (Figure 9). However, the predomination of disinfection stage to freshwater eutrophication (98.77%) could be related to the untreated nutrients, which cause eutrophication, as mentioned by Remy *et al.* (2014).

3.2. Endpoint damage indicator

At this stage, the results of the mid-point indicators are combined using the ReCiPe Endpoint (H) V.1.11 tool to address three distinct areas of protection (Endpoint indicators); that is, human health, ecosystems and resources. Figure 10 describes the

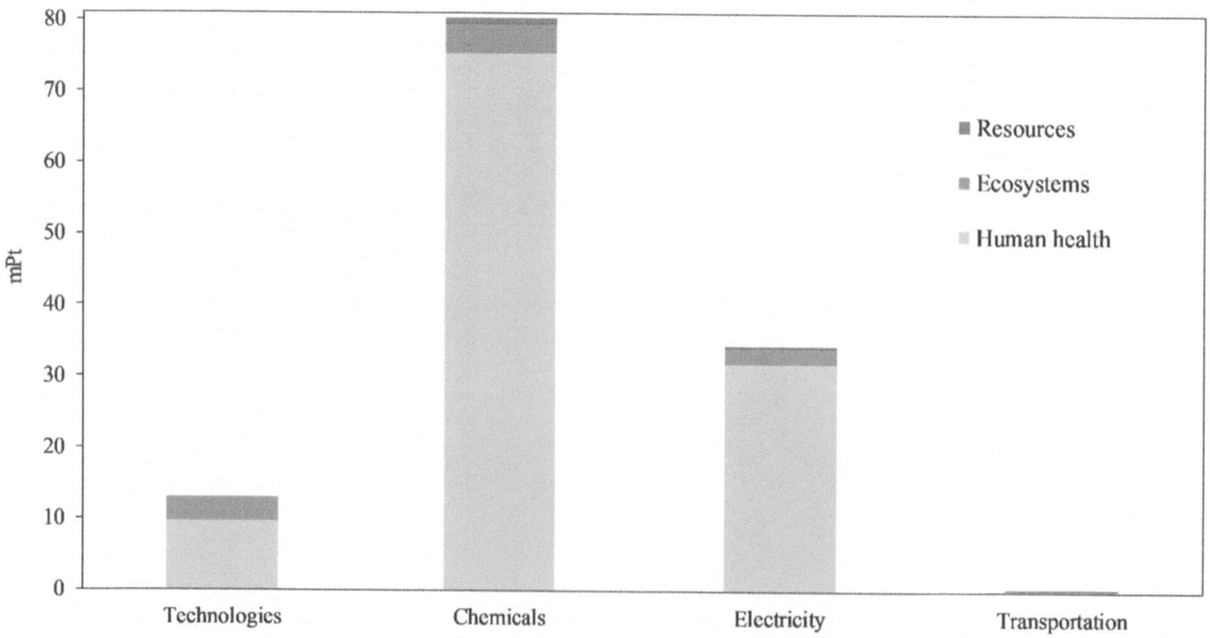

Figure 10 | Endpoint assessment results of chemicals, energy, technology and transportation operating from WWTP.

results of three Endpoint effect indicators for all treatment processes divided into chemicals, electricity, transportation, and self effect in WWTP. As seen in Figure 10, the highest impact contributor in the WWTP was Human health, which accounted for 116.95 mPt, followed by Ecosystems (9.76 mPt) and Resources (0.99 mPt). The dominant impact of Human health is mainly associated with the use of chemicals (75.33 mPt) and electricity (31.9 mPt); the contribution of the technology's characteristics and transportation are insignificant (below 10 mPt).

The highest impact of chemicals and electricity on human health or the environment could be confirmed in Figure 11, which displays the contributions of the different treatment processes in the WWTP regardless of the impact category. Most chemical and electricity consumption is attributed appreciably to DAFs, IC and MBBR, while the disinfection stage has a minor contribution to the overall impact. These results reflect the primary function of the WWTP, which is the eliminating of SS and organic pollutants in the influent. As presented in Fig. S4, as much as 48.62% of chemicals and 62.38% of electricity are attributed to primary DAF and MBBR, respectively. This finding suggests the improvement or change of these techniques could reduce the total environmental impact of WWTP. Moreover, the high share of these environmental impacts (Figure 11) and a moderate contribution to chemicals and electricity consumption (Fig. S4) of the IC phase represents a suitable technique to paper mill WWTP, as stated in many previous studies (Ruan *et al.* 2017; Ha & Ha 2019).

3.3. IC and UASB scenarios environmental analysis

To evaluate how the results changed by using the traditional UASB treatment process, mid-point indicators at two scenarios with electricity generated directly from the biogas, are shown in Table 4 and Fig. S5. According to the table, both IC and UASB have positive contributions (negative values) on five environmental indicators: fine particulate matter formation, terrestrial acidification, human carcinogenic toxicity, human non-carcinogenic toxicity and fossil resource scarcity. These occurred due to the anaerobic bacteria growth stage in both IC and UASB using wastewater, reducing the pollutants and releasing green energy, then decreasing the impacts of these 05 life cycle indicators (Yang *et al.* 2020).

However, both techniques have caused environmental damage (positive values) in stratospheric ozone depletion, ionizing radiation, freshwater eutrophication, terrestrial ecotoxicity, mineral resource scarcity and water consumption due to the required nutrients, minerals for growing bacteria and also release greenhouse gas which impacts toward to these categories. The advancement of IC compared to UASB techniques is clearly evidenced in the seven remaining indicators (global warming, ozone formation- human health, ozone formation- terrestrial ecosystems, marine eutrophication, freshwater ecotoxicity, marine ecotoxicity and land use). IC is characterized by a lower environmental impact (negative values) than the UASB

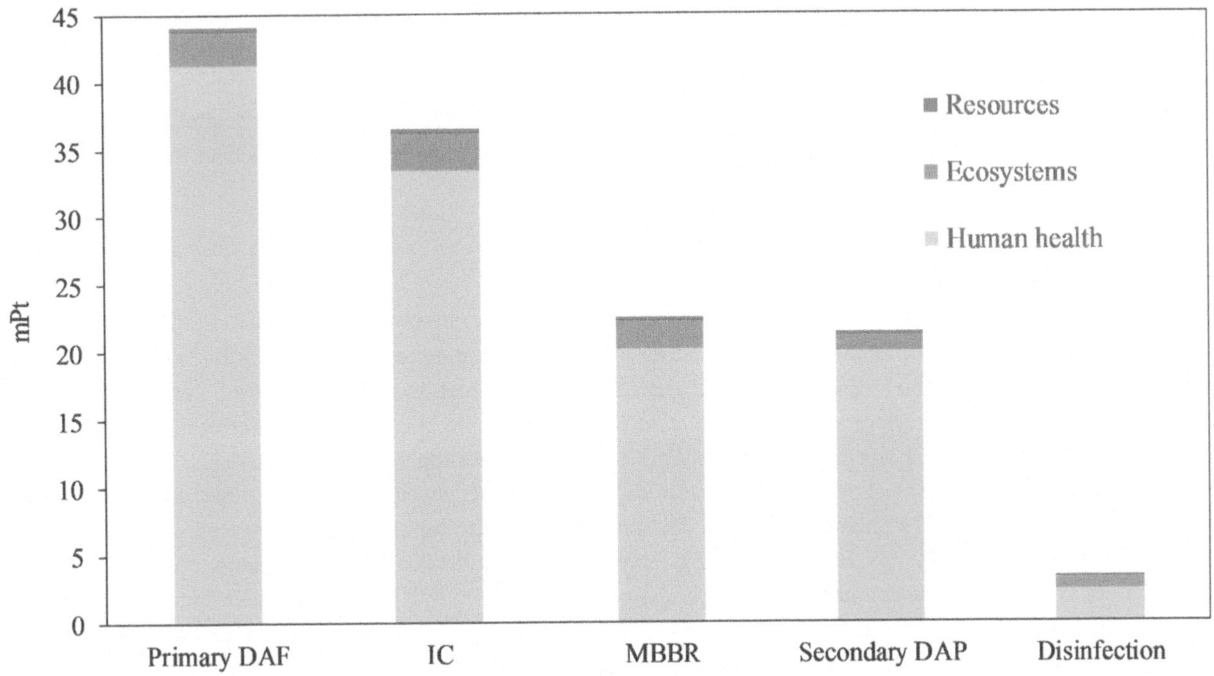

Figure 11 | Endpoint assessment results in the different stages of WWTP.

Table 4 | Mid-point impact indicators per m^3 wastewater for the two scenarios

Impact category	Unit	IC	UASB
Global warming	kg CO_{2eq}	−0.219848	0.101717
Stratospheric ozone depletion	kg $CFC11_{eq}$	86.69×10^{-6}	7.44×10^{-6}
Ionizing radiation	kBq Co-60_{eq}	0.027144	0.035087
Ozone formation, Human health	kg NO_{xeq}	−0.000644	0.000169
Fine particulate matter formation	kg $PM2.5_{eq}$	−0.001493	−0.000412
Ozone formation, Terrestrial ecosystems	kg NO_{xeq}	−0.000642	0.000185
Terrestrial acidification	kg SO_{2eq}	−0.004622	−0.001606
Freshwater eutrophication	kg P_{eq}	0.001947	0.002811
Marine eutrophication	kg N_{eq}	-13×10^{-6}	3×10^{-6}
Terrestrial ecotoxicity	kg 1,4-DCB	1.177477	3.006415
Freshwater ecotoxicity	kg 1,4-DCB	−0.013616	0.011277
Marine ecotoxicity	kg 1,4-DCB	−0.018235	0.014727
Human carcinogenic toxicity	kg 1,4-DCB	−0.021964	−0.007151
Human non-carcinogenic toxicity	kg 1,4-DCB	−1.047671	−0.430230
Land use	m^2a $crop_{eq}$	−0.013326	0.075361
Mineral resource scarcity	kg Cu_{eq}	0.002576	0.005459
Fossil resource scarcity	kg oil_{eq}	−0.247465	−0.066984
Water consumption	m^3	0.014792	0.028623

(positive values) in all these indicators. A possible explanation of this discrepancy finding is two reasons. Firstly, IC has a higher treatment capacity than UASB, as the same conditions of this WWTP, IC generates approximately 7,895 m^3 biogas/day (Driessen *et al.* 2000) as compared with 6,090 m^3 biogas/day of UASB (Metcalf *et al.* 2004); the biogas could become energy (3,088 kWh and 2,382 kWh for IC and UASB, respectively); therefore, IC could hugely reduce the energy or reduce the global warming, ozone formation- human health, ozone formation- terrestrial ecosystems (Driessen *et al.* 2000; Zhang *et al.* 2015). Secondly, both IC and UASB are anaerobic techniques; however, the chemical balance for COD: N: P in IC is 800:5:1 while UASB is 350:5:1, which means the IC process uses fewer chemicals with lower HRT compared to UASB. Arshad *et al.* (2011) and Ali *et al.* (2009) also emphasized that IC has efficient mixing, which reduces the HRT 4 times and with organic loading six times higher than UASB. The above explanations imply less land use and marine eutrophication, freshwater ecotoxicity, and marine ecotoxicity and even give the green effect to these indicators of IC technique.

4. CONCLUSIONS AND FUTURE PERSPECTIVES

This work presented an environmental assessment of a paper mill wastewater treatment plant (WWTP) in Vietnam. Midpoint and Endpoint assessments by ReCiPe 2016 were applied to measure environmental indicators of the WWTP. The obtained Midpoint results indicated that the DAF and IC phase's chemical use and energy consumption were key contributors to the total impact category. Among the categories, terrestrial ecotoxicity carried the majority impact in this 5.786 kg 1,4-DCB/m^3 wastewater, followed by Human non-carcinogenic toxicity and Global warming with 2.61 kg 1,4-DCB/m^3 wastewater and 2.068 kg CO_{2eq}, respectively. The DAF was also the most significant phase impact to three Endpoint categories (44.02 mPt). Moreover, the DAF's chemical use was still the highest contribution to the three Endpoint categories with 80.02 mPt. MBBR was determined as the highest energy consumption (62.38%), which significantly contributed to human health indicators (31.90 mPt). The comparative scenario between IC and UASB (a traditional anaerobic technique) revealed that UASB carries the most significant impact as compared to IC by all the categories. The IC technique even contributes negatively impacts (greener technique) to seven categories within 18 Midpoint categories. These findings highlighted the efficiency of IC on the environmental aspect for this kind of high organic pollutants as paper mill wastewater. The limitation of this study lies in the inventory data; Vietnam does not have the LCIs. The European and OECD data used may not represent 100% LCA cases in other region-specific locations. More system improvement should focus on biological and physical techniques to reduce contaminant and power consumption.

DATA AVAILABILITY STATEMENT

All relevant data are included in the paper or its Supplementary Information.

REFERENCES

Adams, M., Rydpal, K. & Woodfield, M. 2013 EEA air pollutant emission inventory guidebook 2013. Technical guidance to prepare national emission inventories. In: *Report*. EEA, Copenhagen, Denmark.

Afzal, M., Shabir, G., Hussain, I. & Khalid, Z. M. 2008 Paper and board mill effluent treatment with the combined biological–coagulation–filtration pilot scale reactor. *Bioresource Technology* **99** (15), 7383–7387.

Ahmed, M. T. 2010 Life cycle analysis in wastewater: a sustainability perspective. In: *Waste Water Treatment and Reuse in the Mediterranean Region*, D. Barcelo (ed.) Springer, Heidelberg, pp. 125–154.

Akbarjon, N., Huang, F. & Shen, W. 2019 Calculation of N_2O emission in the wastewater treatment process of paper mill. In *The 2nd International Conference on Sustainable Energy, Environment and Information Engineering (SEEIE 2019)*. Atlantis Press, pp. 128–131.

Ali, A., Hashmi, H. N., Querashi, I. A. & Saeed, A. 2009 Treatment feasibility of NSSC pulping effluent using UASB reactor. *HYDRO Nepal: Journal of Water, Energy and Environment* **5**, 46–49.

Arshad, A., Hashim, N. H. & Ghazala, N. 2011 Usage of UASB reactor to assess feasibility of treatment of paper mill effluent. *ASEAN Journal on Science and Technology for Development* **28** (2), 103–114.

Bakraoui, M., Karouach, F., Ouhammou, B., Aggour, M., Essamri, A. & El Bari, H. 2020 Biogas production from recycled paper mill wastewater by UASB digester: optimal and mesophilic conditions. *Biotechnology Reports* **25**, e00402.

Carnevale, E. & Lombardi, L. 2015 Comparison of different possibilities for biogas use by life cycle assessment. *Energy Procedia* **81**, 215–226.

Charalambous, P., Shin, J., Shin, S. G. & Vyrides, I. 2020 Anaerobic digestion of industrial dairy wastewater and cheese whey: performance of internal circulation bioreactor and laboratory batch test at pH 5-6. *Renewable Energy* **147**, 1–10.

Chatzisymeon, E., Foteinis, S., Mantzavinos, D. & Tsoutsos, T. 2013 Life cycle assessment of advanced oxidation processes for olive mill wastewater treatment. *Journal of Cleaner Production* **54**, 229–234.

Chen, H., Wei, Y., Liang, P., Wang, C., Hu, Y., Xie, M., Wang, Y., Xiao, B., Du, C. & Tian, H. 2020a Performance and microbial community variations of a upflow anaerobic sludge blanket (UASB) reactor for treating monosodium glutamate wastewater: effects of organic loading rate. *Journal of Environmental Management* **253**, 109691.

Chen, W., Oldfield, T. L., Patsios, S. I. & Holden, N. M. 2020b Hybrid life cycle assessment of agro-industrial wastewater valorisation. *Water Research* **170**, 115275.

Chen, J., Liu, Y., Liu, K., Hu, L., Yang, J., Wang, X., Song, Z.-L., Yang, Y., Tang, M. & Wang, R. 2021 Bacterial community composition of internal circulation reactor at different heights for large-scale brewery wastewater treatment. *Bioresource Technology* **331**, 125027.

Corominas, L., Byrne, D. M., Guest, J. S., Hospido, A., Roux, P., Shaw, A. & Short, M. D. 2020 The application of life cycle assessment (LCA) to wastewater treatment: a best practice guide and critical review. *Water Research* **184**, 116058.

de Oliveira Schwaickhardt, R., Machado, Ê. L. & Lutterbeck, C. A. 2017 Combined use of VUV and UVC photoreactors for the treatment of hospital laundry wastewaters: reduction of load parameters, detoxification and life cycle assessment of different configurations. *Science of The Total Environment* **590–591**, 233–241.

Driessen, W., Tielbaard, M., Habets, L. & Yspeert, P. 2000 Anaerobic treatment of evaporator condensates from the chemical pulp industry. In *VI Latin American IWA Workshop and Seminar on Anaerobic Digestion*, Recife, Brazil, pp. 1–9.

Dubey, A. K., Gupta, P. K., Garg, N. & Naithani, S. 2012 Bioethanol production from waste paper acid pretreated hydrolyzate with xylose fermenting *pichia stipitis*. *Carbohydrate Polymers* **88** (3), 825–829.

Farooqi, I. H. & Basheer, F. 2017 Treatment of Adsorbable Organic Halide (AOX) from pulp and paper industry wastewater using aerobic granules in pilot scale SBR. *Journal of Water Process Engineering* **19**, 60–66.

Gallego-Schmid, A. & Tarpani, R. R. Z. 2019 Life cycle assessment of wastewater treatment in developing countries: a review. *Water Research* **153**, 63–79.

Ha, N. T. & Ha, M. B. 2019 Applying internal circulation anaerobic reactor for wastewater treatment: a case study in Saigon paper mill wastewater treatment plant. *Architecture Civil Engineering Environment* **12** (3), 145–151.

Hansen, E., Zadura, L., Frankowski, S. & Wachowicz, M. 1999 Upgrading of an activated sludge plant with floating biofilm carriers at Frantschach Swiecie S.A. to meet the new demands of year 2000. *Water Science and Technology* **40** (11), 207–214.

Hauschild, M. Z., Rosenbaum, R. K. & Olsen, S. I. 2018 *Life Cycle Assessment*. Springer, Cham, Switzerland.

Hubbe, M. A. 2007 Water and papermaking. 2. White water components. *Paper Technology* **48** (2), 31–40.

Huijbregts, M. A. J., Steinmann, Z. J. N., Elshout, P. M. F., Stam, G., Verones, F., Vieira, M., Zijp, M., Hollander, A. & van Zelm, R. 2017 ReCiPe2016: a harmonised life cycle impact assessment method at midpoint and endpoint level. *The International Journal of Life Cycle Assessment* **22** (2), 138–147.

Klöpffer, W. 2012 The critical review of life cycle assessment studies according to ISO 14040 and 14044. *The International Journal of Life Cycle Assessment* **17** (9), 1087–1093.

Larsson, M., Truong, X.-B., Björn, A., Ejlertsson, J., Bastviken, D., Svensson, B. H. & Karlsson, A. 2015 Anaerobic digestion of alkaline bleaching wastewater from a kraft pulp and paper mill using UASB technique. *Environmental Technology* **36** (12), 1489–1498.

Li, Q. & Li, Y. 2019 Coproduction of hydrogen and methane in a CSTR-IC two-stage anaerobic digestion system from molasses wastewater. *Water Science and Technology* **79** (2), 270–277.

Liang, J., Mai, W., Wang, J., Li, X., Su, M., Du, J., Wu, Y., Dai, J., Tang, Q., Gao, J., Liu, Y., Tang, J. & Wei, Y. 2021 Performance and microbial communities of a novel integrated industrial-scale pulp and paper wastewater treatment plant. *Journal of Cleaner Production* **278**, 123896.

Liu, J., Wang, C., Wu, K., Huang, L., Tang, Z., Zhang, C., Wang, C., Zhao, X., Yin, F., Yang, B., Liu, J., Yang, H. & Zhang, W. 2020 Novel start-up process for the efficient degradation of high COD wastewater with up-flow anaerobic sludge blanket technology and a modified internal circulation reactor. *Bioresource Technology* **308**, 123300.

Lopes, T. A. D. S., Queiroz, L. M. & Kiperstok, A. 2018 Environmental performance of a full-scale wastewater treatment plant applying Life Cycle Assessment. *Revista Ambiente & Água* **13** (4), 1–10.

Lutterbeck, C. A., Kist, L. T., Lopez, D. R., Zerwes, F. V. & Machado, Ê. L. 2017 Life cycle assessment of integrated wastewater treatment systems with constructed wetlands in rural areas. *Journal of Cleaner Production* **148**, 527–536.

Metcalf, L., Eddy, H. P., Burton, F. L., Stensel, H. D. & Tchobanoglous, G. 2004 *Wastewater Engineering: Treatment and Reuse*. McGraw Hill, New York.

Naushad, M. 2018 *Life Cycle Assessment of Wastewater Treatment*. CRC Press, Boca Raton, FL.

Pennington, D. W., Potting, J., Finnveden, G., Lindeijer, E., Jolliet, O., Rydberg, T. & Rebitzer, G. 2004 Life cycle assessment part 2: current impact assessment practice. *Environment International* **30** (5), 721–739.

Raghuvanshi, S., Bhakar, V., Sowmya, C. & Sangwan, K. S. 2017 Waste water treatment plant life cycle assessment: treatment process to reuse of water. *Procedia CIRP* **61**, 761–766.

Rebello, T. A., Roque, R. P., Gonçalves, R. F., Calmon, J. L. & Queiroz, L. M. 2021 Life cycle assessment of urban wastewater treatment plants: a critical analysis and guideline proposal. *Water Science and Technology* **83** (3), 501–514.

Remy, C., Miehe, U., Lesjean, B. & Bartholomäus, C. 2014 Comparing environmental impacts of tertiary wastewater treatment technologies for advanced phosphorus removal and disinfection with life cycle assessment. *Water Science and Technology* **69** (8), 1742–1750.

ReportLinker 2021 *Vietnam Paper Packaging Market – Growth, Trends, COVID-19 Impact, and Forecasts (2021–2026)*. Available from: https://www.reportlinker.com/p06129761/Vietnam-Paper-Packaging-Market-Growth-Trends-COVID-19-Impact-and-Forecasts.html?utm_source=GNW (accessed 21/09/02 2021).

Revell, L. E., Bodeker, G. E., Smale, D., Lehmann, R., Huck, P. E., Williamson, B. E., Rozanov, E. & Struthers, H. 2012 The effectiveness of N_2O in depleting stratospheric ozone. *Geophysical Research Letters* **39** (15), 1–6.

Ruan, J., Chen, X., Huang, M. & Zhang, T. 2017 Application of fuzzy neural networks for modeling of biodegradation and biogas production in a full-scale internal circulation anaerobic reactor. *Journal of Environmental Science and Health, Part A* **52** (1), 7–14.

Sabba, F., Terada, A., Wells, G., Smets, B. F. & Nerenberg, R. 2018 Nitrous oxide emissions from biofilm processes for wastewater treatment. *Applied Microbiology and Biotechnology* **102** (22), 9815–9829.

Sharaai, A. H., Mahmood, N. Z. & Sulaiman, A. H. 2009 Life cycle impact assessment (LCIA) of potable water treatment process in Malaysia: comparison between dissolved air flotation (DAF) and ultrafiltration (UF) technology. *Australian Journal of Basic and Applied Sciences* **3** (4), 3625–3632.

Tabesh, M., Masooleh, M. F., Roghani, B. & Motevallian, S. S. 2019 Life-cycle assessment (LCA) of wastewater treatment plants: a case study of Tehran, Iran. *International Journal of Civil Engineering* **17** (7), 1155–1169.

Teow, Y. H., Chong, M. T., Ho, K. C. & Mohammad, A. W. 2021 Comparative environmental impact evaluation using life cycle assessment approach: a case study of integrated membrane-filtration system for the treatment of aerobically-digested palm oil mill effluent. *Sustainable Environment Research* **31** (1), 1–14.

Tiku, D. K., Kumar, A., Sawhney, S., Singh, V. P. & Kumar, R. 2007 Effectiveness of treatment technologies for wastewater pollution generated by Indian pulp mills. *Environmental Monitoring and Assessment* **132** (1), 453–466.

Vlasopoulos, N., Memon, F. A., Butler, D. & Murphy, R. 2006 Life cycle assessment of wastewater treatment technologies treating petroleum process waters. *Science of the Total Environment* **367** (1), 58–70.

Wang, C., Peng, Z., Feng, K., Chen, Z. & Liu, H. 2020 A study on the treatment efficiency of internal circulation biological aerated filters for refinery wastewater and the transformation of main organic pollutants. *Environmental Science and Pollution Research* **27** (18), 22902–22912.

Yang, X., López-Grimau, V., Vilaseca, M. & Crespi, M. 2020 Treatment of textile wastewater by CAS, MBR, and MBBR: a comparative study from technical, economic, and environmental perspectives. *Water* **12** (5), 1306.

Zhang, A., Shen, J. & Ni, Y. 2015 Anaerobic digestion for use in the pulp and paper industry and other sectors: an introductory mini-review. *BioResources* **10** (4), 8750–8769.

First received 26 October 2021; accepted in revised form 28 January 2022. Available online 8 February 2022

doi: 10.2166/wst.2022.040

Analysis of the variation of costs for sewage sludge transport, recovery and disposal in Northern Italy: a recent survey (2015–2021)

M. Domini [a,*], A. Abbà [b] and G. Bertanza [b]

[a] Department of Mechanical and Industrial Engineering (DIMI), University of Brescia, via Branze, 43, Brescia 25123, Italy
[b] Department of Civil, Environmental, Architectural Engineering and Mathematics (DICATAM), University of Brescia, via Branze, 43, Brescia 25123, Italy
*Corresponding author. E-mail: marta.domini@unibs.it

MD, 0000-0002-0733-6581; AA, 0000-0002-1377-9741; GB, 0000-0002-2965-023X

ABSTRACT

The cost for the recovery/disposal of residues produced by wastewater treatment plants (WWTPs) represents a relevant aspect in the economics of WWTP management. We elaborated and applied a desk-based methodology to analyze tender processes and to investigate quantitatively and qualitatively the variability of unit costs for sewage sludge (EWC 190805) management in Northern Italy from 2015 to 2021. We conducted a qualitative survey for operators of the sector. Unit costs for sludge management increased over time. The unit cost was mostly affected by the type of sludge recovery/disposal being the cost for recovery in agriculture (direct or by composting) lower than incineration and landfilling. Other variables influencing the cost were the distance and discount offered by the contractor. Regulatory and judiciary events determined a sudden increase of sludge disposal costs in Italy between 2017 and 2019. Recovery in agriculture, when practicable, remains the preferable option. Results can orient and support the optimization of sludge management costs and be of interest for future studies at national or European level.

Key words: biosolids, management cost, recovery/disposal routes, tender process, wastewater treatment plant

HIGHLIGHTS

- A systematic methodology based on the analysis of tender processes to investigate trends in sewage sludge disposal costs was developed.
- The cost of sludge recovery/disposal increased over time in Northern Italy in 2015–2021.
- Variables most influencing the cost were sludge quality type of recovery/disposal.
- Recovery/disposal only influenced costs more than transport.

GRAPHICAL ABSTRACT

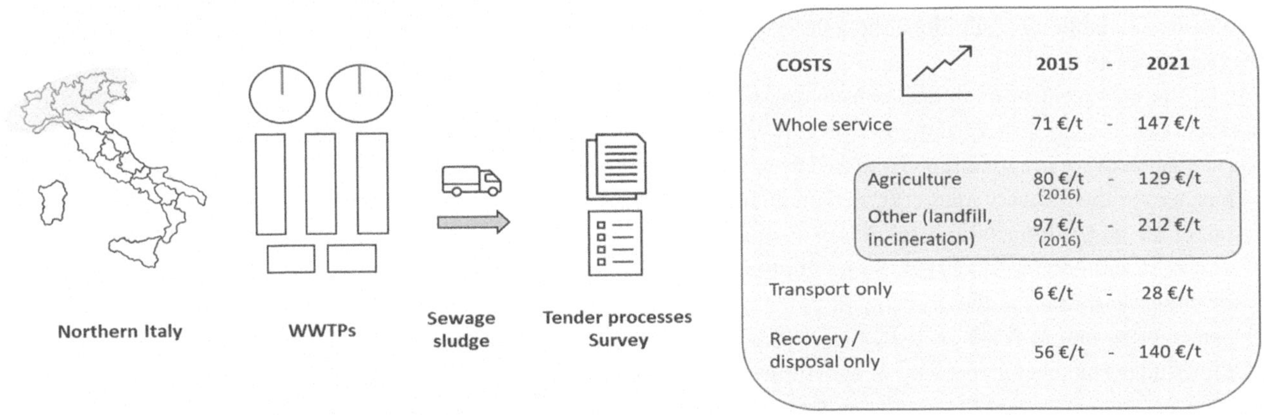

INTRODUCTION

The generation of sewage sludge is an unavoidable consequence of wastewater treatment. Wastewater treatment plants (WWTPs) produce on average 0.3 kg of dry matter (DM) per kg of chemical oxygen demand (COD) removed, with 75–85% water content (Ginestet & Camacho 2007; Canziani 2016), and approximately 20–25 kg DM per person per year in Europe (EU-27) (EUREAU 2021).

Sewage sludge derived from urban WWTPs is a waste labelled under the European Waste Catalogue (EWC) with the code 190805. Once treated, sludge is managed mainly by recovery and reuse in agriculture or for energy production by means of incineration, or disposed of by landfilling (Cieślik *et al.* 2015; Shaddel *et al.* 2019). The most practiced route for sludge management in Europe is the reuse on land (Eurostat 2021), regulated by the Sewage Sludge Directive 86/278/EEC (Bianchini *et al.* 2016; Collivignarelli *et al.* 2020; Campo *et al.* 2021). The flexibility in national implementation provided by the Directive resulted in a variety of approaches and limit values adopted by each member State (Hudcová *et al.* 2019; Gianico *et al.* 2021). In 2018, countries mainly recovering sewage sludge in agriculture, directly or by composting, were Ireland, Lithuania, Hungary, Spain, Czech Republic and Norway, while in northern countries such as Netherlands, Belgium Germany and Switzerland incineration was predominant, and landfilling was mainly practiced in Malta, Bosnia Herzegovina and Romania (Eurostat 2021).

In Italy, a decrease of disposal of sewage sludge in landfills was observed, from 40% in 2010 (Eurostat 2021) to 7% in 2019 (ISPRA 2021). In 2019, in Italy, 56% of sewage sludge was disposed of (of which 13.2% was in landfills and 7.7% by incineration) and 41% was recovered, of which 67.9% was by activities of recovery/reuse of organic substances (ISPRA 2021).

There is a great variability in the cost of alternatives for sludge disposal in European countries and the costs can vary according to local conditions (Foladori *et al.* 2010). The costs for sludge management and disposal may reach 40–60% of the total costs for managing the entire WWTP, including personnel, maintenance, energy and sludge treatment and disposal (Foladori *et al.* 2010). A publication from the EU (EC 2002) confirmed lower cost for sludge addressed to landspreading of solid and semi-solid sludge (160–210 €/t DM), whilst landfilling, mono-incineration and co-incineration of sludge with other wastes entail the highest costs (210–330 €/t DM) considering operational and investment costs. Foladori *et al.* (2010) reported the average cost of wet sludge disposal in Europe as 63.4 €/t (in the range of 0–120 €/t), including transport, being data from 2003 to 2007. It corresponded to an average cost of 317 €/t DM, with 80% water content, varying from 359 €/t DM for landfilling and 139 €/t DM for recovery in agriculture, with higher cost associated with landfill and incineration. Kacprzak *et al.* (2017) estimated an average cost for sewage sludge disposal in selected countries in EU (data up to 2013) varying between 160 and 310 €/t DM, with variation among countries and type of recovery/disposal.

At the European level, main factors influencing sludge management and leading to very different costs for disposal were: climatic conditions, land resources, transport, local regulations and fertilizer limits, amount of sludge production, disposal options and ethical factors (values and priorities) related to the acceptability of specific practices or technologies (Foladori *et al.* 2010; Collivignarelli *et al.* 2020). Also, increased consideration to climate changes and mitigation of greenhouse gas emissions focus attention on additional benefits of sludge applications to soils and resource recovery from sludge as phosphates (Kacprzak *et al.* 2017).

In Italy, the management of sludge, considering loading, transport, analysis and recovery or disposal, was estimated to impact between 15 and 40% of the costs of a WWTP (ATIA ISWA Italia 2019). In Lombardy region, the unit cost for sewage sludge recovery in agriculture decreased from 90 €/t to about 50–70 €/t in 2015 due to the increasing competitiveness of the business in the recovery market (Canziani 2016). In 2016, the cost for sewage sludge recovery by incineration was in the range of 80–90 €/t (Canziani 2016). Visigalli *et al.* (2019) reported typical unit cost for sludge disposal as 70–250 €/t, while transport counted for 15–20 €/t. Some utilities of Integrated Water Service (IWS) carried out an estimation of unit costs for sludge disposal for the Investment plans of Optimal Territorial Basins (OTB) ('Piano d'Ambito')[1]. One utility, which manages several WWTPs in the Milan province, observed an increase of costs from 2015 to 2019 from 55 €/t to 103 €/t for sludge suitable for recovery in agriculture, from 81 to 120 €/t for incineration and from 104 to 202 €/t for landfilling. Another one, managing the two large plants serving the city of Milan, registered an increase (2015–2020) from 68 to

[1] In Italy, the water services are organized in Optimal Territorial Basins (OTB), defined by the Regions and individualized through the territorial and functional integration of the different activities of the integrated cycle. A single utility is in charge of the whole water cycle, from catchment and distributing domestic water to collecting and treating wastewater (Integrated Water Service).

122 €/t for sludge recovery in agriculture and from 43 to 115 €/t for dried sludge recovered by cement factories. These increases were attributed mainly to market externalities and regulatory uncertainty that occurred between 2015 and 2018, and to the increase in sludge production due to improved performance of treatment plants. In Italy, over time, regulations posed more restrictive limits for the reuse of sludge in agriculture – such as the recent Law 109/2018 (Italian Parliament 2018) that introduced new limits for hydrocarbons, some organic micropollutants and restricted limit values for some heavy metals. However, improved technologies related to digestion, dewatering and thickening in the sludge line of WWTPs enhanced the potential of its recovery such as for energy production, together with higher competition and increased demand which affected the cost of sludge management (Campo *et al.* 2021).

As for many public services, the entrusting of services for the management of sludge produced by WWTPs in Italy is administered through tender processes. This study proposes a methodology and shows results on the variation of costs for sludge management in the last six years in Northern Italy, with the aim of supporting the sector in the optimization of costs and sludge management. Systematic and comprehensive studies and published data on sludge disposal costs for European countries, specifically Italy, are scarce. In this study, we contribute to the sector by researching sludge disposal costs and the variables affecting their variation in the North of Italy. This area, with seven regions representing about 30% of its territory, hosts most Italian treatment plants: 56% of landfills for special waste, 57% of co-incineration plants, 57% of composting plants (ISPRA), with Lombardy being one of the Italian regions with the highest reuse of fertilizer in agriculture derived by sludge (Ref Ricerche 2018).

The understanding and investigation of unit costs for sludge recovery/disposal and their variability are of great relevance due to their incidence on WWTP management costs and their relationship with waste destination, from the perspective of a circular economy, both at national and European level.

METHODS

A desk review of publicly available tender processes for the management (collection, transport, and recovery/disposal) of sewage sludge (European Waste Code EWC 190805) treated by urban wastewater treatment plants (WWTPs) in Northern Italy was conducted. In Italy, water and wastewater treatment plants are mostly managed by public bodies (97%) which can outsource services, such as waste management, through public tender processes. The study covered the following seven regions in the North of Italy: Piedmont, Lombardy, Liguria, Aosta Valley, Veneto, Trentino Alto Adige and Friuli Venezia Giulia, and a time frame of about seven years (Jan 2015–Sept 2021). Tender documents were collected from websites of contracting authorities and water management companies in the regions. Additionally, a combination of keywords was searched on Italian tendering websites 'smaltimento', 'fanghi', 'rifiuti speciali', '190805' (recovery/disposal, sludge, special waste, 190805). Inclusion criteria were: the date (2015–2021), the EWC (190805), the location (selected regions), the indication of the tender base price and/or the awarded price (or the data to calculate it) for at least one of the management activities for sludge produced by WWTPs (collection, transport, reuse/recovery/disposal, analysis).

Collected data for each tender process were: company name, year, region and province, tender base price and/or awarded price, raw sludge amount (tons), number of WWTPs served, type of recovery/disposal, document source, discount offered, total value of the bid, duration of the service requested. Data were collected, summarized and analysed by using Microsoft Excel (2010).

We conducted a quantitative and a qualitative analysis.

For the quantitative analysis, data were grouped by tender processes that reported (or for which was possible to calculate) the base unit costs (€/t) and the awarded unit cost (€/t), intended for the dewatered sludge exiting the WWTP raw. For each group, data on costs were divided in three subcategories based on the following sludge management activities: (1) whole service (loading, transport, recovery/disposal and analysis), (2) transport only, (3) recovery/disposal only. In tender processes not indicating the unit cost, it was calculated by dividing the total cost by the tons of sludge to be managed. The awarded price, when not indicated, was calculated by applying the discount offered by the awarded company to the base tender price. When bids were divided into lots, meaning that the total amount of the bid was divided into small segments independent from each other, the cost indicated for each lot was extracted separately.

For the unit costs, we calculated the mean, median, minimum and maximum, 1st and 3rd quartile and the standard deviation. Results were summarized and compared over time.

A qualitative analysis of the variation of costs of sludge management was conducted by a desk review and a survey. For the desk review, we carried out: (1) identification and selection of the variables that may influence the price, (2) identification and grouping of tender processes with information on selected variables, (3) qualitative analysis of data identified variables influencing the price variability.

Based on results of the qualitative analysis from the desk review, we prepared a survey to be submitted to utilities involved in sludge management tendering process (Supplementary Material S1). The questionnaire, to be filled online, was sent to utilities in the area of study of which we had a direct contact. The questionnaire was composed of 15 questions about variables which can affect the cost of sludge recovery and disposal. The questionnaire was aimed at confirming the desk review findings and clarify aspects that were not easily emerging from the desk review. The questionnaire was prepared by Google Modules (Google LLC) and answers were collected and analysed by Microsoft Excel (2010).

Results were summarized and compared.

RESULTS

We analysed 131 websites and identified 49 reporting tender processes meeting inclusion criteria. The desk review resulted in the identification of 194 tender processes; for 100 of them it was possible to extract both the base and awarded unit costs, for 22 the awarded unit cost only, and for 72 the base unit cost only. The most represented region was Lombardy with 42% of bids, followed by Veneto and Piedmont. Fifty-nine per cent of tender processes referred to the period 2019–2021. A summary of extracted data is presented in Table 1. Each tender process could report one or more base and/or awarded unit cost, as an example, when the service to be contracted was divided into lots (intended as parts in which the bid can be divided into that maintain their functionality and feasibility): the same tender process can indicate a base price for lot n. 1 and a different price for lot n. 2.

Quantitative analysis

In Northern Italy, the tender base unit cost for the whole service (loading, transport, recovery/disposal and analysis) (n=208) increased by 125% on average over the studied period, with the median value starting at 78.6 €/t in 2015 up to 176.9 €/t in 2021 (Figure 1). The median values of tender base unit costs for transport only (n=25) ranged from 13 €/t in 2015 to 30 €/t in 2021; the median value for recovery/disposal only (n=21) ranged from 59 €/t in 2015 to 120 €/t in 2021. See Table S2 and Figure S2.1 in the Supplementary Material for detailed results.

The awarded unit costs for the whole service (loading, transport, recovery/disposal and analysis) (n=179) increased by 95% over the last seven years, with the median value starting at 62.4 €/t in 2015 up to 122.2 €/t in 2021 (Figure 2). The median values for awarded unit costs for transport only (n=26), ranged from 5.7 €/t in 2015 to 22.8 €/t in 2021; the median value for awarded unit costs for recovery/disposal only (n=21) ranged from 55.7 €/t in 2015 to 145.0 €/t in 2021. See Table S2 and Figure S2.2 in the Supplementary Material for detailed results.

Even if the components of the price by service, i.e. transport vs recovery/disposal, were indicated (or were possible to be calculated) in few tender processes, results showed that the activity with more influence on the total price is the recovery/disposal more than the transport (Figure 3). The unit cost for transport represented on average 16 and 18% (awarded and

Table 1 | Summary of tender processes analyzed per year and data extracted from tender processes

		Number of data extracted in tender processes					
		Tender base price			Awarded price		
Year	Number of tender processes	Whole service	Transport only	Recovery/disposal only	Whole service	Transport only	Recovery/disposal only
2015	14	12	4	3	9	2	2
2016	16	22	3	0	13	0	1
2017	22	19	2	2	23	1	1
2018	27	25	0	4	18	1	3
2019	41	35	4	5	28	5	0
2020	52	63	6	6	68	11	7
2021	22	32	6	1	20	6	7
Total	**194**	**208**	**25**	**21**	**179**	**26**	**21**

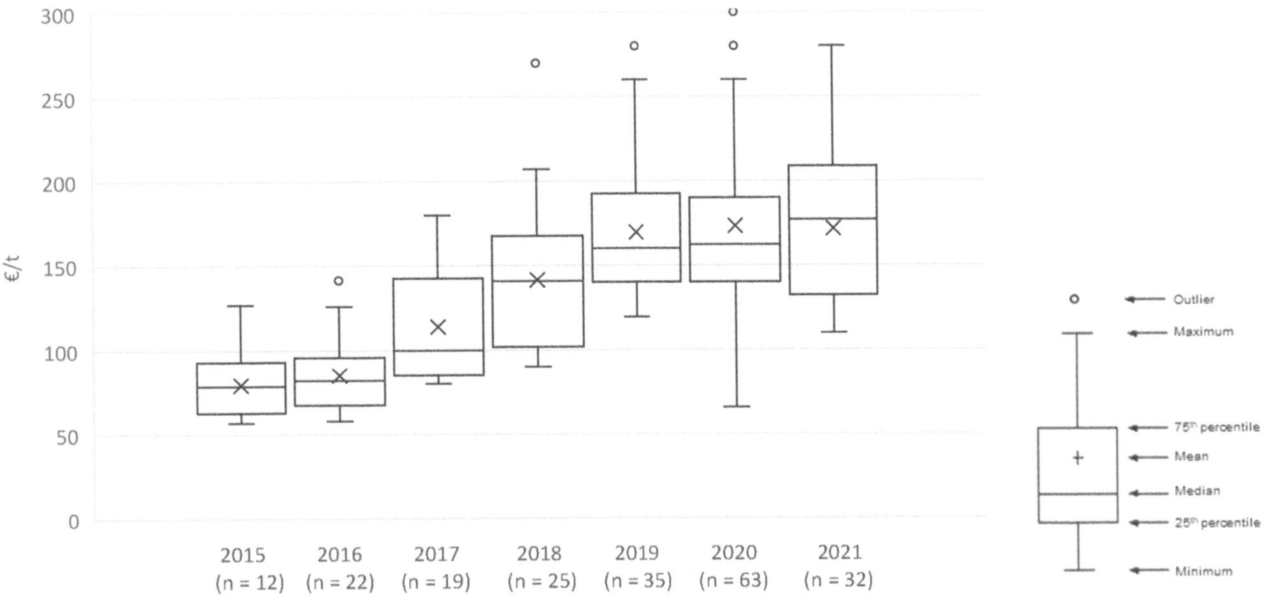

Figure 1 | Results of elaborations for tender base unit costs for sludge management activities (loading, transport, recovery/disposal and analysis) in Northern Italy (2015–2021). The amount of data (base unit cost) per year *n* is indicated in parenthesis.

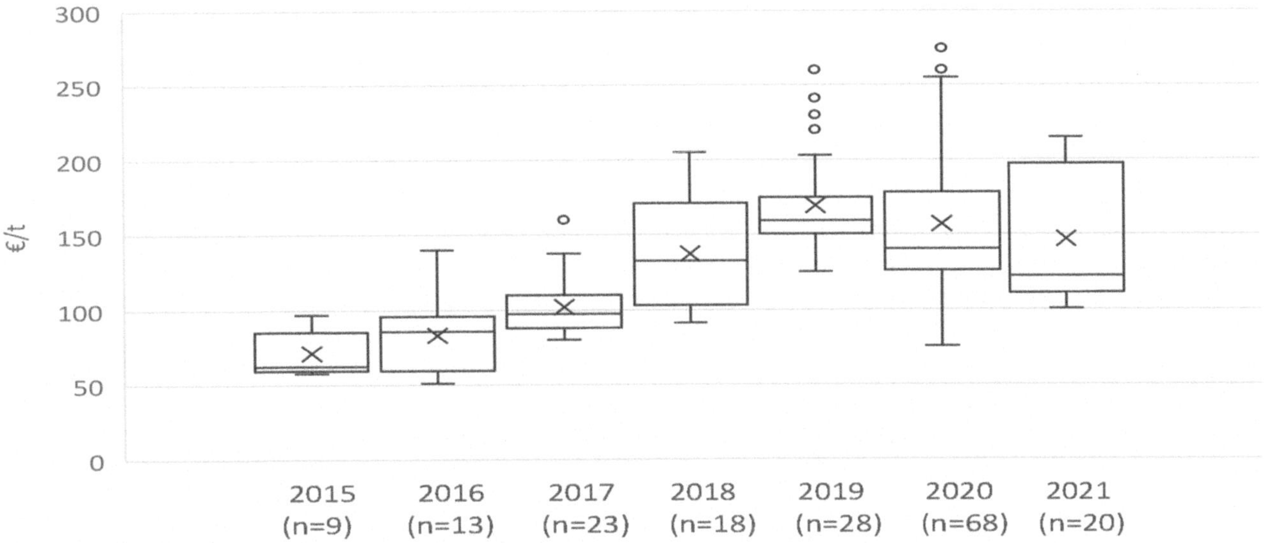

Figure 2 | Results of elaborations for awarded unit costs for sludge management activities (loading, transport, recovery/disposal and analysis) in Northern Italy (2015–2021). The amount of data (awarded unit cost) per year *n* is indicated in parenthesis.

tender base price respectively) of the total cost, and was not observed to have a relevant effect over time on the overall cost of the management of sludge exiting the WWTPs (Table S2, Figures S2.1 and 2.2 in Supplementary Material).

Qualitative analysis

Qualitatively, we identified and analysed, by the desk review, the following variables that could influence the unit cost of sludge management: type of recovery/disposal (agriculture, incineration, landfill), distance to be covered for transport to recovery/disposal facilities, area (region or province), type of sludge (dewatered/dried), discount offered by the contractor, number of plants included in the bid, and quantity of sludge to be managed.

The unit cost was influenced by the type of recovery/disposal the sludge could undergo. The cost for recovery or reuse was lower than the cost for disposal, with the recovery in agriculture being lower and the disposal in landfill the highest. We could identify 22 tender procedures explicitly indicating the possible destination for the sludge: we extracted the unit prices for

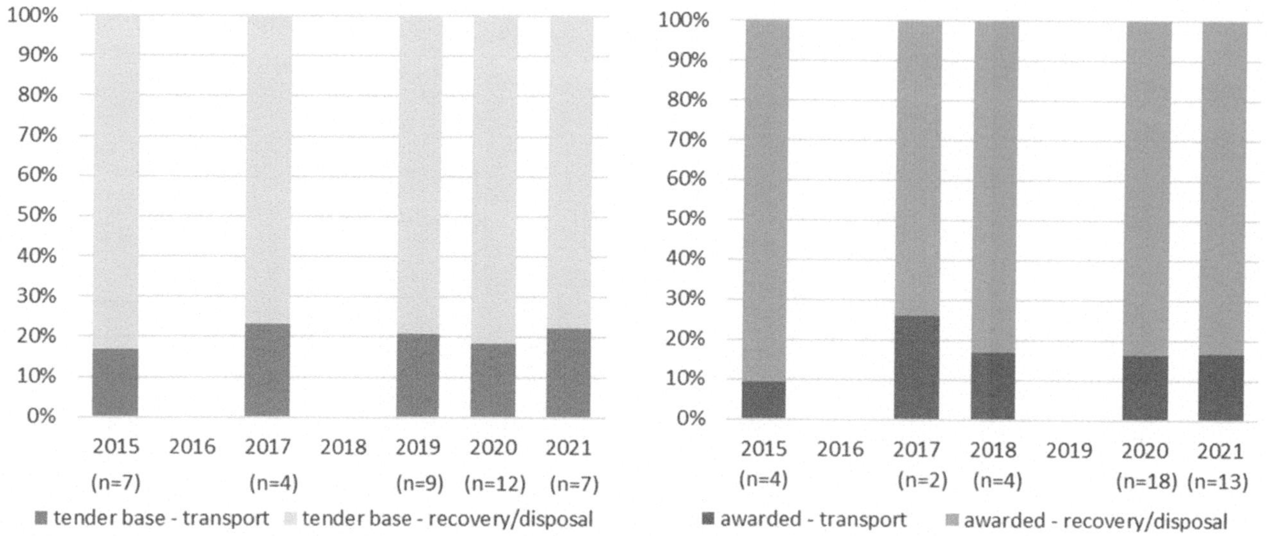

Figure 3 | Weight, in percentage, calculated on the average of tender-base and awarded unit costs for transport only and recovery/disposal only over time (2015–2021) in Northern Italy. The amount of data per year *n* is indicated in parenthesis.

managing the sludge to be recovered in agriculture and to undergo other types of recovery/disposal, e.g. incineration, landfill. On average the price for the recovery of sludge in agriculture is lower than other recovery or disposal (Figure 4).

The distance of WWTPs from recipients also emerged as having an influence on the unit cost: in a few calls for tender, a slightly higher unit cost was indicated for collecting sludge from WWTPs further from recipients.

A few calls for tender fractioned the total amount of the bid into lots to facilitate the participation of small and medium enterprises. Among criteria for the partition into lots, we identified the partition by zone, which grouped plants in a nearby area, and the partition by quality of sludge which consequently involved their possible destination. The difference among unit costs related to lots can be traced back to variables such as distance or type of sludge or type of recovery/disposal. Moreover, the awarded price was influenced by the discount offered by the contractor and can vary within the same tender process by lots when they were assigned to different contractors.

The analysis also suggested that the price can be influenced by the type of sludge (dewatered/dried) but available information was not sufficient to determine the extent of impact on the price.

Finally, there were three cases in which tender documents specified different base prices related to the need for special means of transports due to the access or the structure of WWTPs; and one tender process indicated a higher price in relation with or to the urgency of sludge collection (within 24 hours being more expensive than within 3 days).

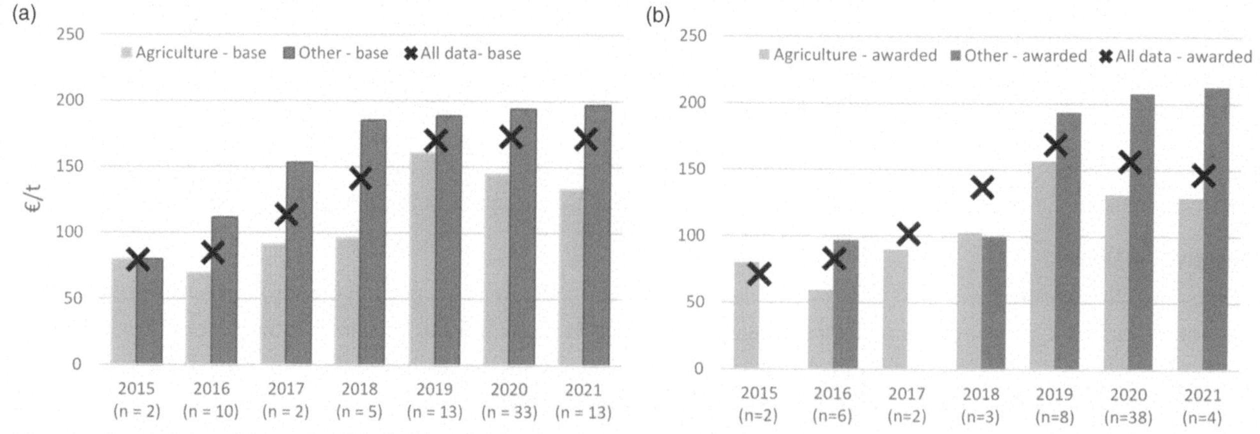

Figure 4 | Average values for the unit cost per year of sludge recovery in agriculture, other recovery/disposal of sludge, all data, base (a) and awarded (b) price. The amount of data per year (*n*) is indicated in parenthesis.

Qualitatively, no relation was observed between the unit cost and the number of plants to be served, or the quantity of sludge to manage, or the area (region or province).

The survey was submitted to 31 utilities of IWS, 11 of which responded. Sludge management activities covered the whole service, whilst it was pointed out that in some plants analysis or loading could be not included in the bid. Respondents estimated that sludge management costs impacted on the overall WWTPs management cost of a percentage comprised of between 11 and 40% in 2020.

The main variable affecting sludge management cost resulted in the sludge quality and type of recovery/disposal, followed by the distance from the destination plan. The quantity of sludge (three respondents), the discount applied by the contractor (two respondents), the loading/transport, the area, the number of plants (one respondent) were less mentioned as relevant in the definition of the price. All respondents affirmed to have noticed an increase of sludge costs over the last six years, the increase being mostly related to the activity of recovery/disposal. Respondents confirmed that recovery of sludge for agricultural reuse is less expensive than energy recovery, and that the most expensive destination is the disposal of in landfill, which could be up to twice the cost of recovery (*'landfilling costs up to twice than other destinations'*, ' *disposal in other plants costs about twice than agriculture recovery'*), some affirming *'The possibility of recovery in agriculture lower considerably the cost'*, *'The sludge managing cost is mainly linked to the possibility of recovery in agriculture'*, *'Incineration and landfill are more expensive'*.

One respondent pointed out that recovery at cement factories is still the cheapest destination for dried sludge, even compared to recovery in agriculture (dewatered sludge). Respondents declared to be aware of the possible destination of sludge according to its quality when preparing the bid and to consider it for the definition of the tender process base price.

When asked about reasons for the increasing trend of the unit cost for sludge management, respondents attributed it to more restrictive regulations and normative uncertainty, one mentioning the lowered acceptability limits for sludge to be recovered in agriculture; the diminished and limited availability of sites receiving or treating sludge; the limited competition among operators of the recovery/disposal sector; and, the increase of sludge quantities and the demand for recovery/disposal sites.

DISCUSSION

Study results showed that in the North of Italy, in the last seven years, average unit prices for sludge management increased. The increase affected tender base prices more than awarded unit prices. The increase was mainly due to the component of the price representing the recovery/disposal of sludge than the transport. Variables affecting the price were principally the type of recovery/disposal, then the distance, the discount offered, while the influence of the area or number of plants in the bid was not noticeable. Our study confirmed the trend of costs as well variables influencing it already identified in previous studies and reports at regional and European level.

The type of recovery/disposal of sludge represents an interesting variable because, in its turn, it is influenced by other factors such as the availability of sites as lands or treatment plants, the number and type of operators in the market, the competitiveness, and the regulatory framework, in particular for sludge recovery in agriculture.

We observed an increase of prices especially between 2016 and 2019, followed by a stabilization or even a little decrease. It is to be noted that in Lombardy, since 2015, the introduction of the regional regulation 'DGR 2031/X/2014' (Lombardy Region 2014), distinguishing limits for suitable and high-quality sludge to be recovered in agriculture, had a crucial impact on the regulatory framework. In the following years (2017–2019) continuous normative changes and judiciary issues, especially on parameters and limits for agriculture recovery, led to a great uncertainty (Ricerche 2018). The situation resulted in a slowdown of sludge recovery as fertilizer, especially in Lombardy, and a strong public opposition, and consequent cost increase (Utilitalia 2017). This generated in parallel an increase request for cross-border disposal, or disposal from other regions (ARERA 2019). At the same time, technological improvements and process optimizations resulted internationally to an improvement of WWTPs performance with a consequent increase of sludge production, and an increased demand for recovery and disposal options. These events, as underlined by utilities representatives responding to the survey, were the main aspects causing the abrupt increase of sludge disposal costs from 2016 to 2019. Also, since 2017, we observed an increased variability of prices, especially of tender base prices.

Sludge recovery in agriculture, when practicable, emerged to be economically advantageous and preferable.

As the variable area intended as the geographical location (region or province) did not emerge as relevant in directly influencing the price from the qualitative analysis of bids, we can hypothesize that other variables, mentioned in the survey, were affecting the cost, such as WWTPs being located in areas with less availability of land for sludge recovery, or treatment plants as incinerators, cement factories, or landfills for special waste.

Moreover, strategies implemented by utilities of IWS, such as the possibility and balance of sludge quantities to dewatering or dry, affected sludge type of recovery/disposal and their management costs. It was not possible to determine the extent of this influence from our results.

The developed methodology, based on public data from tender processes, has the potential to be applied to investigate price trends of activities related to WWTPs undergoing public tender processes, such as waste management and procurement, as an example, of reagents, or services.

Limitations of the study were: tender processes analyzed were limited to those accessible and available online; information and data of interest for the study were reported inconsistently in tender documents; disaggregated data per activity (transport and recovery/disposal) were limited. Despite these limitations, we consider that the study shows a reliable picture of the economics related to activities for the recovery or final disposal of sludge from WWTPs in Northern Italy.

CONCLUSIONS

This study clearly shows an increase of the costs for the management of sludge that occurred in Northern Italy in the last seven years (2015–2021). This is mainly related to the recovery/disposal services rather than transport. The study also evidences as the recovery in agriculture is less expensive than other forms of recovery or disposal. Moreover, regulatory framework changes also affected the variation of costs. It emerges that, from an economic point of view, and despite the increasing trend of costs over time, agricultural recovery of sludge should be prioritized when possible.

We consider the methodology applied was simple and useful and it was successful to investigate the trend of costs of activities related to WWTPs that undergo public tender processes. Results can orient and support the optimisation of sludge management costs.

This systematic analysis could be of interest for studies or analysis at national or European level in future.

DATA AVAILABILITY STATEMENT

All relevant data are included in the paper or its Supplementary Information.

REFERENCES

Arera 2019 *Avvio di indagine conoscitiva sulle modalità di gestione e di valorizzazione dei fanghi derivanti dalla depurazione delle acque reflue [Launch of a Fact-Finding Investigation on the Methods of Management and Valorisation of the Sludge Deriving From the Wastewater Tretment]*. Deliberazione 22 gennaio 2019 – 20/2019/R/IDR. Available from: https://www.arera.it/allegati/docs/19/020-19.pdf (accessed 25 October 2021).

ATIA ISWA Italia 2019 *Tavolo tecnico 'Fanghi di depurazione' [Technical Group 'Sewage Sludge']*. Available from: http://www.atiaiswa.it/wp-content/uploads/2019/07/Tavolo-tecnico-fanghi-ATIA_ISWA_finale.pdf (accessed 11/01/2022).

Bianchini, A., Bonfiglioli, L., Pellegrini, M. & Saccani, C. 2016 Sewage sludge management in Europe: a critical analysis of data quality. *International Journal of Environment and Waste Management* **18** (3), 226. doi:10.1504/IJEWM.2016.10001645.

Campo, G., Cerutti, A., Lastella, C., Leo, A., Panepinto, D., Zanetti, M. & Ruffino, B. 2021 Production and destination of sewage sludge in the Piemonte Region (Italy): the results of a survey for a future sustainable management. *International Journal of Environmental Research and Public Health* **18** (7), 3556. https://doi.org/10.3390/ijerph18073556.

Canziani, R. 2016 I fanghi, inevitabili prodotti della depurazione delle acque di rifiuto [Sludge, unavoidable products of wastewater treatment]. *Ingegneria dell' Ambiente* **3** (3), 177–178. https://doi.org/10.14672/ida.v3i3.743.

Cieślik, B. M., Namieśnik, J. & Konieczka, P. 2015 Review of sewage sludge management: standards, regulations and analytical methods. *Journal of Cleaner Production* **90**, 1–15. https://doi.org/10.1016/j.jclepro.2014.11.031.

Collivignarelli, M. C., Abbà, A. & Benigna, I. 2020 The reuse of biosolids on agricultural land: critical issues and perspective. *Water Environment Research* **92** (1), 11–25. doi:10.1002/wer.1196.

Eureau 2021 *Waste Water Treatment – Sludge Management – Briefing Note*. Available from: https://www.eureau.org/resources/briefing-notes/5629-briefing-note-on-sludge-management/file (accessed 25 October 2021).

European Commission 2002 *Disposal and Recycling Routes for Sewage Sludge part 4 – Economic Report*. Office for Official Publications of the European Communities, Luxembourg. ISBN 92-894-1801-X. Available from: https://ec.europa.eu/environment/archives/waste/sludge/pdf/sludge_disposal4.pdf (accessed 22 October 2021).

Eurostat 2021. Available from: https://ec.europa.eu/eurostat (accessed 8 January 2022).

Foladori, P., Andreottola, G. & Ziglio, G. 2010 *Sludge Reduction Technologies in Wastewater Treatment Plants*. IWA Publishing Ltd, London, UK. ISBN13: 9781843392781.

Gianico, A., Braguglia, C. M., Gallipoli, A., Montecchio, D. & Mininni, G. 2021 Land application of biosolids in Europe: possibilities, constraints and future perspectives. *Water* **13**, 103. https://doi.org/10.3390/w13010103.

Ginestet, P. & Camacho, P. 2007 *Technical Evaluation of Sludge Production and Reduction. Comparative Evaluation of Sludge Reduction Routes*. IWA Publishing Ltd, London, UK, pp. 1–15. https://doi.org/10.2166/9781780402352.

Hudcová, H., Vymazal, J. & Rozkošný, M. 2019 Present restrictions of sewage sludge application in agriculture within the European Union. *Soil and Water Research* **14**, 104–120. https://doi.org/10.17221/36/2018-SWR.

ISPRA 2021 *Rapporto rifiuti speciali [Report on Special Waste]*. Edizione 2021. Rapporti 344 - 2021, Roma. ISBN 978-88-448-1052-8.

Italian Parliament 2018 *Decreto Legge 28 settembre 2018 – Disposizioni urgenti per la città di Genova, la sicurezza della rete nazionale delle infrastrutture e dei trasporti, gli eventi sismici del 2016 e 2017, il lavoro e le altre emergenze [Law 28 September 2018 – Urgent Provisions for the City of Genova, the Safety of the National Infrastructure and Transport Network, the Seismic Events of 2016 and 2017, the Work and the Other Emergencies]*. Italian Official Gazette, September 28, 2018, n. 226, Ordinary Supplement n. 25.

Kacprzak, M., Neczaj, E., Fijałkowski, K., Grobelak, A., Grosser, A., Worwag, M., Rorata, A., Brattebob, H., Almåsc, Å. & Singh, B. R. 2017 Sewage sludge disposal strategies for sustainable development. *Environmental Research* **156**, 39–46. doi:10.1016/j.envres.2017.03.010.

Lombardy Region 2014 *Deliberazione del Consiglio Regionale X/2031, 1 luglio 2014 – Disposizioni regionali per il trattamento e l'utilizzo, a beneficio dell'agricoltura, dei fanghi di depurazione delle acque reflue di impianti civili ed industriali in attuazione dell'art. 8, comma 8, della Legge regionale 12 luglio 2007, n. 12. Conseguente integrazione del punto 7.4.2, comma 6, n. 2) della D.G.R. 18 aprile 2012, n. IX/3298, riguardante le linee guida regionali per l'autorizzazione degli impianti per la produzione di energia elettrica da fonti energetiche rinnovabili [Deliberation of Regional Council X/2031 July 1, 2014 Containing Regional Dispositions for Treatment and use, for the Benefit of Agriculture, of Sewage Sludge From Urban and Industrial Wastewater Treatment Plants in Implementation of art. 8, Paragraph 8 of Regional Law July 12, 2007, n. 12. Consequent Integration of Point 7.4.2, Paragraph 6, n. 2 of D.G.R. April 18, 2012, n. IX/3298, Regarding Regional Guidelines for the Authorization of Plants for the Production of Electricity From Renewable Energy Sources]*.

Ref Ricerche 2018 *I fanghi della depurazione: l'acqua entra nell'economia circolare [Sewage Sludge: Water Enters Into the Circular Economy]. Acqua* 107, ISSN 2531-3215. Available from: https://www.sipotra.it/wp-content/uploads/2018/10/I-fanghi-della-depurazione-l%e2%80%99acqua-entra-nell%e2%80%99economia-circolare.pdf (accessed 27 October 2021).

Shaddel, S., Bakhtiary-Davijany, H., Kabbe, C., Dadgar, F. & Østerhus, S. W. 2019 Sustainable sewage sludge management: from current practices to emerging nutrient recovery technologies. *Sustainability* **11** (12), 3435. https://doi.org/10.3390/su11123435.

Utilitalia 2017 *Utilizzo dei fanghi di depurazione in agricoltura. Indagine Utilitalia sui fanghi prodotti dal trattamento delle acque reflue urbane [Use of Sewage Sludge in Agriculture. Study by Utlitalia on Sewage Sludge From Urban Wastewater Treatment]*. Available from authors.

Visigalli, S., Spinosa, L. & Canziani, R. 2019 Tecnologie di disidratazione dei fanghi [Technologies for sludge dewatering]. *Ingegneria dell' Ambiente* **6** (4). https://doi.org/10.32024/ida.v6i4.235.

First received 8 November 2021; accepted in revised form 23 January 2022. Available online 3 February 2022

doi: 10.2166/wst.2022.037

Comparing statistical process control charts for fault detection in wastewater treatment

H. L. Marais *, V. Zaccaria and M. Odlare
Future Energy Center, Mälardalen University, Västerås SE-721 23, Sweden
*Corresponding author. E-mail: heidi.marais@mdh.se

HLM, 0000-0002-3097-459X; MO, 0000-0001-5480-0167

ABSTRACT

Fault detection is an important part of process supervision, especially in processes where there are strict requirements on the process outputs like in wastewater treatment. Statistical control charts such as Shewhart charts, cumulative sum (CUSUM) charts, and exponentially weighted moving average (EWMA) charts are common univariate fault detection methods. These methods have different strengths and weaknesses that are dependent on the characteristics of the fault. To account for this the methods in their base forms were tested with drift and bias sensor faults of different sizes to determine the overall performance of each method. Additionally, the faults were detected using two different sensors in the system to see how the presence of active process control influenced fault detectability. The EWMA method performed best for both fault types, specifically the drift faults, with a low false alarm rate and good detection time in comparison to the other methods. It was shown that decreasing the detection time can effectively reduce excess energy consumption caused by sensor faults. Additionally, it was shown that monitoring a manipulated variable has advantages over monitoring a controlled variable as set-point tracking hides faults on controlled variables; lower missed detection rates are observed using manipulated variables.

Key words: fault detection, process efficiency, process monitoring, statistical control charts, univariate, wastewater treatment

HIGHLIGHTS

- The best fault detection performance was obtained with the EWMA chart.
- Manipulated variable monitoring improves controlled variable sensor fault detection.
- Fault detection in wastewater treatment processes can improve the energy efficiency.

GRAPHICAL ABSTRACT

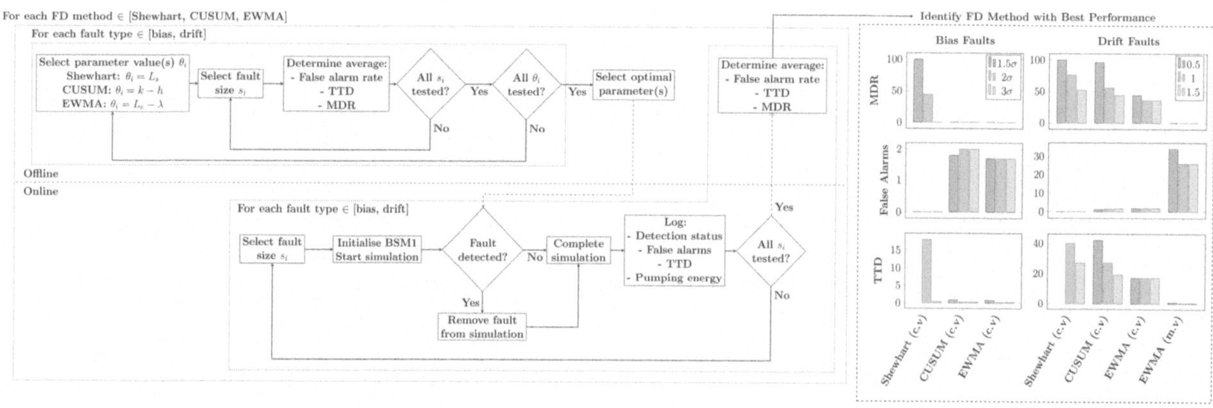

INTRODUCTION

Process monitoring is an important part of all industrial processes. Often process outputs are constrained by quality or safety requirements and continuous monitoring is necessary to ensure that those standards are met (Isermann 1997). In wastewater treatment plants (WWTPs) this is becoming more relevant as environmental regulations result in more stringent requirements on the quality of the effluent (Olsson *et al.* 2005). An important part of process monitoring is fault detection (FD), as it involves identifying whether a process has deviated from the normal operating state (Rosen *et al.* 2003). The development and implementation of reliable FD methods has the potential to improve process control by ensuring that the control system has access to reliable data (Olsson & Newell 1999). This is vital for WWTPs that have historically lagged behind other industrial processes with regards to control and automation due to the push to move from treatment facilities to water resource recovery facilities; more precise and flexible control of the process will be required.

Available FD methods vary greatly in complexity (Venkatasubramanian *et al.* 2003), however, a common technique is to place limits on certain variables and declare a fault when the variable exceeds the limit (Chiang *et al.* 2000). This is called limit sensing and is a traditional process monitoring method. Limit sensing is not particularly sensitive or robust because it ignores interactions between different variables, this makes it a univariate technique (Chiang *et al.* 2000). When considering FD in an industrial process, like the WWTP, there are advantages to investigating simple univariate FD methods despite the high level of complexity of the system (Marais *et al.* 2020). Some of the primary benefits of evaluating univariate methods are to obtain a benchmark to evaluate the performance of more advanced and costly FD methods against, as well as in their potential to be combined with advanced methods. Furthermore, the use of automated FD in industrial WWTPs is limited; optimising and demonstrating the capabilities of simple univariate methods is an important step to increase the use of FD methods in industrial processes due to the slow acceptance rate of more advanced methods.

Common univariate FD methods are often based on statistical process control and include, but are not limited to, Shewhart charts, cumulative sum (CUSUM) charts, and exponentially weighted moving average (EWMA) charts (Chiang *et al.* 2000). While numerous modifications have been suggested to these chart-based methods, their base forms are more commonly seen in applications to WWTPs. Some of the earliest investigations of these methods applied to WWTPs include that of Thomann *et al.* (2002) and Thomann (2008), who suggested the use of the Shewhart control chart to give warnings and alarms for out-of-control behaviour in a WWTP for an ammonium sensor and a suspended solids sensor, as well as a nitrate and phosphate sensor. Schraa *et al.* (2006), however, claimed that the Shewhart and CUSUM charts should perform poorly for the WWTP due to seasonal effects and non-constant variance, and advocated for the use of the EWMA control chart due to its good performance for a slow varying mean.

A more thorough comparison is presented by Corominas *et al.* (2011), comparing the Shewhart chart, standard EWMA, EWMA performed on residuals, and other EWMA variants, utilising the Benchmark Simulation Model No.1 long-term (BSM1_LT) platform. They investigated both bias and drift sensor faults on dissolved oxygen sensors, one in a control loop and one not. Using a proposed evaluation index based on several factors (such as the number of false alarms, the missed detection rate (MDR), and the time to detection (TTD)) they found that the EWMA chart was prone to high false alarm counts and that better performance was achieved with the EWMA calculated on residuals and when monitoring the actuator instead of the controlled variable.

The use of control charts on residuals has been a popular application of the methods, with Baklouti *et al.* (2017), Baklouti *et al.* (2018b), and Baklouti *et al.* (2021) utilising EWMA charts, or variations there of (maximum adaptive or maximum double adaptive), on residuals calculated from state estimation. They used the Benchmark Simulation Model No.1 (BSM1) process and tested bias type sensor faults on a chemical oxygen demand sensor (Baklouti *et al.* 2017) and drift and bias faults on a dissolved oxygen sensor (Baklouti *et al.* 2018b, 2021). They too compared the results based on the MDR, the number of false alarms, and the TTD. Using the same BSM1 layout and sensors, Baklouti *et al.* (2018a) investigated Shewhart charts, Pareto optimised EWMA, and multiscale optimised EWMA also on calculated residuals. Their results showed high MDRs obtained with the Shewhart method coupled with fewer false alarms, and overall good results for the EWMA charts.

Sánchez-Fernández *et al.* (2018) considered dissolved oxygen sensor faults in the Benchmark Simulation Model No.2 (BSM2) and found that the univariate EWMA chart on residuals performed as well, or better, than multivariate monitoring using principal component analysis (PCA) with T^2 and Q statistics. Similarly, Kazemi *et al.* (2021) reached the same conclusion but for the univariate CUSUM chart when monitoring residuals; they tested a process fault, as well as bias and drift faults on a pH sensor, and showed that the CUSUM chart outperformed the multivariate monitoring for small fault sizes, and in the cases with missing data. Finally, Riss *et al.* (2021) used a modified CUSUM chart, based on the median, and combined it with the random forest method to successfully perform fault detection on real data from a WWTP.

Evidently these univariate chart-based methods have potential to be used in WWTPs, however, none of the aforementioned studies provide a rigorous comparison of the different methods for a range of fault types and sizes. These methods are known to perform differently depending on the type of fault that is occurring; for example, theoretically, CUSUM and EWMA charts are good at detecting small and recurring changes in the data, while Shewhart charts are better for detecting large and sudden shifts in the data (Montgomery 2009). As a result, it is difficult to say which of these methods is best suited for use in process monitoring and FD in a WWTP.

In this study the goal was to evaluate which of these simple FD strategies performs best for specific faults, as well as for the general case. To achieve this a range of faults and fault sizes were tested, using the BSM1 platform, to get an idea of the overall performance of each method. The three methods mentioned (Shewhart, CUSUM, and EWMA charts) were compared for two types of sensor faults, bias and drift, on the nitrate sensor in the second bioreactor in the BSM1 process. Bias and drift faults are common forms of sensor faults, where the former represents a sudden and sustained deviation from the true value while the latter is a gradual deviation and is generally harder to detect. Both the nitrate sensor in the second bioreactor (a controlled variable) and the flow rate of the internal recycle (a manipulated variable) were considered for process monitoring. In this investigation no modifications were made to the base methods, however, a systematic approach to optimal parameter selection for the methods was proposed. This work demonstrates the effectiveness of a simple univariate FD method and identifies potential areas for improvements and future work.

METHODS

Description of fault detection methods

In univariate FD, based on statistical process control charts, the effectiveness of the method depends on the upper and lower threshold values, or control limits, that are selected. These are chosen to minimise the rate of false alarms, to reduce the TTD, and to minimise the MDR. Naturally, there is a trade-off that must be made between these criteria. Each method (Shewhart, CUSUM, EWMA) has parameters that determine the threshold limits, the behaviour of the statistic being monitored. In some cases, there is also a dependence on historical data or knowledge about the expected values of the samples. These control chart methods perform best when the data is stationary; WWTPs are not completely stationary and have diurnal and seasonal variations. Diurnal variations can be taken into consideration with the variance, but addressing seasonal variations may require utilising different chart parameters at different times of the year (Olsson & Newell 1999). This is shown in the work of Corominas *et al.* (2011) who suggest adapting the control limits based on the temperature, or simply adopting summer and winter control limits. In this work only diurnal variations are considered as considering limits for different seasons should not affect the analysis of which method performs best given different fault types and sizes. The data required to determine optimal parameter sets is not extensive and should only be representative of the normal operating state of the process.

When considering the Shewhart control chart, the form for individual measurements is appropriate for the sensor monitoring problem. The control chart is constructed on the sensor output, x_i, has a centreline at the mean, μ_0, and has upper and lower control limits (UCL, LCL) calculated as per Equations (1) and (2) (Montgomery 2009), where $\overline{MR}$ is the average moving range for the data of interest, L_s is the limit size, and d_2 is a function of the sample size which, for the current application, is always equal to 1.128 (Montgomery 2009).

$$UCL = \mu_0 + L_s \frac{\overline{MR}}{d_2} \tag{1}$$

$$LCL = \mu_0 - L_s \frac{\overline{MR}}{d_2} \tag{2}$$

The mean and average moving range are based on the expected behaviour of the data, while the limit size can be adjusted to influence the number of false alarms and the MDR.

The CUSUM chart is particularly well suited to applications involving individual measurements, and also performs well when detecting small shifts in the variable of interest (Montgomery 2009). The tabular CUSUM (one of the more common forms) involves the summation of deviations of the samples from the target, both above and below a certain threshold. Because of the cumulative nature it incorporates more information about the behaviour of the sample than the Shewhart method. The deviations, which are the statistics being monitored, are calculated in Equations (3) and (4) (Montgomery 2009),

$$C_i^+ = max\ [0, x_i - (\mu_0 + K) + C_{i-1}^+] \tag{3}$$

$$C_i^- = max\ [0, (\mu_0 - K) - x_i + C_{i-1}^-] \tag{4}$$

where K is called the reference, or allowance, value and relates to the size deviation that is desirable to detect, and $C_0^+ = C_0^- = 0$. The threshold value, H, defines the control limits. It is common practice to define both K and H as multiples of the standard deviation (σ) of the data: $K = k\sigma$ and $H = h\sigma$(Montgomery 2009). Both C_i^+ and C_i^- are monitored which results in two lines on the CUSUM control chart; the UCL (H) can be violated by C_i^+, or the LCL ($-H$) by C_i^-.

The final method is the EWMA method, this control chart is good for detecting small shifts and it is typically used with individual observations. The threshold values are defined according to Equations (5) and (6) (Montgomery 2009),

$$UCL = \mu_0 + L_e\sigma\sqrt{\left(\frac{\lambda}{2-\lambda}\right)[1-(1-\lambda)^{2i}]} \tag{5}$$

$$LCL = \mu_0 - L_e\sigma\sqrt{\left(\frac{\lambda}{2-\lambda}\right)[1-(1-\lambda)^{2i}]} \tag{6}$$

where λ, a smoothing parameter, is $\in (0, 1)$, and L_e represents the width of the control limits. The monitored statistic is shown in Equation (7),

$$z_i = \lambda x_i + (1-\lambda)z_{i-1} \tag{7}$$

where $z_0 = \mu_0$ (Montgomery 2009). The parameters λ and L_e are tuned to determine the performance, with small values of λ detecting smaller deviations in the variable of interest but requiring narrower control limits (Montgomery 2009).

Performance evaluation criteria

As mentioned, the various control chart parameters affect the number of false alarms that occur, the MDR, and the TTD. It is necessary to define these criteria. Consider the case where a single fault occurs during a simulation, we define the interval within the control limits as $\text{CL} = [LCL, UCL]$, and the set of points outside the control limits as its complement $\overline{CL}$. When a control limit is exceeded by a monitored statistic, J_i, where J_i is x_i, C_i^+, C_i^-, or z_i, this is equivalent to $J_i \in \overline{CL}$. We define v_i which evaluates if a control limit is violated by the monitored statistic at timestep

$$v_i = \mathbb{1}_{\overline{CL}}(J_i) \tag{8}$$

where $\mathbb{1}$ is the indicator function. We also define f_i which shows if a fault is occurring, and c_i which tracks if a new false alarm has occurred:

$$f_i = \begin{cases} 1 & \text{if fault occurring} \\ 0 & \text{otherwise} \end{cases} \tag{9}$$

$$c_i = \begin{cases} i & \text{if } v_i = 1 \text{ and } f_i = 0 \\ c_{i-1} & \text{otherwise} \end{cases} \tag{10}$$

where $c_1 = 0$. We define a false alarm as a control limit violation in the absence of a fault. The number of false alarms is the number of times a threshold is incorrectly crossed, two successive erroneous violations are thus considered a single false alarm. The number of false alarms for a simulation period can be calculated as

$$\text{Number of False Alarms} = \sum_{i=1} v_{i+1}\,(1 - \mathbb{1}_i(c_i)) \tag{11}$$

Since each simulation contains a single fault, as previously stated, any control limit violation within the duration of the fault indicates the detection of the fault. This corresponds to the set of correct detections, $F = \{i{:}v_i f_i = 1\}$, being non-empty. If this set is empty, it indicates that the monitored statistic failed to violate a control limit during a fault event and the fault was undetected. The MDR is determined for several simulations as the percentage of simulations where the fault was undetected. Finally, the TTD is the time between the fault event occurring (which in a real fault case can be hard to pinpoint) and the time at which the monitored statistic correctly violates a control limit. It can be calculated as

$$\text{TTD} = \underset{i \in F}{\text{argmin}}\, i - i_{fault\ start} \tag{12}$$

The TTD in the case where a fault is undetected is undefined.

WWTP used for simulations

The MATLAB Simulink implementation of the BSM1 was used to simulate the WWTP. This setup consists of five biological reactors placed in series and modelled with the Activated Sludge Model No. 1 (ASM1). Two of the reactors are treated as anoxic and three as aerated, resulting in a total anoxic volume of 2,000 m^3 and a total aerated volume of 3,999 m^3. Following the reactors is a non-reactive ten-layer settler, modelled with the Takács model, which has a volume of 6,000 m^3. The model assumes a constant temperature of 15 °C. A detailed description of the BSM1 can be found in Gernaey *et al.* (2014). The layout is shown in Figure 1.

The primary sensor of interest in this study was the nitrate sensor ($S_{NO,2}$) in the second biological reactor, as shown in Figure 1. This is a standard sensor in the BSM1 layout and has a setpoint of 1 $gN{\cdot}m^{-3}$ in the control loop. All sensor faults investigated were introduced to this sensor, $S_{NO,2}$. An additional sensor was added to the BSM1 to measure the flow rate of the internal recycle, Q_a. This was a class A sensor with a measurement range of 0–1,00,000 $m^3{\cdot}day^{-1}$, based on the recommendations in Gernaey *et al.* (2014). Sensor noise was obtained from the IWA Modelling & Integrated Assessment Specialist Group (2020). Both the $S_{NO,2}$ and the Q_a sensors were used for process monitoring, with the focus being on the former.

All simulations were run using the provided dry-weather influent which simulates for a 14-day period. The simulations were initialised as per Gernaey *et al.* (2014).

Determination of optimal parameters

The simulation work was divided into two parts, where the first involved the calibration and optimisation of the parameters used in the control charts. For this, the BSM1 was simulated under normal operating conditions with no added faults, and the response of the $S_{NO,2}$ sensor recorded. Bias and drift faults were then added in post-processing to this pre-obtained sensor signal to generate a wide range of fault conditions in the $S_{NO,2}$ sensor that could be used to determine the parameters of the control chart FD methods.

The faults were added with start times between days 7 and 12 of the simulation, with durations between 4 and 48 h. Latin hypercube sampling was used to generate start times and durations within these limits for each fault size of interest. The bias

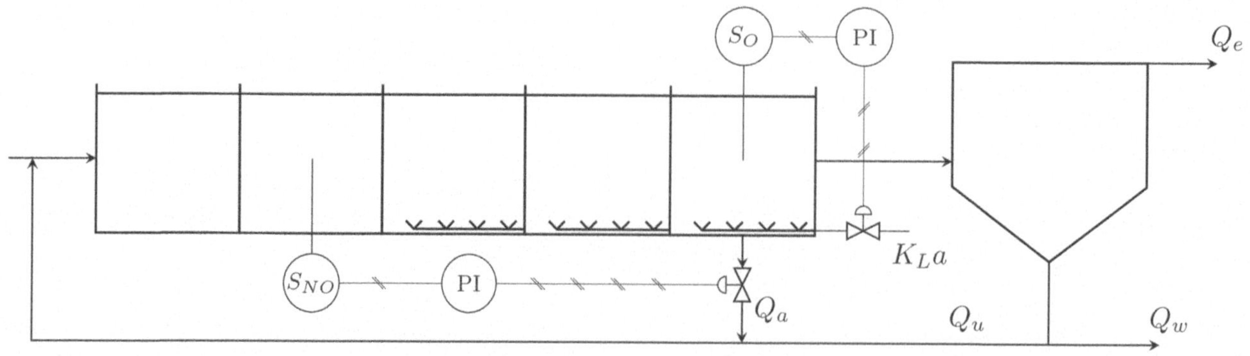

Figure 1 | The layout of the BSM1 WWTP, adapted from Gernaey *et al.* (2014), showing the two standard control loops.

fault sizes ranged from 1σ gN·m^{-3} to 5σ gN·m^{-3}, where $\sigma = 0.565$ gN·m^{-3}, while the drift fault slopes were between 0.25 gN·m^{-3}·day^{-1} and 3 gN·m^{-3}·day^{-1}.

The procedure was as follows and is shown in Figure 2:

(1) For a given FD method, the relevant parameters were selected from a range.
(2) A fault of specific type and size was added to the sensor data, with characteristics described in the preceding paragraph.
(3) The FD method was used to detect faults on the generated faulty data.
(4) The number of false alarms, TTD, and MDR were logged.
(5) Steps 2–4 were repeated 20 times for the same fault type and size (different start times and durations).
(6) The false alarms, MDR, and TTD were averaged for the given fault type, size, and parameter combination.
(7) Fault size was adjusted and steps 2–6 repeated.
(8) After iterating through all fault sizes and types, the parameters were changed and steps 1–7 repeated.

This was done for all three FD methods, where parameter ranges were selected based on recommended values from theory (Montgomery 2009):

- Shewhart: $1 \leq L_s \leq 5$ with 100 samples.
- CUSUM: $0.5 \leq k \leq 2$ and $1.5 \leq h \leq 5$ with 25 samples in each range.
- EWMA: $0.5 \leq L_e \leq 4.5$ and $0.05 \leq \lambda \leq 0.5$ with 25 samples in each range.

The results from this were used to select an optimal parameter combination for each FD method.

Online FD tests

The BSM1 was used for online FD tests with the control chart methods and a schematic of the procedure is given in Figure 3. Each simulation was initialised from the same starting point, and a unique fault added during the simulation. The faults were added with start times between days 7 and 12 of the simulation and with durations between 4 and 48 h, where again Latin hypercube sampling was used to generate sample points within these limits. The bias fault sizes tested were 1.5σ gN·m^{-3}, 2σ gN·m^{-3}, and 3σ gN·m^{-3}, and the drift fault slopes were 0.5 gN·m^{-3}·day^{-1}, 1 gN·m^{-3}·day^{-1}, and 1.5 gN·m^{-3}·day^{-1}.

Blocks were added into the Simulink model to perform the FD, according to the descriptions of the methods provided previously, with the calibrated parameters. If the fault was correctly detected the fault was removed from the system at the time of detection to evaluate the effect which improved FD can have on the performance of the WWTP.

The dependent variables, or criteria for evaluation, were the number of false alarms that occurred, the MDR, and the TTD, as well as the signals from the sensors being considered and the pumping energy evaluation index built into the BSM1 platform. The pumping energy in BSM1 is calculated based on the flow rates of the external recycle, the internal recycle, and the waste sludge (Gernaey *et al.* 2014), it is given in kWh·day^{-1} and makes up part of the operational cost index in the BSM1 performance evaluation (Gernaey *et al.* 2014). The pumping energy is selected as an evaluation criterion due to its dependence on the internal recycle, Q_a, which is directly affected by faults in the nitrate sensor in the second biological reactor.

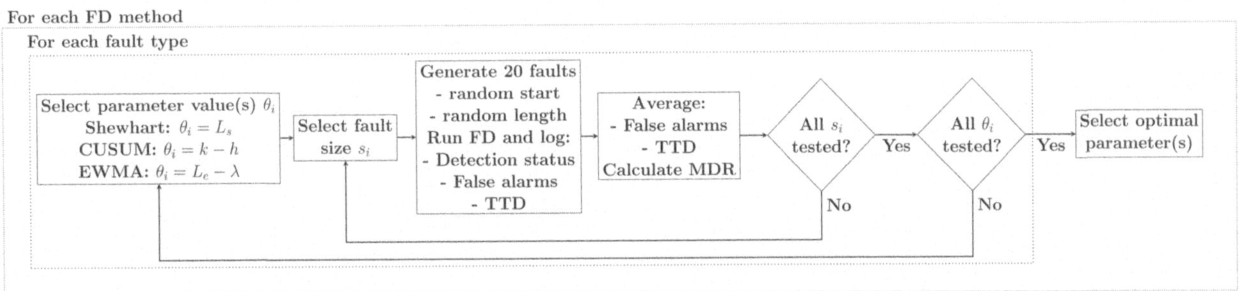

Figure 2 | Procedure for optimal parameter selection, performed for each FD method (Shewhart, CUSUM, EWMA) with two fault types (bias and drift).

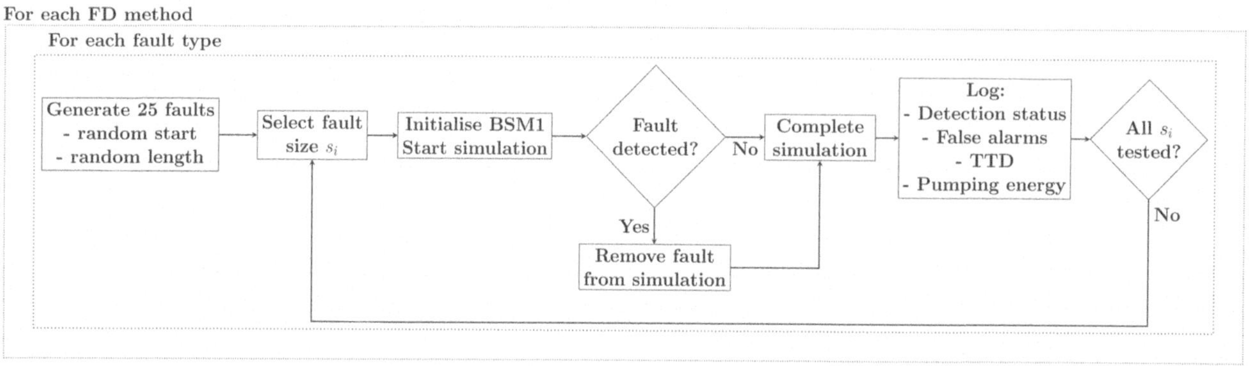

Figure 3 | Procedure for online testing of FD methods to determine an average performance for each fault size – fault type – FD method combination.

RESULTS AND DISCUSSION

Determination of optimal parameters

As mentioned, the determination of optimal parameters was done using the offline artificially generated faulty data. The faults were introduced to the $S_{NO,2}$ sensor in the second biological reactor. Figure 4 shows the results from the offline bias, and drift, fault tests using the Shewhart control chart for FD.

What is immediately evident is that attempting to minimise the TTD results in a rapid increase in the number of false alarms. As the aim is to minimise these two criteria, and the MDR, any point on the Pareto front with a low MDR is a candidate optimal solution for some trade-off between the criteria. It can also be noted that the drift faults have significantly longer TTD values (unnormalised bounds of 0.45–20.2 h compared to 0.045–3.7 h for bias faults), as well as higher MDRs. The false alarms for both methods vary between 0 and 135 when unnormalised.

Three points are marked on Figure 4: the points which have the smallest distance to the origin for the bias faults (upward facing triangle) and the drift faults (downward facing triangle) respectively, the right facing triangle is the point using the standard limit size. It is perhaps not unexpected that the different fault types have different optimal parameters. The limit sizes (L_S) that correspond to these three points are 2.37, 2.05, and 3, respectively. While the smaller limit sizes give shorter detection times, it was decided to proceed with the limit size of 3 due to the lower false alarm rate without sacrificing significantly on the MDR (0% for bias faults, 4% for drift faults). The Shewhart chart is well established and commonly used with a limit size of 3 which, by selecting this limit size, makes it a baseline method that can be used for comparison.

A similar line of reasoning can be followed for selecting the optimal parameters for the CUSUM chart, which has results shown in Figure 5. The limits of the unnormalised data are 0–27 and 0–30 false alarms for bias and drift faults

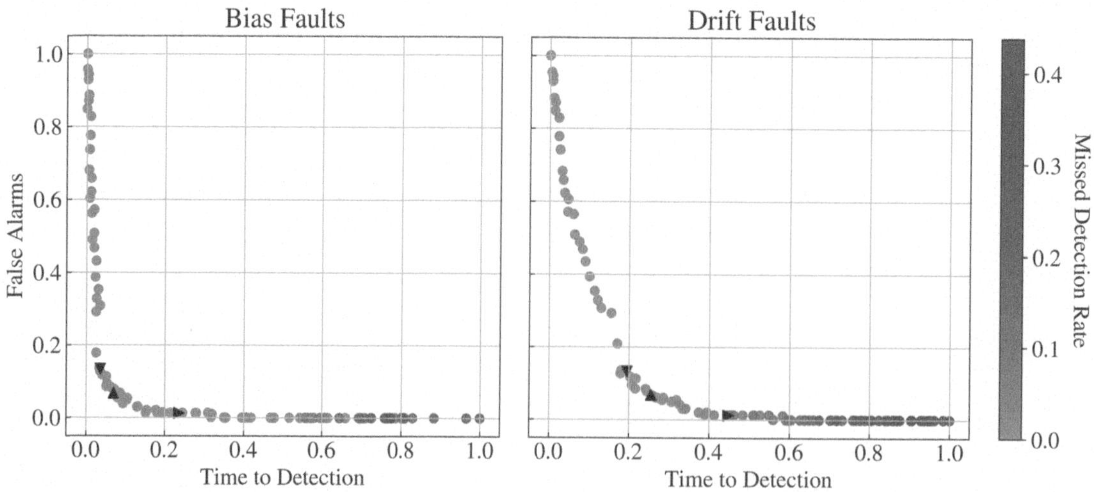

Figure 4 | Normalised performance of offline Shewhart FD.

respectively, and 0.21–3.1 h and 1.2–17.7 h for the detection times for the bias and drift faults. The upward and downward facing triangles have the same meaning as before, while the right facing triangle represents a third parameter set. The reference and threshold values, k and h, corresponding to these points are 1–2.38, 0.62–4.85, and 1.5–1.6, respectively (written as k–h).

An interesting observation is the significantly higher MDR rate for the drift faults when compared to the bias faults; decreasing the size of the threshold value (h) appears to have the largest effect on the MDR considering the three highlighted points. This suggests that knowing what sizes of faults are dangerous or costly is important to optimal calibration of the control chart. However, as before, reducing the number of false alarms clearly comes at the cost of increasing the MDR. The point closest to the origin in the case of the drift faults has a 2.5% MDR, with 4 false alarms on average (unnormalized), while the third parameter set has a 6% MDR with on average 0.85 false alarms (unnormalized); the false alarms decreased by a factor of 4.7 while the MDR increased by a factor of 2.4. For this reason, the third parameter set, 1.5–1.6, was selected for this method.

Finally, the EWMA chart results are shown in Figure 6. An interesting difference can be observed compared to Figures 4 and 5; the TTD appears to be able to reach a minimum while the number of false alarms remains acceptably low. This is slightly deceptive, however, due to the normalisation. In Figure 5 the absolute values for maximum number of false alarms were 27 and 30 for bias and drift respectively, while in Figure 6 the absolute value is 125 for both fault cases, which means that the 0.3 point in Figure 6 corresponds to 37.5 false alarms, and is significantly higher than what was obtained for the CUSUM approach. The unnormalised detection times have bounds of 0.018–2.5 h for bias faults, and 0.13–14.6 h for drift faults.

The three parameter sets for the upward, downward, and right facing triangles correspond to limit size – smoothing factor pairs (L_e–λ) of 2.5–0.48, 1.33–0.05, and 3.05–0.4. The effects that the parameters have on the performance for detecting bias faults is significantly lower than on the drift faults. Due to the significantly higher rate of false alarms when compared to the previous methods the third parameter set was chosen as it reduced the false alarms without reaching high MDRs or unacceptable TTDs. In all three cases, it was prioritised to reduce the number of false alarms over minimising the TTD. This choice was motivated by the relatively slow dynamics in a WWTP, while false alarms can disrupt operations and cause unnecessary cost to the process.

To summarise the parameter selection and the performance metrics obtained:

- Shewhart: $L_s = 3$, False alarms = 1.75 (bias and drift), TTD=0.825 h (bias), 10.2 h (drift), MDR=0% (bias), 4.25% (drift).
- CUSUM: $k = 1.5$, $h = 1.6$, False alarms=0.830 (bias), 0.845 (drift), TTD=0.903 h (bias), 10.5 h (drift), MDR=0% (bias), 6% (drift).
- EWMA: $L_e = 3.05$, $\lambda = 0.4$, False alarms=4.72 (bias), 4.75 (drift), TTD=0.569 h (bias), 8.30 h (drift), MDR=0% (bias), 2.25% (drift).

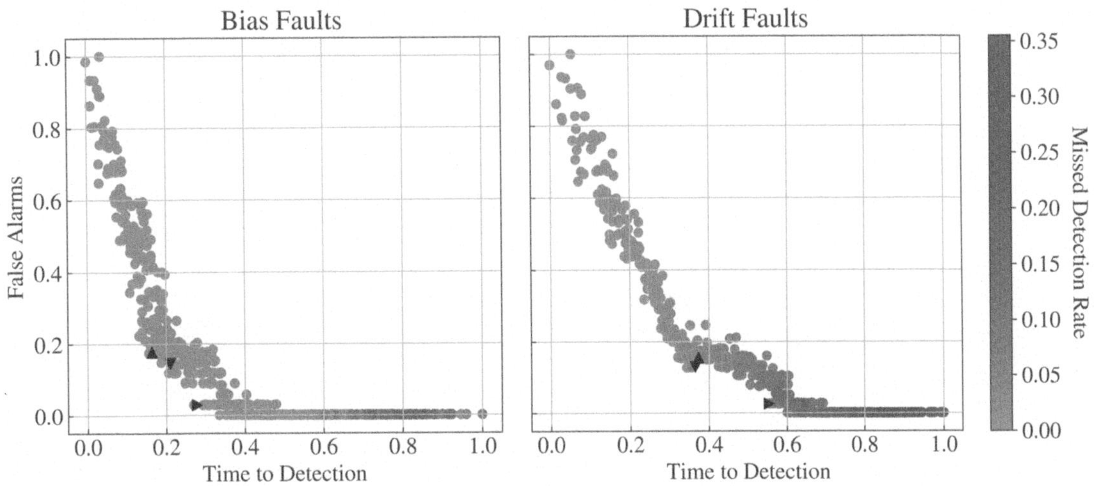

Figure 5 | Normalised performance of offline CUSUM FD.

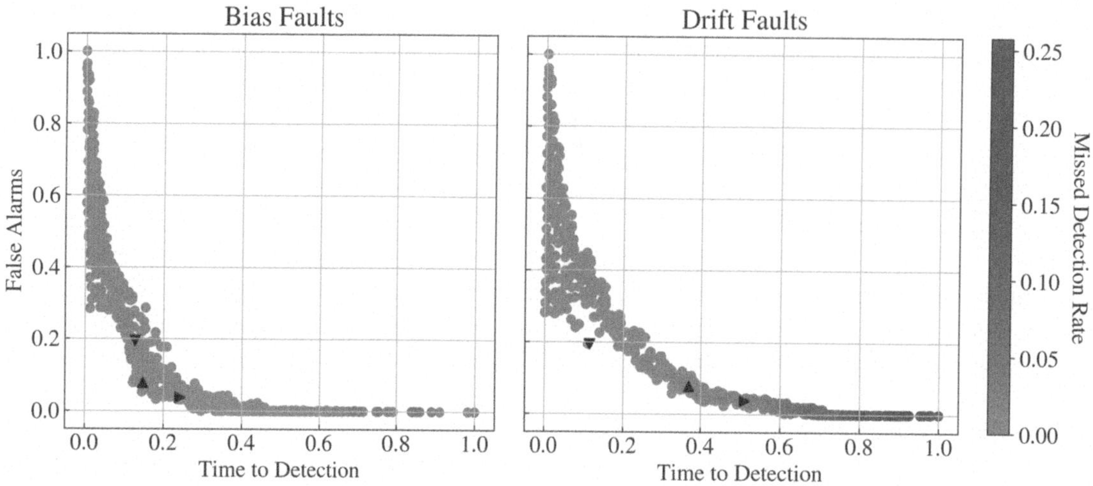

Figure 6 | Normalised performance of offline EWMA FD.

In these offline tests the EWMA method had the shortest detection times and lowest missed detection rates, while the CUSUM method had the fewest false alarms.

Online FD: monitoring the nitrate sensor

The parameters selected in the prior section were used in the online FD tests for monitoring the nitrate sensor. In Figure 7 examples of the control charts are shown for the two types of faults. Note that the faults are different due to the random start and end times. In the Shewhart method the statistic that is monitored is the sensor output itself, as discussed previously; this results in a noisy output shown in the Shewhart charts in Figure 7 due to the absence of additional filtering of the data. The CUSUM chart shows both the positive and negative statistics (C_i^+ and C_i^-) on the same plot and is difficult to interpret due to sudden spikes in the statistic when faults are detected, note that the y-axis is bounded as the absolute values of the statistics are not of interest, only the period for which it is above or below the threshold values. In the CUSUM chart the size and severity of the fault is unclear from the chart, however, the simplicity and the few numbers of false alarms produced is appealing. The EWMA method appears to be a good compromise between the aforementioned two charts, due to the ability to adjust the noise and rate of response with the smoothing factor but being able to visually judge the severity of the fault from the chart.

The analysis of the results will focus on the three performance evaluation criteria. As mentioned in the descriptions of the simulations, three fault sizes were tested for each fault type. Table 1 shows the results for the bias faults.

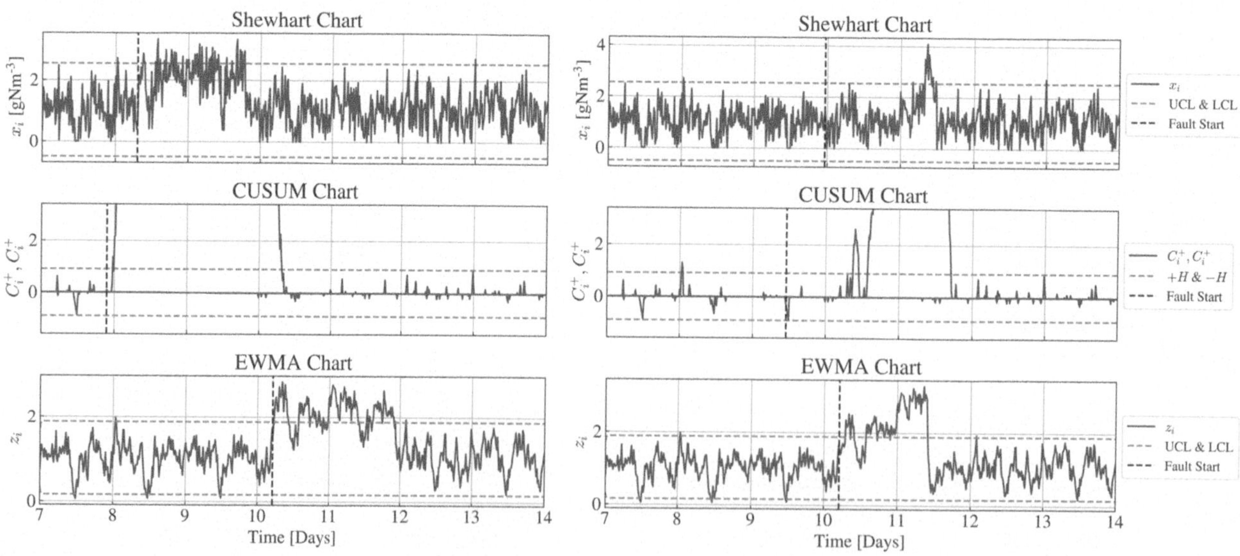

Figure 7 | Examples of control charts for online FD of left, bias faults and right, drift faults.

Table 1 | Average performance for online FD tests with bias faults

	Shewhart			CUSUM			EWMA		
Fault size [gN·m^{-3}]	1.5σ	2σ	3σ	1.5σ	2σ	3σ	1.5σ	2σ	3σ
MDR [%]	100	44	0	0	0	0	0	0	0
False alarms	0.0	0.0	0.0	1.8	2.0	2.0	1.7	1.7	1.7
TTD [hours]	NA	18	0.4	0.85	0.25	0.25	0.8	0.25	0.25

The Shewhart chart, despite its good performance in the offline tests, had extremely large MDRs for the two smaller fault sizes, failing to detect any faults with a bias size of 1.5σ gN·m^{-3}. However, the CUSUM and EWMA methods had no missed detections for the bias faults, as expected based on the results of the offline tests, and were able to successfully detect all the faults.

The method with the highest MDR corresponds to that with the lowest number of false alarms and indicates that the limit selected for the Shewhart chart was perhaps too conservative for the actual behaviour of the data. The false alarms of the CUSUM and EWMA charts are satisfactorily low. What is surprising, however, is that the EWMA had fewer false alarms than the CUSUM method, when operating online. This is contradictory to the expected behaviour based on the offline tests. Finally, to address the TTD, it should be noted that 0.25 h is the sampling rate in the BSM1 simulation platform and therefore the reported TTD cannot be lower than this. It is likely that the CUSUM and EWMA charts would successfully detect the larger bias faults in a shorter time were a higher sampling rate used. However, due to the relatively slow dynamics in a WWTP and the appreciable sensor noise, a higher sampling rate would not provide much benefit to the control chart analysis as currently presented. The benefits of a higher sampling rate would be in the potential to apply filtering and smoothing algorithms to the sensor signal.

The results from the drift FD tests are shown in Table 2, and look drastically different to those of the bias faults. The main difference is seen in the high MDRs across all chart types, along with the long TTDs. In the offline FD tests the highest MDR was 6%, whereas in Table 2 the lowest value is 36%. This can be explained by the presence of active online control that was present in the simulations. The nitrate sensor that was monitored, and was the faulty sensor, is the controlled variable in one of the two default BSM1 control loops. This results in the sensor output being pushed back to the set-point value and, in the case of the slow drift faults, this happens before a control limit is crossed and an alarm triggered.

Based on what is seen in Table 2, the EWMA method appears best suited to the drift faults as it can consistently detect more than 50% of the fault cases, a claim none of the other methods can make. It is not unexpected that the number of false alarms are similar in magnitude to those seen in the bias fault tests – these are dependent on the parameters of the FD methods and the variations in the data, not the faults that are occurring. The long TTDs are likely due to the faults reaching a size where the controller can no longer compensate for them, and a limit is then violated. This is suboptimal as at this stage the fault has already had a large effect, likely negative, on the process. Quicker detection of these drift faults is essential. An interesting observation, however, is the apparent lack of dependence on fault size that the EWMA has in its TTD. This is a desirable characteristic of a FD method.

From the above discussion it can be concluded that for the selected parameters used in the control charts, the EWMA FD method outperforms the other two methods for the tested faults, bias and drift, and sizes.

Online FD: effect of control

As mentioned, the presence of online control had a large effect on the ability to detect faults in the sensor measuring the controlled variable. Figure 8 illustrates the controller concealing such a sensor fault; the top plot shows the sensor output, and

Table 2 | Average performance for online FD tests with drift faults

	Shewhart			CUSUM			EWMA		
Fault size [gN·m^{-3}·day^{-1}]	0.5	1	1.5	0.5	1	1.5	0.5	1	1.5
MDR [%]	100	76	52	96	56	44	44	36	36
False alarms	0.0	0.0	0.0	1.1	1.4	1.6	1.7	1.8	1.7
TTD [hours]	NA	40	27	42	27	19	17	17	17

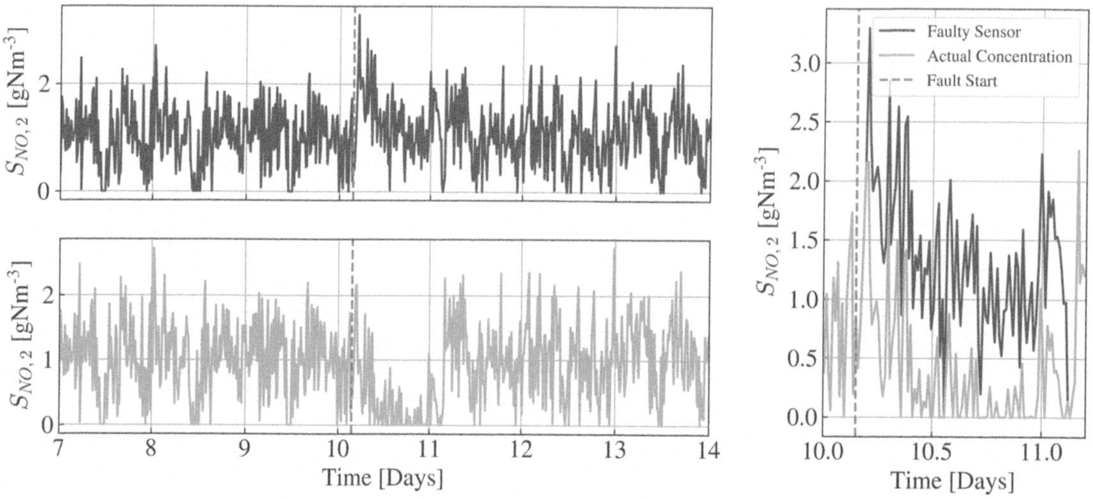

Figure 8 | Example of the controller concealing the sensor fault. Top: faulty sensor output, bottom: actual concentration in reactor, right: both outputs for the duration of the fault.

while there is a small spike, the controller quickly addresses the deviation from the set-point. The bottom plot shows the actual concentration in the reactor, which has deviated from the desired value due to the sensor fault.

An obvious step to take, based on these results, is to consider how the results change if the manipulated variable is monitored, with the goal of detecting concealed faults in the controlled variable sensor. The manipulated variable, in this case, is the internal recycle flow rate (Q_a). The behaviour of this variable is more cyclic in nature than that of the $S_{NO,2}$ sensor, and so it is less suited for use with control charts. However, to illustrate the potential improvement in MDR that can be gained, simulations were performed for the drift faults monitoring Q_a with the EWMA method. The EWMA method was not re-optimised with the same meticulousness for the new variable, due to the reliance on active process control for this FD scheme to work the offline optimisation could not be done in the same way. However, for deviations in the manipulated variable itself it was observed that the same parameter set, L_e–λ of 3.05–0.4, provided a reasonable trade-off between the false alarms and TTD while minimising the MDR. This supported the use of this parameter set for these preliminary tests. Table 3 shows the results of the online FD tests, comparing to the prior results from monitoring the nitrate concentration.

It is clear from Table 3 that, despite the higher number of false alarms, monitoring the manipulated variable is significantly more effective at detecting drift faults in the controlled variable sensor. All the faults were successfully detected, and the TTD was significantly lower than when monitoring the controlled variable.

Effect of faults on process performance

To motivate the need for good FD, the effect of the $S_{NO,2}$ sensor faults on the pumping energy (built in BSM1 performance evaluation index) is considered. This index was selected as the internal recycle flow rate is significantly affected by sensor faults in the $S_{NO,2}$ sensor and is a major factor in the pumping energy index. The changes in the pumping energy from the baseline value are shown versus the time taken to detect a particular fault, for varying fault sizes, in Figure 9.

From the Figure it can be seen that the duration of the fault, rather than its size, determines the effect that it will have on the process. Of course, this is not always the case, but for the sensor faults considered it is the dominant factor. The smaller fault

Table 3 | Average performance for online FD tests with drift faults using the EWMA method, comparing monitoring the manipulated variable (Q_a) to the controlled variable ($S_{NO,2}$)

	Monitor Q_a			Monitor $S_{NO,2}$		
Fault size [gN·m^{-3}·day^{-1}]	0.5	1	1.5	0.5	1	1.5
MDR [%]	0	0	0	44	36	36
False alarms	34	26	26	1.7	1.8	1.7
TTD [hours]	0.80	0.35	0.35	17	17	17

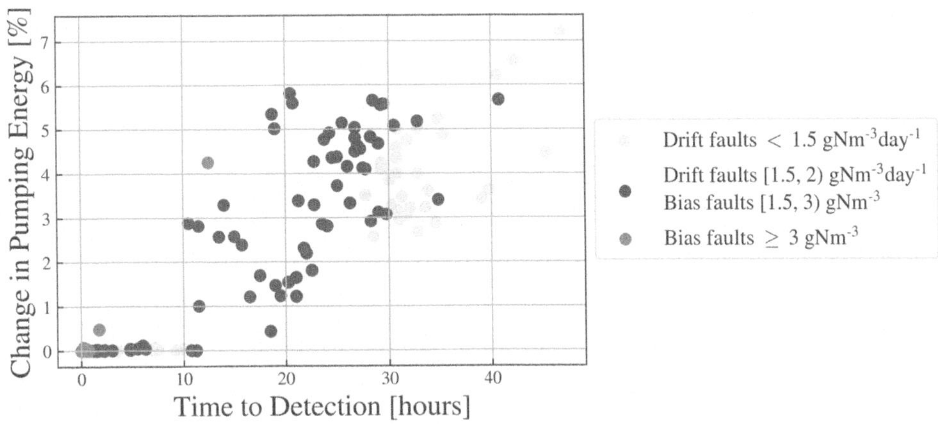

Figure 9 | The change in pumping energy plotted against the time taken to detect a fault.

sizes have the largest impact on the pumping energy requirements due to their extended presence in the process as a result of being more difficult to detect than larger faults. Another observation is that the first 10 h do not have a significant effect on the pumping energy; this suggests that accepting slower detection times in order to lower the number of false alarms is not expected to influence the process significantly. This is expected due to the relatively slow dynamics present in the WWTP.

CONCLUSIONS

It is clear that faults, even small faults, can have a significant effect on the performance and efficiency of the process, as seen in the up to 7% change in pumping energy. Rapid detection, and correction, of faults is crucial to optimising the operation of the process and ensuring that the quality limits placed on the effluent are not compromised. The three univariate control chart-based FD methods that were tested and compared are well established and have proved valuable in the field of FD applied to WWTPs. It was shown, however, that diligent calibration of the methods is vital due to the large effect that their respective parameters have on their performance. An approach for determining optimal parameters was developed and it was seen to be highly dependent on the criteria considered.

When the optimised chart methods were applied to online FD using the BSM1 the EWMA method outperformed the other methods for both bias and drift faults. The Shewhart method performed poorly in comparison to the CUSUM and EWMA methods for both fault types. When considering the bias faults, the CUSUM chart performed well, similarly to the EWMA chart. However, the CUSUM chart failed on the drift faults with high MDRs and TTDs. The EWMA chart was able to consistently detect more than 50% of the fault cases. The discrepancies between the results of the offline and online tests are explained by the active process control which was able to hide many of the drift sensor faults through the controller compensating for the apparent change in the controlled variable. It was shown that monitoring a manipulated variable can be a solution to identifying these faults.

The EWMA method should be further investigated, considering modifications that have been suggested to the base method, to further reduce the MDR and TTD in the case of drift faults. This method may be improved further through the use of filtering methods on the sensor signals, such as moving average widows, due to the apparent lack of urgency in detecting these faults shown in Figure 9.

ACKNOWLEDGEMENT

This study has been done within the international project Control4Reuse with partners from Sweden, France and Brazil. The project is part of the IC4WATER programme, in the frame of the collaborative international consortium of the 2017 call of the Water Challenges for a Changing World Joint Programme Initiative (Water JPI). The authors would like to thank Formas (Project No 2018-02213) for funding the Swedish part of this project, within the above mentioned initiative.

The authors also acknowledge and express gratitude to Dr Ulf Jeppsson and the other IWA Task Group members for making the BSM1 code available.

Dr Eva Nordlander is also thanked for her contributions in the form of supervision and valuable discussions related to the work.

CONFLICT OF INTEREST STATEMENT

The authors declare that they have no known competing interests or personal relationships that could have influenced the work reported in this paper.

DATA AVAILABILITY STATEMENT

Data cannot be made publicly available; readers should contact the corresponding author for details.

REFERENCES

Baklouti, I., Mansouri, M., Nounou, H., Slima, M. B. & Hamida, A. B. 2017 Fault detection in a wastewater treatment plant. In: *Proc of the 2017 International Conference on Advanced Technologies for Signal and Image Processing (ATSIP)*. pp. 1–5. https://doi.org/10.1109/ATSIP.2017.8075523.

Baklouti, I., Mansouri, M., Hamida, A. B., Nounou, H. & Nounou, M. 2018a Monitoring of wastewater treatment plants using improved univariate statistical technique. *Process Safety and Environmental Protection* **116**, 287–300. https://doi.org/10.1016/j.psep.2018.02.006.

Baklouti, I., Mansouri, M., Hamida, A. B., Nounou, H. & Nounou, M. 2018b Max-double adaptive EWMA for fault detection of wastewater treatment plants. In: *Proc of the 2018 IEEE/ACS 15th International Conference on Computer Systems and Applications (AICCSA)*. pp. 1–6. https://doi.org/10.1109/AICCSA.2018.8612891.

Baklouti, I., Mansouri, M., Hamida, A. B., Nounou, H. N. & Nounou, M. 2021 Enhanced operation of wastewater treatment plant using state estimation-based fault detection strategies. *International Journal of Control* **94** (2), 300–311. https://doi.org/10.1080/00207179.2019.1590735.

Chiang, L. H., Russell, E. L. & Braatz, R. D. 2000 *Fault Detection and Diagnosis in Industrial Systems*. Springer Science & Business Media, London, UK.

Corominas, L., Villez, K., Aguado, D., Rieger, L., Rosén, C. & Vanrolleghem, P. A. 2011 Performance evaluation of fault detection methods for wastewater treatment processes. *Biotechnology and Bioengineering* **108** (2), 333–344. https://doi.org/10.1002/bit.22953.

Gernaey, K. V., Jeppsson, U., Vanrolleghem, P. A. & Copp, J. B. 2014 *Benchmarking of Control Strategies for Wastewater Treatment Plants*. IWA Publishing, London, UK. https://doi.org/10.2166/9781780401171.

Isermann, R. 1997 Supervision, fault-detection and fault-diagnosis methods – an introduction. *Control Engineering Practice* **5** (5), 639–652. https://doi.org/10.1016/S0967-0661(97)00046-4.

IWA Modelling & Integrated Assessment Specialist Group 2020 *Benchmarking – Modelling & Integrated Assessment*. Available from: http://iwa-mia.org/benchmarking/ (accessed 23 March 2020).

Kazemi, P., Bengoa, C., Steyer, J.-P. & Giralt, J. 2021 Data-driven techniques for fault detection in anaerobic digestion process. *Process Safety and Environmental Protection* **146**, 905–915. https://doi.org/10.1016/j.psep.2020.12.016.

Marais, H. L., Nordlander, E., Thorin, E., Wallin, C., Dahlquist, E. & Odlare, M. 2020 Outlining process monitoring and fault detection in a wastewater treatment and reuse system. In: *Proc of the 2020 European Control Conference (ECC)*, pp. 558–563.

Montgomery, D. C. 2009 *Introduction to Statistical Quality Control*, 6th edn. John Wiley & Sons, Inc., Hoboken, NJ, USA.

Olsson, G. & Newell, B. 1999 *Wastewater Treatment Systems*. IWA Publishing, London, UK.

Olsson, G., Nielsen, M., Yuan, Z., Lynggaard-Jensen, A. & Steyer, J.-P. 2005 *Instrumentation, Control and Automation in Wastewater Systems*. IWA Publishing, London, UK.

Riss, G., Romano, M., Memon, F. A. & Kapelan, Z. 2021 Detection of water quality failure events at treatment works using a hybrid two-stage method with CUSUM and random forest algorithms. *Water Supply* **21**, 3011–3026. https://doi.org/10.2166/ws.2021.062.

Rosen, C., Röttorp, J. & Jeppsson, U. 2003 Multivariate on-line monitoring: challenges and solutions for modern wastewater treatment operation. *Water Science and Technology* **47** (2), 171–179. https://doi.org/10.2166/wst.2003.0113.

Sánchez-Fernández, A., Baldán, F. J., Sainz-Palmero, G. I., Benítez, J. M. & Fuente, M. J. 2018 Fault detection based on time series modeling and multivariate statistical process control. *Chemometrics and Intelligent Laboratory Systems* **182**, 57–69. https://doi.org/10.1016/j.chemolab.2018.08.003.

Schraa, O., Tole, B. & Copp, J. B. 2006 Fault detection for control of wastewater treatment plants. *Water Science and Technology* **53** (4–5), 375–382. https://doi.org/10.2166/wst.2006.143.

Thomann, M. 2008 Quality evaluation methods for wastewater treatment plant data. *Water Science and Technology* **57** (10), 1601–1609. https://doi.org/10.2166/wst.2008.151.

Thomann, M., Rieger, L., Frommhold, S., Siegrist, H. & Gujer, W. 2002 An efficient monitoring concept with control charts for on-line sensors. *Water Science and Technology* **46** (4–5), 107–116. https://doi.org/10.2166/wst.2002.0563.

Venkatasubramanian, V., Rengaswamy, R., Yin, K. & Kavuri, S. N. 2003 A review of process fault detection and diagnosis: part I: quantitative model-based methods. *Computers & Chemical Engineering* **27** (3), 293–311. https://doi.org/10.1016/S0098-1354(02)00160-6.

First received 14 December 2021; accepted in revised form 18 January 2022. Available online 2 February 2022

doi: 10.2166/wst.2021.608

Recovery profile of anaerobic ammonium oxidation (anammox) bacteria inhibited by ZnO nanoparticles

Safiye Can [a], Tugba Sari [b] and Deniz Akgul [a,*]

[a] Department of Environmental Engineering, Marmara University, Goztepe 34722, Istanbul, Turkey
[b] Department of Bioengineering, Marmara University, Goztepe 34722, Istanbul, Turkey
*Corresponding author. E-mail: deniz.tufan@marmara.edu.tr

SC, 0000-0002-7854-2294; TS, 0000-0001-6347-1238; DA, 0000-0002-8095-5645

ABSTRACT

The potential effects of nanoparticles (NPs) on biological treatment processes have become significant due to their increasing industrial applications. The purpose of this research was to investigate the self-recovery ability of anammox bacteria following acute ZnO NPs toxicity. In this context, a 2-liter lab-scale anammox reactor was operated for 550 days to enrich the biomass required to the batch exposure tests. Anammox culture was firstly exposed to four different doses of ZnO NPs (50, 75, 100 and 200 mg/L) for 24 h. Then, the ZnO NPs were removed and self-recovery performance of the anammox bacteria was assessed by evaluating the nitrogen removal capacities for 72 h. Besides the nitrogen removal performance, extracellular polymeric substances (EPS) production was also detected to deeply understand the response of the enriched anammox culture against ZnO NPs exposure. The results revealed that sudden and high load of ZnO NPs (100 and 200 mg/L) resulted in persistent impairment to the nitrogen removal performance of the enriched anammox culture. However, relatively lower doses (50 and 75 mg/L) caused deceleration of the nitrogen removal performance during the recovery period. In addition, EPS content in the reactor decreased along with escalating load of ZnO NPs.

Key words: acute inhibition, anammox bacteria, extracellular polymeric substances, nanoparticle toxicity, self-recovery, zinc oxide nanoparticles

HIGHLIGHTS

- Enriched anammox cultures were exposed to different ZnO NPs doses.
- Total nitrogen consumption per VSS decreased by the escalating burden of ZnO NPs.
- This study takes attention to the self-recovery ability of anammox bacteria against shock ZnO NPs loads.
- Anammox bacteria could not recover their activity following the acute inhibition of high ZnO NPs load.

GRAPHICAL ABSTRACT

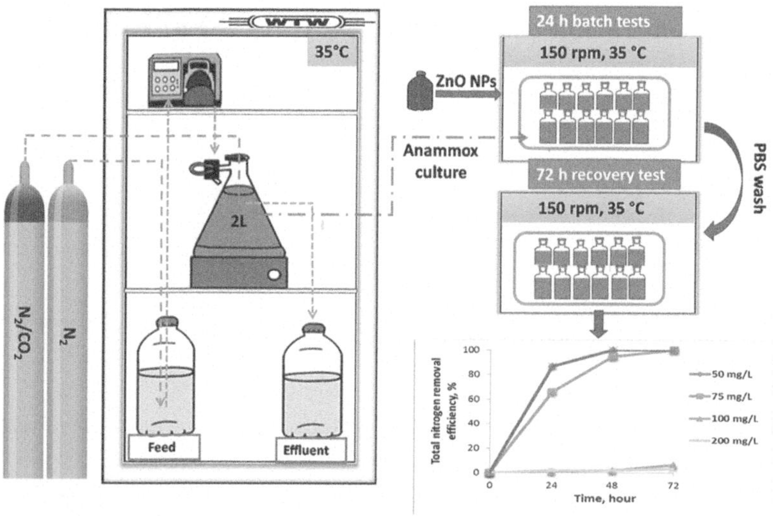

INTRODUCTION

Living things need nitrogen element in order to synthesize many crucial biomolecules such as DNA, chlorophyll, ATP and to maintain their life. Dinitrogen gas (N_2) is transformed into ammonium compound which is the available form of nitrogen for the utilization of (micro) organisms. The fixed ammonia in the soil is then released to the atmosphere by consecutive nitrification and denitrification processes. Basically, ammonia is converted firstly to nitrite and subsequently to nitrate by the nitrification process. In denitrification, nitrate is turned into nitrite and then N_2 gas. Therefore, the global nitrogen cycle represents the natural nitrogen flow in the earth and provides the continuity of life (Stein & Klotz 2016).

Nitrogen-containing compounds such as ammonia and nitrate, are two of the main pollutants in wastewaters. Along with the increase in anthropogenic activities, the removal of these contaminants from wastewaters has become a critical issue. If ammonia is discharged to the ecosystem without treatment, it causes serious environmental problems. The excess amount of ammonia in the water bodies is used by aquatic plants and algae as nutrient and accelerates their growth. The excessive growth of these plants and algae is named eutrophication, which affects the water quality and results in oxygen depletion in water and eventually kills the aquatic organisms (Heisler *et al.* 2008; Luo *et al.* 2015; Wurtsbaugh *et al.* 2019). Therefore, many countries apply stringent nitrogen discharge standards. In European Union, discharge limit of total nitrogen in treated wastewaters is 15 mg/L (10,000–100,000 population equivalent (pe)) and 10 mg/L (>100,000 pe) for areas sensitive to eutrophication (EC 2019). Besides, free nitrous acid can be formed when nitrate accumulates in the environment. Even if ammonia and free nitrous acid exist at low concentrations, they inhibit many microorganisms. Nitrification products may also adversely affect human health (Weralupitiya *et al.* 2021). Drinking water contaminated with nitrate causes blue baby syndrome, which can be fatal in newborn babies (Knobeloch *et al.* 2000). Hence, various chemical, physical, and biological treatment processes are applied to remove nitrogen compounds from wastewaters. Inspired by the global nitrogen cycle, conventional nitrification/denitrification processes are mostly applied to treat nitrogen compounds in wastewater (Weralupitiya *et al.* 2021). However, the nitrification process carried out under an oxic environment, leads to high energy requirements due to the aeration cost of the process. Additionally, the need for external organic carbon addition in the denitrification process, excess sludge production and its disposal policy result in high costs (Cao *et al.* 2017; McCarty 2018).

Anaerobic ammonium oxidation (Anammox) enables the direct conversion of ammonium into dinitrogen gas under anoxic conditions. The central metabolism of the anammox reaction has been proposed as three main steps (Equations (1)–(3)) (Strous *et al.* 2006). Therefore, it can be thought of as a shortcut to the classical nitrification/denitrification processes

(Weralupitiya *et al.* 2021).

$$NO_2^- + 2H^+ + e^- \rightarrow NO + H_2O \ (E_0' = +0.38V) \quad (1)$$

$$NO + NH_4^+ + 2H^+ + 3e^- \rightarrow N_2H_4 + H_2O \ (E_0' = +0.66V) \quad (2)$$

$$N_2H_4 \rightarrow N_2 + 4H^+ + 4e^- \ (E_0' = -0.75V) \quad (3)$$

Anammox bacteria belong to the phylum Planctomycetes, order Brocadiales. Up to now, seven genera of Anammox bacteria have been identified: Kuenenia, Brocadia, Anammoxoglobus, Jettenia, Brasilis, Scalindua, and Anammoximicrobium. Even though they are diversely distributed in the environment such as marine and freshwater ecosystems, the wastewater treatment systems are the main habitat of these bacteria (Pereira *et al.* 2017). Along with the discovery of Anammox at the end of the '90s, this process has been rapidly implemented in different industries. At present, the Anammox process has been applied more than 200 wastewater treatment facilities. The most prominent advantages of this process compared to conventional biological processes are that anammox bacteria use inorganic carbon as a carbon source, reduce the energy requirement by 60%, and produce 80% less waste sludge due to their slow growth rate (Cao *et al.* 2017). Furthermore, it is a more sustainable process, decreasing greenhouse gas emissions by 90% (Manonmani & Joseph 2018). Based on the overall consideration of these factors, the Anammox process has become an alternative treatment technology to conventional nitrification/denitrification processes. Today, the Anammox process has been successfully integrated for ammonia removal from high-strength wastewaters; domestic sewage, sludge liquor, sludge supernatant, sludge digestate, slaughterhouse wastewater, piggery wastewater, synthetic coke-oven wastewater, monosodium glutamate wastewater, semi-conductor factory wastewater, pharmaceutical wastewater, livestock wastewater, tannery wastewater, seafood processing wastewater, digested fish canning effluent, and municipal wastewaters (Adams *et al.* 2020; Aransiola *et al.* 2021). Despite all advantages, certain drawbacks of the process restrict its applications in the industry. Slow-growing anammox bacteria are very vulnerable to not only environmental factors but also several inhibitory compounds including antibiotics, aromatic and organic pollutants, heavy metals, nanoparticles, and so on (Jin *et al.* 2012; Madeira & De Araújo 2021).

Nanotechnology is an emerging technology that encompasses various disciplines including chemistry, physics, material science, engineering, biology, etc. It deals with the matter at a nanoscopic scale (Porter & Youtie 2009). Nanotechnology has gained great attention because the physicochemical properties of nanoparticles (NPs) can easily be altered for a specific purpose (Bhushan 2017). In 2022, the global nanotechnology market value is expected to rise up to $16.8 billion (Liu *et al.* 2019). Zinc oxide (ZnO) NPs are one of the most commercially used NPs (Liu *et al.* 2019). Paints, ointments, rubber, cement, glass, plastics, adhesives, foods, pigments, fire retardants, ferrites, sealants, batteries, are examples of products containing ZnO NPs (Sabir *et al.* 2014). Because it improves UV protection capability, and acting as anti-acne and aging agent, ZnO NPs have been immensely used in the cosmetic sector (Werkneh *et al.* 2018). ZnO NPs are also gathering unique properties to the semiconductors such as wide band gap, high electron mobility, and good transparency (Sabir *et al.* 2014).

Increasing use of NPs in many consumer products and different industries, and unintentional release of them into the ecosystem at every stage of their life cycles, cause a great concern about their potential toxicological risks on biological systems (Sari *et al.* 2020b). Up to now, the potential toxicity of NPs on biological wastewater treatment processes has been reported by several studies in the current literature (Chen *et al.* 2021). Among the NPs examined by Song *et al.* (2018), ZnO NPs were found to be the most toxic one due to their inhibition mechanisms on bacterial cells (Gutwiński *et al.* 2021). ZnO NPs are among the top 3 most-produced NPs worldwide (Chen & Chen 2017). In the current literature, there are a few studies related to the impacts of ZnO NPs on the Anammox process (Sari *et al.* 2020a). These studies mostly focused on short- and long-term effects of ZnO NPs on the Anammox process, but not on its recovery profile following acute exposure.

Due to the extremely long doubling time of anammox bacteria (7–20 days) (Kartal *et al.* 2012), challenging start-up period of the process is one of the main obstacles. Therefore, in order to shorten the start-up time of the Anammox process, many researchers have focused on optimizing key parameters such as seeding sludge, operational strategy, biomass immobilization, and so on (Verma *et al.* 2021), and the addition of chemical reagents into the system for stimulating the bacteria (Ganesan & Vadivelu 2019; Miodoński *et al.* 2019). Therefore, in full-scale applications of the Anammox process, in case of a sudden toxicity, recovery would be a better way than restarting it (Ni & Meng 2011). In addition, prior to exploring state-of-art recovery or control strategy in real time, the response of the system itself without any attempt should be well-understood (Arora *et al.* 2021). In this context, this study aimed to figure out the self-recovery profile of the Anammox process following 24 h acute exposure of ZnO NPs at four different doses in order to assess whether the inhibition of anammox bacteria is persistent or not. In this respect, the current study is the first in the literature to the best of our knowledge. Therefore, this study's primary purpose is investigating acute impacts of the ZnO NPs on the nitrogen treatment capacity of the Anammox process and evaluating the self-recovery ability of the Anammox process immediately after a short-term inhibition period.

MATERIALS AND METHODS

Preparation of synthetic wastewater and ZnO NPs stock solution

Synthetic wastewater was prepared as previously described by Yapsakli *et al.* (2017). Ammonium nitrogen (NH_4^+-N) and nitrite nitrogen (NO_2^--N) was prepared in the forms of $(NH_4)_2SO_4$ and $NaNO_2$, respectively. The composition of synthetic wastewater is given in Table 1 in detail. The anammox bioreactor was fed daily with synthetic wastewater. Prior to feeding, the synthetic wastewater was purged via dinitrogen gas to remove dissolved oxygen.

Commercially produced ZnO NPs (<100 nm) was prepared as a suspension of 20% ZnO by wt (Sigma Aldrich, USA) by following the procedure described by Mu *et al.* (2012). 0.1 mM sodium dodecylbenzene sulfonate (SDBS) was used to dilute the ZnO suspension in order to procure 1 g/L ZnO stock solution. SDBS, a dispersing agent, assists in the stabilization of NPs in the stock solution (Mu *et al.* 2012; Zhang *et al.* 2017; Song *et al.* 2018). Then, 1 h ultrasonication was employed to the stock solution (25 °C, 40 kHz, 250 W) by Sonopuls Ultrasonic Homogenizer (Bandelin, Germany). Particle size distribution and zeta potential were controlled after 24 h using a Zetasizer (Malvern Panalytical, UK).

Enrichment of anammox bacteria and reactor operation

The experimental setup of the anammox bioreactor was established as previously described by Sari *et al.* (2020a). A lab-scale anammox bioreactor with 2 L working volume was established in order to enrich anammox bacteria for batch tests. The anammox seeding sludge was taken from the up-flow anammox bioreactor that has been operated for more than 10 years in our laboratory. The bioreactor was placed in a TS 606-G/4-i incubator (WTW, Germany) to maintain mesophilic conditions (35 ± 0.5 °C) and operated in sequencing batch reactor (SBR) mode for a 24 h cycle with a 20-min fill time, 22.67-h reaction time, 30-min settling time, and 20-min effluent withdrawal time. Hydraulic retention time (HRT) was set to be 2 days. Therefore, volumetric exchange ratio per cycle was 0.5 during the enrichment period. N_2/CO_2 (90/10%) gas mixture was also introduced to the bioreactor in order to provide anoxic conditions and supply the inorganic carbon requirement. pH of the reactor was kept at 7.5 ± 0.3. During the enrichment, influent substrate concentrations were kept low at first to help

Table 1 | Composition of synthetic wastewater

Constituent	Quantities in synthetic wastewater
NH_4^+-N and NO_2^--N	1:1.15 ratio
$NaHCO_3$	1.04 g/L
K_2HPO_4	174.2 mg/L
$CaCl_2$	73.5 mg/L
$MgCl_2$	102 mg/L
Trace Element Solution 1	1 ml/L
$Na_2EDTA{\cdot}2H_2O$: 10 g/L	
$FeSO_4$: 5 g/L	
Trace Element Solution 2	1 ml/L
$Na_2EDTA{\cdot}2H_2O$: 10 g/L	
$ZnSO_4.7H_2O$: 0.43 g/L	
$CoCl_2.6H_2O$: 0.24 g/L	
$MnCl_2.4H_2O$: 0.99 g/L	
$CuSO_4.5H_2O$: 0.25 g/L	
$NiCl_2.6H_2O$: 0.19 g/L	
H_3BO_4: 0.014 g/L	

adaptation of anammox bacteria to their new environment. Thereafter, the substrate concentrations were gradually increased considering the stable nitrogen treatment capacity. Finally, NH_4^+-N and NO_2^--N concentrations were adjusted to be 100 and 115 mg N/L, respectively, which were the same concentrations used in the short-term exposure tests.

Batch exposure tests

Acute exposure tests were conducted using amber serum flasks having an effective volume of 50 mL. Synthetic wastewater containing 100 mg N/L NH_4^+-N and 115 mg N/L NO_2^--N, ZnO NPs and biomass were transposed to serum flasks in this order. The volatile suspended solids (VSS) in each serum flask was 1.5 ± 0.5 g/L. The pH was set to 7.5 ± 0.2. In the batch exposure tests, anammox bacteria were subjected to four different ZnO NPs dosages of 50, 75, 100 and 200 mg/L. N_2 gas was used to purge the serum flasks for 3 min to prevent oxygen suppression before starting the experiment. Afterwards, each flask was sealed with a rubber stopper and aluminum crimp. All flasks were kept in an Ecotron incubation shaker (INFORS HT, Sweden) at 35 °C and 150 rpm for 24 h incubation period. Each condition was conducted in triplicate. A 0.5 mL sample was collected from each serum flask at 3 h intervals in order to analyze total nitrogen concentrations and determine total nitrogen removal efficiencies.

Recovery tests

The recovery ability of the anammox bacteria exposed to the ZnO NPs was observed during recovery tests. ZnO NPs attached to the bacteria were removed by applying the washing procedure described by Zhang *et al.* (2017) with some modifications. Biomass in serum flasks was transferred into 50 mL centrifuge tubes and centrifuged at 35 °C and 3,000 g for 10 min using a Universal 320R centrifuge (Hettich, Germany). After removing the supernatant, 30 mL of 10 mM phosphate buffer solution (PBS) was added onto the pellets, and tubes were shaken at 35 °C and 150 rpm for 30 minutes using an Ecotron incubation shaker (INFORS HT, Sweden). The centrifugation step was repeated twice more. Finally, the washed anammox sludge was transferred into new serum flasks containing fresh synthetic wastewater. For the preparation of serum flasks in the recovery tests, the same procedure was followed with that of exposure tests, except with ZnO NPs addition. Besides, the operation time in the recovery test was 72 h, and the liquid samples were collected from the flasks at 24 h intervals.

Analytical methods

pH was daily tested by HQ40D digital portable multimeter kit (HACH, USA). VSS (SM 2540E), ammonia nitrogen (SM 4500-NH_3-A), and nitrite nitrogen (SM 4500-NO_2-B) were determined according to Standard Methods (APHA/AWWA/WEF 2005). The Anthrone method and modified Lowry method was applied to detect polysaccharide (PS) and protein (PN) concentrations in EPS content, respectively. During the EPS analysis, bovine serum albumin and glucose were used as standard solutions for the PN and PS analysis, respectively. A modified heat extraction method was performed as previously described by Morgan *et al.* (1990) for EPS extraction. Subsequently, supernatant parts were filtered through 0.45 μM membrane filter and stored at −20 °C. A UV-2450 Spectrophotometer (Shimadzu, Japan) was used for all EPS quantification experiments. Each measurement was performed with two independent samples and each sample was tested in duplicate.

Statistical analysis

The results of the experiments were demonstrated as the mean ± standard error using Microsoft Excel.

RESULTS AND DISCUSSION

Enrichment of anammox bacteria

In the current study, the established anammox bioreactor was operated for 550 days, and start-up period of the bioreactor was divided into three phases. Figure 1 represents the nitrogen components and the treatment capacity of the bioreactor during the start-up and enrichment period. Following the inoculation (Phase I: Day 1–53), mild conditions were provided by adding relatively lower substrate concentrations in order to adapt anammox bacteria to their new environment. Numerically, NH_4^+-N and NO_2^--N concentrations within the bioreactor were kept at 39.32 ± 12.36 mg N/L and 41.92 ± 12.54 mg N/L, respectively, during the start-up period. In the initial days of Phase I, the anammox bioreactor showed poor nitrogen treatment capacity with inconsistent low nitrogen removal efficiencies. Along with observing satisfactory nitrogen removal performances (>80%), phase II (Day 54–255) was initiated. In Phase II, initial NH_4^+-N and NO_2^--N concentrations were increased stepwise to enhance the microbial activity and enrich the anammox culture. On days 166–234, total nitrogen concentration within the bioreactor reached its maximum level (358.51 ± 35.36 mg N/L). Furthermore, the final phase (Phase III)

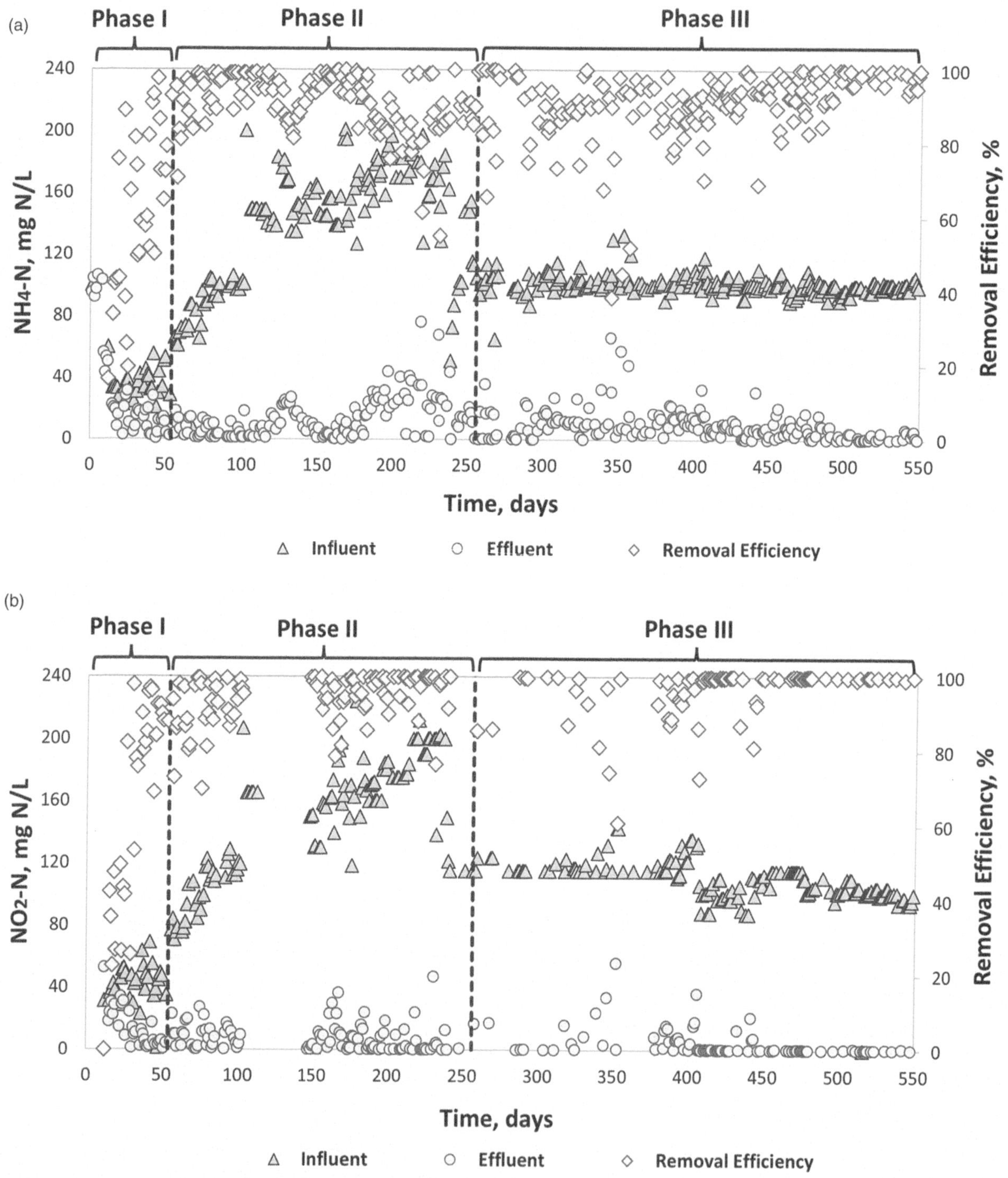

Figure 1 | Nitrogen treatment capacity of the anammox bioreactor. (a) ammonium; (b) nitrite removal during the start-up period. Phase I: Days 1–53; Phase II: Days 54–255; Phase III: Days 256–550.

comprises the days from 256 to 550. In this phase, the main purpose was to create the conditions to be applied in the short-term exposure tests and acclimatize the enriched anammox culture to the conditions. Therefore, influent NH_4^+-N and NO_2^--N concentrations were adjusted to be 100 and 115 mg N/L, respectively, in the final phase. From day 458, satisfying nitrogen treatment performances were observed. During this period, the removal efficiencies of NH_4^+-N and NO_2^--N was obtained as 95.87 ± 4.21% and 99.79 ± 0.22%, respectively.

Recovery profile of anammox bacteria following the acute exposure of ZnO NPs

The ZnO NPs toxicity on anammox activity in a short-term manner has already been reported by a few studies in the current literature. However, the observed inhibition effects of ZnO NPs show differences in these studies due to the different experimental setups, operational strategies, microbial dynamics, and the sludge type (Sari *et al.* 2020a). Besides, to the best of our knowledge, no study has assessed whether the acute toxicity of ZnO NPs on anammox bacteria is persistent or not. Therefore, this study has focused on enlightening the recovery profile of the enriched anammox culture following the acute ZnO NPs toxicity. In order to achieve this, anammox bacteria were exposed to 50, 75, 100, and 200 mg/L ZnO NPs for 24 h. During the 24 h incubation period, the changes in substrate concentrations were observed, and the total nitrogen (the sum of ammonium and nitrite) removal efficiencies were calculated.

During the incubation period, total nitrogen concentration diminished over time in all experiment groups (Figure 2(a)). In control groups, which were not exposed to any NPs dosage, 90.91 $\pm$ 7.44% of TN had been removed from the system within

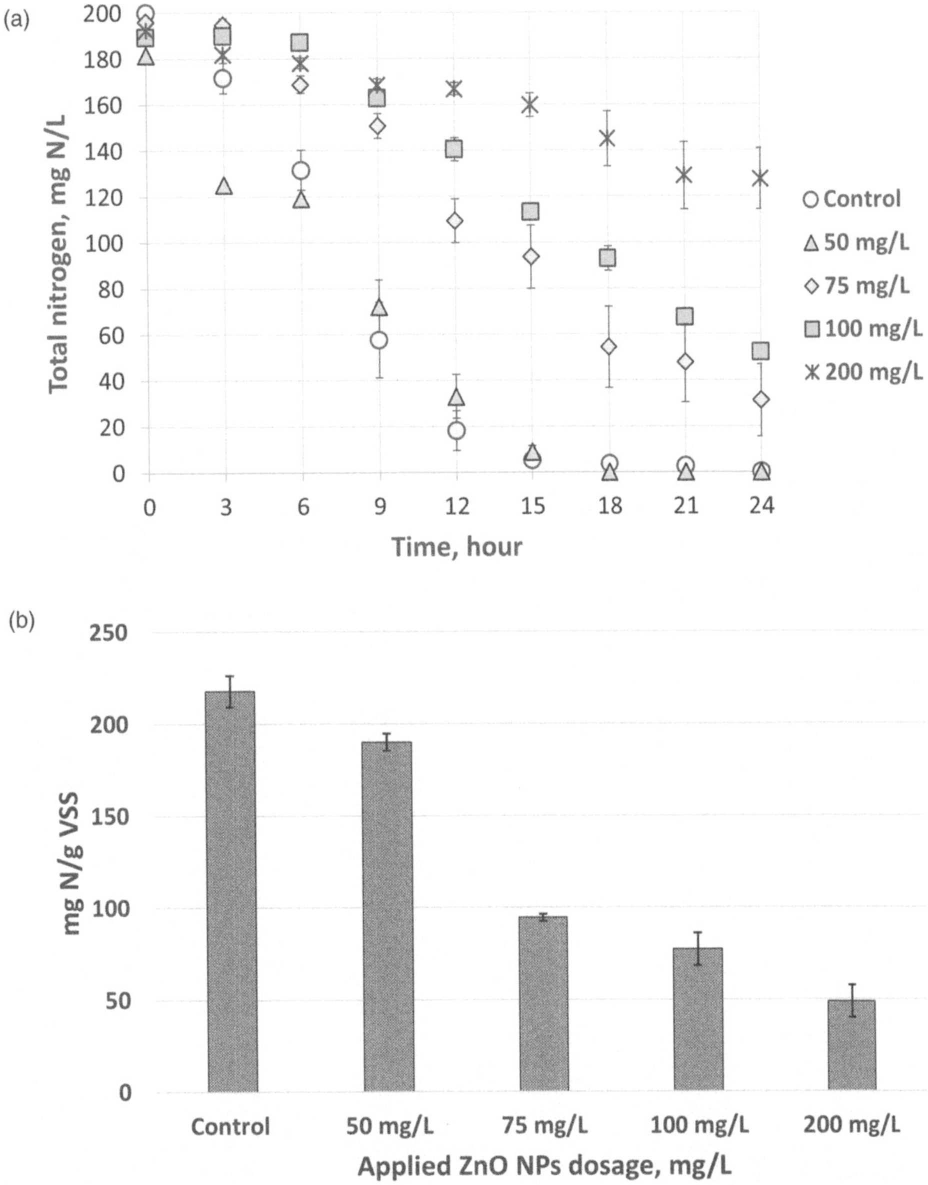

Figure 2 | Impacts of 50, 75, 100, 200 mg/L ZnO NPs on anammox bacteria during 24 h incubation period. (a) total nitrogen concentrations; (b) the consumed total nitrogen per VSS (mg N/g VSS) after 24 h. Data indicate average, and error bars represent standard error of the results from three independent samplings.

12 h. The experimental groups exposed to 50 mg/L ZnO NPs dose showed similar nitrogen removal trends with the control groups. Numerically, the TNRE was detected to be 81.83 ± 9.10% in 12 h, and all nitrogen was treated within 18 h. As the applied ZnO NPs dosages increased, the nitrogen treatment capacity of the enriched anammox culture considerably decreased. During 75 mg/L ZnO NPs exposure, 44.04 ± 9.50% of TN was removed from the system in 12 h, and the TNRE was found to be 83.70 ± 14.65% at the end of the experiment. Moreover, the TNRE was detected to be 72.53 ± 1.86% as a result of the load of 100 mg/L ZnO NPs in 24 h. As for 200 mg/L ZnO NPs exposure, only 33.96 ± 10.49% of the total nitrogen was removed by the enriched anammox culture within 24 h. In order to be able to comment more accurately, the consumed amount of TN per VSS was calculated for each group (Figure 2(b)). Even though similar nitrogen removal performances were observed in both control and the 50 mg/L ZnO experiment groups, the TN consumed per VSS was lower in the 50 mg/L exposure group (190.00 ± 8.26 mg N/g VSS) than that of the control group (217.79 ± 12.03 mg N/g VSS). Additionally, a sharp decrease in TN consumption per VSS was observed when the exposure dose of ZnO NPs increased. In the acute exposures of 75, 100, and 200 mg/L ZnO NPs doses, the consumed TN per VSS was found to be 94.37 ± 3.31, 77.19 ± 15.34, 48.70 ± 15.24 mg N/g VSS, respectively.

Prior to the self-recovery test, the anammox bacteria were firstly washed to get rid of ZnO NPs from the serum flasks as described in the materials and method part. Following the washing procedure, all serum flasks were fed with 100 mg N/L NH_4^+-N and 115 mg N/L NO_2^--N concentrations. Thereafter, the nitrogen treatment capacities of the anammox systems were examined. During the 72 h recovery period, TN concentrations in the serum flasks were measured at 24 h time intervals and total nitrogen removal efficiencies were determined (Figure 3). The inhibition tests demonstrated that the application of 50 mg/L ZnO NPs did not have any impairment on the nitrogen removal performance of the anammox bacteria. Even all ammonium and nitrite nitrogen had been removed within 18 h. However, after removing ZnO NPs from the anammox system, the treatment of total nitrogen took more than 24 h during the recovery period. In addition, it has already been demonstrated that, 24 h exposure time of 75 mg/L ZnO NPs caused deterioration of nitrogen removal performance of anammox bacteria by the loss of 16.30 ± 14.65% nitrogen removal capacity. As for the recovery period following the 75 mg/L ZnO NPs application, it is much longer than that of the 50 mg/L ZnO NPs application. During the recovery experiment, 65.93% of TN has been treated in the first 24 h, and the TN removal efficiency reached 94.51% in 48 h. This indicates that increasing dose of NPs worsens the nitrogen removal performance. As the NPs dose rose to 100 mg/L, 27.47 ± 1.86% of TN was not removed from the system in the exposure period. However, it was observed that only 6.32% of TN was removed by enriched anammox culture during the recovery period. When experiment groups of 75 and 100 mg/L are compared, although there was a slightly similar nitrogen removal performance between them during the exposure period, anammox bacteria couldn't recover their activities even after the 72 hour recovery period following the 100 mg/L ZnO NPs exposure. The 200 mg/L ZnO NPs load to the anammox system substantially impacted the nitrogen removal performance,

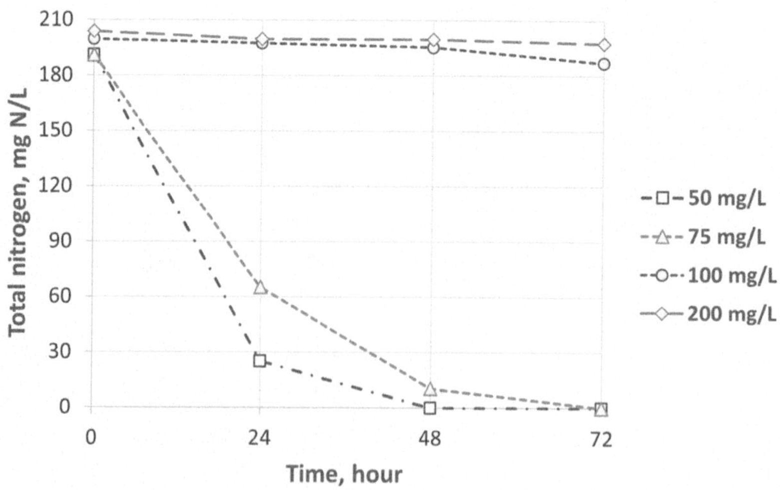

Figure 3 | Recovery profile of anammox bacteria following the acute exposure tests.

and the TNRE of the anammox bacteria was detected to be 33.96 ± 10.49%. After 200 mg/L NPs exposure, anammox bacteria showed almost no nitrogen removal performance (3.09%) during its recovery period.

The recovery profiles following the exposure tests strongly exhibited that sudden and high load of ZnO NPs (100 and 200 mg/L) resulted in persistent impairment on the nitrogen removal performance of the enriched anammox culture. However, relatively lower doses (50 and 75 mg/L) caused deceleration of nitrogen removal performance during the recovery period. The most likely reason for this may be the amount of Zn^{2+} released from the ZnO NPs (Liu *et al.* 2011; Mu & Chen 2011; Zheng *et al.* 2011). It has already been reported that free metal ions dissolved from the surface of NPs are one of the major inhibition mechanisms (Ee *et al.* 2018; Song *et al.* 2018). In Zn-overloaded environments, the binding capacity of EPS, which is a defense mechanism of microorganisms (Tang *et al.* 2018), might be exceeded and Zn ions might penetrate into the cells (Zhang *et al.* 2018a). In such case, the zinc ions may chelate to sulfhydryl groups and disrupt heme c and protein bonds. Eventually, free Zn ions may lead to disorder in the cell metabolism by disrupting the function of intracellular proteins (Zhang *et al.* 2019). Therefore, an increment in the released Zn cations corresponding to the applied ZnO NPs doses may gradually increase inhibitive effects on anammox systems. Additionally, not only dissolved ion toxicity, but also the undissolved NPs might be the reason for the loss of nitrogen removal capacity, because undissolved NPs may lead to pits forming on the cell membrane. Smaller NPs may also penetrate to the cells and increase the permeability of the cell membrane. Additionally, reactive oxygen species (ROS) may be produced at the ZnO NPs surface owing to the electrostatic properties. ROS generation causes oxidative stress and the viability of anammox bacteria might have decreased (Zhao *et al.* 2019). It has also been reported that, when the ZnO NPs dosage increases, more NPs accumulate on the biomass (Sari *et al.* 2020a).

In the current literature, there is only a study investigating the reversibility of ZnO NPs inhibition on the Anammox process following long-term NPs addition (524 days) (Sari *et al.* 2020a). Contrary to the findings in the current study, the previously published study reported that anammox bacteria recovered themselves in 120 days. The most possible reason for this difference is the slow metabolism of the anammox bacteria. In this study, anammox bacteria were not allowed to adapt ZnO NPs and the bacteria were suddenly exposed to a huge amount of ZnO NPs dosage. Afterward, the recovery period only lasted for 72 h. Hence, anammox bacteria might not be able to secrete EPS to protect themselves and could not protect and/or recover their cell integrity under these conditions (Song *et al.* 2018). Therefore, it is noteworthy to mention that the Anammox process without any additional recovery strategy is vulnerable in the face of a sudden and massive amount of ZnO NPs loads to the wastewater treatment facilities. Therefore, the discharge limits of wastewaters containing ZnO NPs should be carefully determined considering the worldwide ZnO NPs utilization.

EPS response of the enriched anammox culture after 24 h acute exposure of ZnO NPs

EPS, an organic matrix, is an important defense mechanism of microorganisms against many inhibitors including nanoparticles (Tang *et al.* 2018). Plenty of studies have reported an increment in the EPS production when environmental stress increases (He *et al.* 2017; You *et al.* 2017). Therefore, the EPS production may be an adaptation strategy developed by the microbial community (Zhang *et al.* 2018c). On the other hand, some studies have revealed opposite results (Mu *et al.* 2012; Zhao *et al.* 2019; Sari *et al.* 2020b). Therefore, there are many different outcomes related to EPS secretion in the current literature. In this study, the EPS content was examined to understand the response of the enriched anammox culture exposed to several ZnO NPs dosages.

The EPS composition varies depending on the microbial dynamics. Besides, proteins (PN) and polysaccharides (PS) have mainly been found in EPS content (Tsuneda *et al.* 2003). It has been pointed out that in complex microbial communities, such as enriched anammox culture, PN secretion is considerably higher than the PS secretion (Tang *et al.* 2018). In the current study, far more PN content was produced than PS content. Moreover, both PN and PS production decreased as the applied NPs dosage increased (Figure 4). This phenomenon may have two possible reasons. Firstly, anammox bacteria are slow-growing bacteria, and they may need a longer lag phase for the EPS secretion to protect themselves (Song *et al.* 2018). However, in this study, the bacteria had been shocked with a high load of ZnO NPs in only 24 h incubation period. Therefore, they may not have been able to produce enough EPS to adapt to changing environmental stress. Secondly, the migration and absorption of Zn^{2+} ions released from ZnO NPs might play a role in the potential inhibition of ZnO NPs on the anammox process (Zhao *et al.* 2019). In this case, cells may utilize their energy to provide cell integrity rather than EPS secretion (Zhang *et al.* 2018b). Thereby, it might lead to a decrement in EPS secretion.

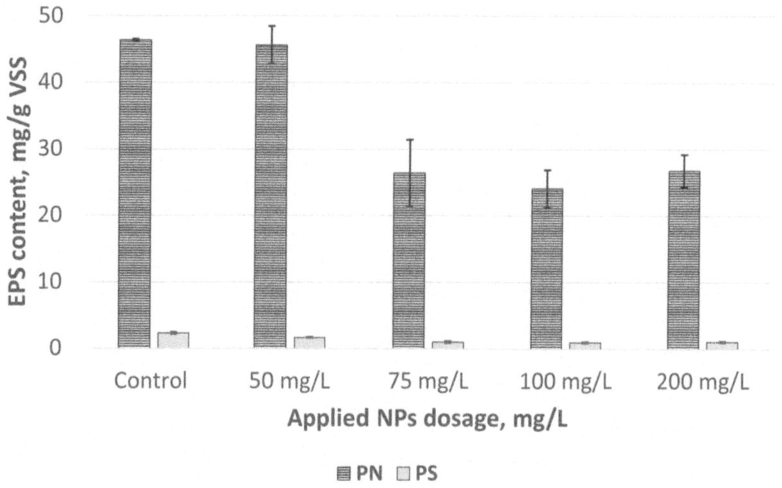

Figure 4 | EPS response of the enriched anammox culture after 24 h exposure of ZnO NPs. Data indicate average, and error bars represent standard error of the results from two independent samplings, each tested in duplicate.

CONCLUSION

This study was carried out to exhibit recovery performance of the anammox system in case of shock ZnO NPs loads to the wastewater treatment facilities. Acute exposure tests revealed that TNRE of the anammox bacteria and the consumed TN per VSS decreased as the applied dosage of ZnO NPs increased. The anammox bacteria could not recover their nitrogen removal performance even in a 72 h recovery period in consequence of the acute exposure of 100 and 200 mg/L ZnO NPs. The most likely reason for this may be the amount of Zn released from the ZnO NPs. In other words, as the dose of ZnO NPs increases, the dissolved toxic Zn ion concentration increases, which may lead to more deteriorative effects on the nitrogen removal capacity. Furthermore, the adsorption of undissolved NPs and even their penetration to the cells might be another reason. Besides, escalating the burden of applied ZnO NPs dosage resulted in a decline in the EPS content that is proportional to the deterioration of nitrogen removal performance of the anammox bacteria.

ACKNOWLEDGEMENTS

This study was funded by the Scientific and Technological Research Council of Turkey (TUBITAK – Project No: 117Y214) and Marmara University Scientific Research Committee (BAPKO- Project No: FEN-C-YLP-090217-0053).

DATA AVAILABILITY STATEMENT

All relevant data are included in the paper or its Supplementary Information.

REFERENCES

Adams, M., Xie, J., Kabore, A. w. J., Chang, Y., Xie, J., Guo, M. & Chen, C. 2020 Research advances in anammox granular sludge: a review. *Critical Reviews in Environmental Science and Technology*, 1–44.

APHA/AWWA/WEF 2005 *Standard Methods for the Examination of Water and Wastewater*. American Public Health Association (APHA), American Water Works Association (AWWA), Water Environment Federation (WEF), Washington, DC, USA.

Aransiola, S. A., Ijah, U. J. J., Abioye, O. P. & Victor-Ekwebelem, M. O. 2021 ANAMMOX application to different wastewaters. In: *Advances in the Domain of Environmental Biotechnology* (Maddela, N. R., García Cruzatty, L. C. & Chakraborty, S., eds). Springer, Singapore, pp. 402–409.

Arora, A. S., Nawaz, A., Qyyum, M. A., Ismail, S., Aslam, M., Tawfik, A., Yun, C. M. & Lee, M. 2021 Energy saving anammox technology-based nitrogen removal and bioenergy recovery from wastewater: inhibition mechanisms, state-of-the-art control strategies, and prospects. *Renewable and Sustainable Energy Reviews* **135**, 110126.

Bhushan, B. 2017 *Springer Handbook of Nanotechnology*. Springer, Berlin, Germany, pp. 39–58.

Cao, Y. S., van Loosdrecht, M. C. M. & Daigger, G. T. 2017 Mainstream partial nitritation-anammox in municipal wastewater treatment: status, bottlenecks, and further studies. *Applied Microbiology and Biotechnology* **101** (4), 1365–1383.

Chen, R. & Chen, C. 2017 Environment, health and safety issues in nanotechnology. In: *Springer Handbook of Nanotechnology*, 4th edn (Bhushan, B., ed.). Springer Nature, Berlin, pp. 1559–1586.

Chen, Y., Zhang, X. & Liu, W. 2021 Effect of metal and metal oxide engineered nano particles on nitrogen bio-conversion and its mechanism: a review. *Chemosphere* **287** (Pt 1), 132097.

EC 2019 *Council Directive 91/271/EEC of 21 May 1991 Concerning Urban Waste-Water Treatment, Commission Staff Working Document, SWD(2019) 700 Final, Part 1/2*. European Commission, Brussels, Belgium.

Ee, B. W. S., Khanna, P., Teng, N. C. & Hun, B. G. 2018 Mechanisms of nanotoxicity to cells, animals, and humans. In: *Nanotoxicology: Toxicity Evaluation, Risk Assessment and Management* (Kumar, V., Dasgupta, N. & Ranjan, S., eds). CRC Press, Boca Raton, Florida, pp. 238–284.

Ganesan, S. & Vadivelu, V. M. 2019 Effect of external hydrazine addition on anammox reactor start-up time. *Chemosphere* **223**, 668–674.

Gutwiński, P., Cema, G., Ziembińska-Buczyńska, A., Wyszyńska, K. & Surmacz-Górska, J. 2021 Long-term effect of heavy metals Cr(III), Zn(II), Cd(II), Cu(II), Ni(II), Pb(II) on the anammox process performance. *Journal of Water Process Engineering* **39**, 101668.

He, Q., Yuan, Z., Zhang, J., Zhang, S., Zhang, W., Zou, Z. & Wang, H. 2017 Insight into the impact of ZnO nanoparticles on aerobic granular sludge under shock loading. *Chemosphere* **173**, 411–416.

Heisler, J., Glibert, P. M., Burkholder, J. M., Anderson, D. M., Cochlan, W., Dennison, W. C., Dortch, Q., Gobler, C. J., Heil, C. A., Humphries, E., Lewitus, A., Magnien, R., Marshall, H. G., Sellner, K., Stockwell, D. A., Stoecker, D. K. & Suddleson, M. 2008 Eutrophication and harmful algal blooms: a scientific consensus. *Harmful Algae* **8** (1), 3–13.

Jin, R. C., Yang, G. F., Yu, J. J. & Zheng, P. 2012 The inhibition of the Anammox process: a review. *Chemical Engineering Journal* **197**, 67–79.

Kartal, B., van Niftrik, L., Keltjens, J. T., Op den Camp, H. J. M. & Jetten, M. S. M. 2012 Chapter 3 - Anammox – growth physiology, cell biology, and metabolism. In: *Advances in Microbial Physiology* (Poole, R. K., ed.). Academic Press, Cambridge, MA, pp. 211–262.

Knobeloch, L., Salna, B., Hogan, A., Postle, J. & Anderson, H. 2000 Blue babies and nitrate-contaminated well water. *Environmental Health Perspectives* **108** (7), 675–678.

Liu, G., Wang, D., Wang, J. & Mendoza, C. 2011 Effect of ZnO particles on activated sludge: role of particle dissolution. *Science of the Total Environment* **409** (14), 2852–2857.

Liu, W. Z., Weng, C. Z., Zheng, J. Y., Peng, X. Q., Zhang, J. & Lin, Z. 2019 Emerging investigator series: treatment and recycling of heavy metals from nanosludge. *Environmental Science-Nano* **6** (6), 1657–1673.

Luo, X., Yan, Q., Wang, C., Luo, C., Zhou, N. & Jian, C. 2015 Treatment of ammonia nitrogen wastewater in low concentration by two-stage Ozonization. *International Journal of Environmental Research and Public Health* **12** (9), 11975–11987.

Madeira, C. L. & De Araújo, J. C. 2021 Inhibition of anammox activity by municipal and industrial wastewater pollutants: a review. *Science of the Total Environment* **799**, 149449.

Manonmani, U. & Joseph, K. 2018 Granulation of anammox microorganisms for autotrophic nitrogen removal from wastewater. *Environmental Chemistry Letters* **16** (3), 881–901.

McCarty, P. L. 2018 What is the best biological process for nitrogen removal: when and why? *Environmental Science & Technology* **52**, 3835–3841.

Miodoński, S., Muszyński-Huhajło, M., Zięba, B., Ratkiewicz, K., Kołbuc, D. & Łagocka, M. 2019 Fast start-up of anammox process with hydrazine addition. *SN Applied Sciences* **1** (6), 523.

Morgan, J. W., Forster, C. F. & Evison, L. 1990 A comparative-study of the nature of biopolymers extracted from anaerobic and activated sludges. *Water Research* **24** (6), 743–750.

Mu, H. & Chen, Y. 2011 Long-term effect of ZnO nanoparticles on waste activated sludge anaerobic digestion. *Water Research* **45** (17), 5612–5620.

Mu, H., Zheng, X., Chen, Y., Chen, H. & Liu, K. 2012 Response of anaerobic granular sludge to a shock load of zinc oxide nanoparticles during biological wastewater treatment. *Water Science & Technology* **46** (11), 5997–6003.

Ni, S. Q. & Meng, J. 2011 Performance and inhibition recovery of anammox reactors seeded with different types of sludge. *Water Science & Technology* **63** (4), 710–718.

Pereira, A. D., Cabezas, A., Etchebehere, C., Chernicharo, C. A. d. L. & de Araújo, J. C. 2017 Microbial communities in anammox reactors: a review. *Environmental Technology Reviews* **6** (1), 74–93.

Porter, A. L. & Youtie, J. 2009 How interdisciplinary is nanotechnology? *Journal of Nanoparticle Research* **11** (5), 1023–1041.

Sabir, S., Arshad, M. & Chaudhari, S. K. 2014 Zinc oxide nanoparticles for revolutionizing agriculture: synthesis and applications. *The Scientific World Journal* **2014**, 925494.

Sari, T., Can, S. & Akgul, D. 2020a Assessment of Anammox process against acute and long-term exposure of ZnO nanoparticles. *Science of the Total Environment* **727**, 138603.

Sari, T., Yar-Bektaş, A. Ö. & Akgül, D. 2020b Short-term inhibitory effects of TiO_2 NPs on Anammox process. *Biotech Studies* **29** (1), 29–37.

Song, Y.-X., Chai, L.-Y., Tang, C.-J., Xiao, R., Li, B.-R., Wu, D. & Min, X.-B. 2018 Influence of ZnO nanoparticles on anammox granules: the inhibition kinetics and mechanism analysis by batch assays. *Biochemical Engineering Journal* **133**, 122–129.

Stein, L. Y. & Klotz, M. G. 2016 The nitrogen cycle. *Current Biology* **26** (3), R94–R98.

Strous, M., Pelletier, E., Mangenot, S., Rattei, T., Lehner, A., Taylor, M. W., Horn, M., Daims, H., Bartol-Mavel, D., Wincker, P., Barbe, V., Fonknechten, N., Vallenet, D., Segurens, B., Schenowitz-Truong, C., Médigue, C., Collingro, A., Snel, B., Dutilh, B. E., Op den Camp, H. J., van der Drift, C., Cirpus, I., van de Pas-Schoonen, K. T., Harhangi, H. R., van Niftrik, L., Schmid, M., Keltjens, J., van de Vossenberg, J.,

Kartal, B., Meier, H., Frishman, D., Huynen, M. A., Mewes, H. W., Weissenbach, J., Jetten, M. S., Wagner, M. & Le Paslier, D. 2006 Deciphering the evolution and metabolism of an anammox bacterium from a community genome. *Nature* **440** (7085), 790–794.

Tang, J., Wu, Y., Esquivel-Elizondo, S., Sorensen, S. J. & Rittmann, B. E. 2018 How microbial aggregates protect against nanoparticle toxicity. *Trends in Biotechnology* **36** (11), 1171–1182.

Tsuneda, S., Aikawa, H., Hayashi, H., Yuasa, A. & Hirata, A. 2003 Extracellular polymeric substances responsible for bacterial adhesion onto solid surface. *FEMS Microbiology Letters* **223** (2), 287–292.

Verma, S., Daverey, A. & Lin, J.-G. 2021 Successful start-up of anammox process from activated sludge and anaerobic sludge in a sequencing batch reactor using an unconventional strategy. *International Biodeterioration & Biodegradation* **156**, 105132.

Weralupitiya, C., Wanigatunge, R., Joseph, S., Athapattu, B. C. L., Lee, T.-H., Kumar Biswas, J., Ginige, M. P., Shiung Lam, S., Senthil Kumar, P. & Vithanage, M. 2021 Anammox bacteria in treating ammonium rich wastewater: recent perspective and appraisal. *Bioresource Technology* **334**, 125240.

Werkneh, A. A., Rene, E. R. & Lens, P. N. L. 2018 Application of nanomaterials in food, cosmetics, and other related process industries. In: *Nanotoxicology: Toxicity Evaluation, Risk Assessment and Management* (Kumar, V., Dasgupta, N. & Ranjan, S., eds). CRC Press, Boca Raton, Florida, pp. 63–80.

Wurtsbaugh, W. A., Paerl, H. W. & Dodds, W. K. 2019 Nutrients, eutrophication and harmful algal blooms along the freshwater to marine continuum. *Wiley Interdisciplinary Reviews-Water* **6** (5), e1373.

Yapsakli, K., Aktan, C. K. & Mertoglu, B. 2017 Anammox-zeolite system acting as buffer to achieve stable effluent nitrogen values. *Biodegradation* **28** (1), 69–79.

You, G. X., Wang, P. F., Hou, J., Wang, C., Xu, Y., Miao, L. Z., Lv, B. W., Yang, Y. Y., Liu, Z. L. & Zhang, F. 2017 Insights into the short-term effects of CeO_2 nanoparticles on sludge dewatering and related mechanism. *Water Research* **118**, 93–103.

Zhang, Z. Z., Xu, J. J., Shi, Z. J., Cheng, Y. F., Ji, Z. Q., Deng, R. & Jin, R. C. 2017 Short-term impacts of Cu, CuO, ZnO and Ag nanoparticles (NPs) on anammox sludge: CuNPs make a difference. *Bioresource Technology* **235**, 281–291.

Zhang, X., Chen, Z., Ma, Y., Zhou, Y., Zhao, S., Wang, L. & Zhai, H. 2018a Influence of elevated Zn (II) on Anammox system: Microbial variation and zinc tolerance. *Bioresource Technology* **251**, 108–113.

Zhang, Z. Z., Cheng, Y. F., Xu, L. Z., Wu, J., Bai, Y. H., Xu, J. J., Shi, Z. J. & Jin, R. C. 2018b Discrepant effects of metal and metal oxide nanoparticles on anammox sludge properties: a comparison between Cu and CuO nanoparticles. *Bioresource Technology* **266**, 507–515.

Zhang, Z. Z., Cheng, Y. F., Xu, L. Z. J., Bai, Y. H., Xu, J. J., Shi, Z. J., Shen, Y. Y. & Jin, R. C. 2018c Evaluating the effects of metal oxide nanoparticles (TiO_2, Al_2O_3, SiO_2 and CeO_2) on anammox process: performance, microflora and sludge properties. *Bioresource Technology* **266**, 11–18.

Zhang, X., Chen, Z., Zhou, Y., Ma, Y., Ma, C., Li, Y., Liang, Y. & Jia, J. 2019 Impacts of the heavy metals Cu (II), Zn (II) and Fe (II) on an anammox system treating synthetic wastewater in low ammonia nitrogen and low temperature: Fe (II) makes a difference. *Science of the Total Environment* **648**, 798–804.

Zhao, J., Zhang, B. & Zuo, J. 2019 Response of anammox granules to ZnO nanoparticles at ambient temperature. *Environmental Technology & Innovation* **13**, 146–152.

Zheng, X., Wu, R. & Chen, Y. 2011 Effects of ZnO nanoparticles on wastewater biological nitrogen and phosphorus removal. *Environmental Science and Technology* **45** (7), 2826–2832.

First received 16 October 2021; accepted in revised form 19 November 2021. Available online 2 December 2021

doi: 10.2166/wst.2021.472

Removal of microplastics from wastewater: available techniques and way forward

Surya Singh [a,*], Madhanraj Kalyanasundaram[b] and Vishal Diwan[a]

[a] Division of Environmental Monitoring and Exposure Assessment (Water & Soil), ICMR – National Institute for Research in Environmental Health, Bhopal 462 030, India

[b] Division of Environmental Health and Epidemiology, ICMR – National Institute for Research in Environmental Health, Bhopal 462 030, India

*Corresponding author. E-mail: suryasingh.nireh@icmr.gov.in

SS, 0000-0001-5981-2343

ABSTRACT

Microplastics are the newly emerged contaminants with a presence in almost every part of the globe. Despite being small in size, microplastic particles have proved to be harmful for plants, animals, humans, and for the ecosystem in general. Water is one of the most important routes through which microplastics transfer from one place to another. Moreover, water is also an important route for the ingestion of microplastics in human, which results in various health issues, such as cancer, mutagenic and teratogenic abnormalities. Thus, microplastics in water is an emerging public health issue which needs attention and, hence, it is important to investigate removal techniques for microplastics in wastewater. Although, there are some biological, chemical/electrochemical, and physical techniques to remove microplastics, their wide scale applicability and cost–effectiveness is an issue. In this review, we have discussed the existing and upcoming treatment technologies for the removal of microplastics from wastewater and also tried to present an overview for the future approaches.

Key words: filtration, microplastic(s), removal techniques, wastewater treatment plants, water

HIGHLIGHTS

- Efficient removal of microplastics from wastewater is inevitable considering its human health aspects.
- Conventional wastewater treatment plants are not fully capable of removing microplastics.
- The need is to have efficient, simple, and low-cost technology for microplastic removal.
- Combination of technologies may provide a suitable option.

GRAPHICAL ABSTRACT

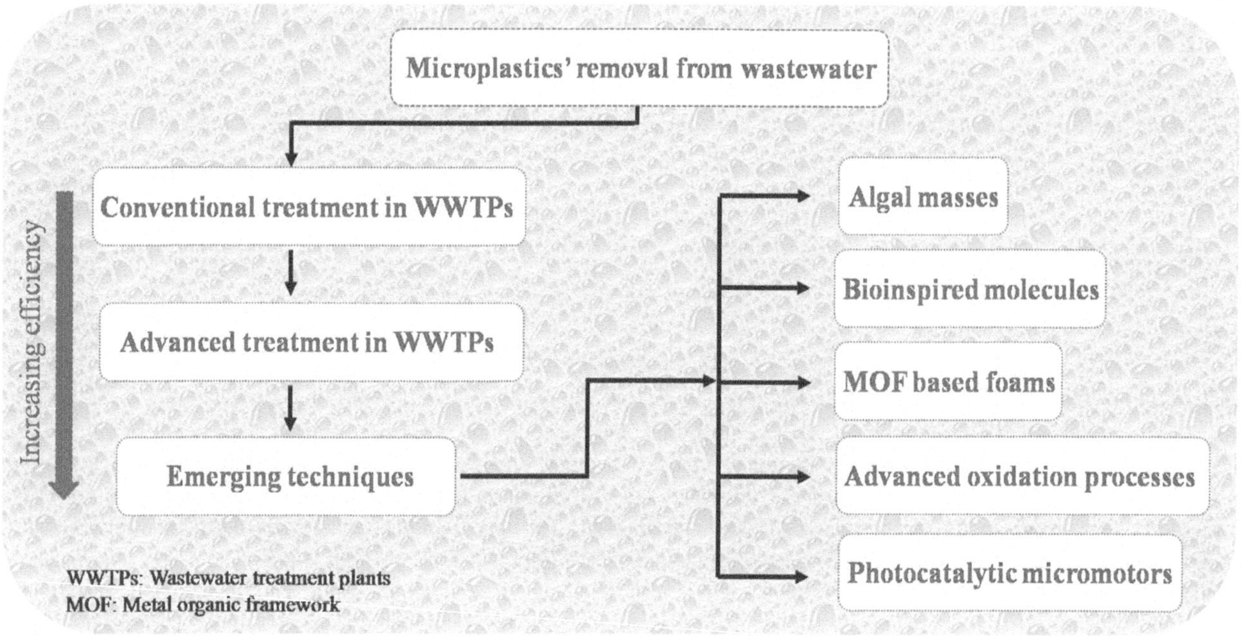

INTRODUCTION

The occurrence of a number of emerging contaminants in various environmental matrices has become a serious concern for the scientific community and policy makers across the globe. Widespread occurrence, bioaccumulation, toxicity, and persistence are the major issues accompanying these emerging contaminants (Richardson & Kimura 2017; Sarkar *et al.* 2021). Microplastic particles are one such contaminant that has gained attention in the recent decade. Small plastic particles, less than 5 mm in size, are termed microplastics and may be in the form of fragment, fibre, pellet, foam, and/or film (Thompson *et al.* 2004; Arthur *et al.* 2009) (Figure 1). These microplastic particles originate either through intentional manufacturing (primary microplastics) or after the breakdown of large-sized plastics (secondary microplastics) due to the action of abiotic and biotic factors in the environment. The occurrence of these microplastic particles has been reported in variety of matrices such as surface water, ground water, marine water, bottled water, soil, air, biotic organisms, food items, honey, sugar (Liebezeit & Liebezeit 2013; Barboza & Gimenez 2015; Galloway & Lewis 2016; Wright & Kelly 2017; Karthik *et al.* 2018; Kosuth *et al.* 2018; Blettler & Wantzen 2019; Conti *et al.* 2020). According to one estimate, approximately 14 million tonnes of microplastics are present on the ocean floor (Barrett *et al.* 2020). The presence of microplastics has serious

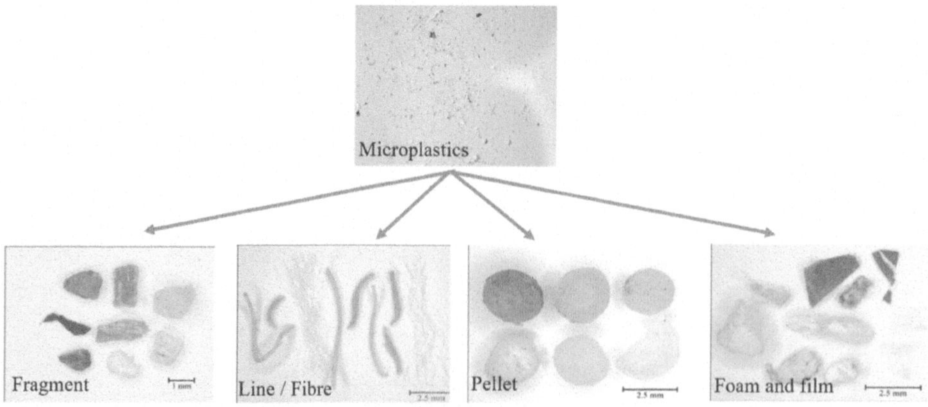

Figure 1 | Types of microplastics in water. (*Adapted from* Karthik *et al.* 2018).

repercussions for the ecosystem as well as for human health. Table 1 summarizes the impact of microplastics over ecosystems and human health.

The occurrence of microplastics (both primary and secondary) in water primarily takes place through the discharge of sewage/wastewater treatment plant effluent and surface run-off (Iyare *et al.* 2020). There are large numbers of industries that use (primary) microplastics for various applications such as medicines or cosmetics (Lechner & Ramler 2015; European Commission Report 2017). After the use of these products at domestic/commercial level, these primary microplastics are washed off and become part of the sewage/wastewater. Conversely, breakdown of large-sized plastic particles into smaller pieces (secondary microplastics) often takes place in landfill sites/dumping grounds/oceans, which further becomes one of the sources of microplastics in the water bodies (Kyrikou & Briassoulis 2007; Galloway & Lewis 2016). As sewage/wastewater treatment plants are not equipped for the complete removal of microplastics, the effluent released from these plants contains substantial quantities of microplastics (Amrutha & Warrier 2020). Upon mixing of this effluent with the freshwater sources, microplastics become part of the fresh/drinking water supply chain as well (Magnusson & Noren 2014; Novotna *et al.* 2019; Okoffo *et al.* 2019). For example, increase in the concentration of microplastics in the Chicago River has been reported to be due to local wastewater treatment plants' effluent (McCormick *et al.* 2014). It is also important to note that many components of water treatment plants and water distribution systems are usually made up of plastic materials such as high density polyethylene, polyvinyl chloride, and polypropylene (Mintenig *et al.* 2019) and, hence, these further contribute towards microplastic generation in the water they carry. The occurrence of microplastics in various water sources is summarized in Table 2.

These plastic particles are of concern owing to their toxicity upon ingestion/inhalation (Table 1). Furthermore, various contaminants, additives/plasticizers, microbes adsorbed onto the surface of microplastics also cause serious health effects (Ensign *et al.* 2012; Deng *et al.* 2017; Schirinzi *et al.* 2017; Smith *et al.* 2018; Prata *et al.* 2020). Preliminary studies have shown that microplastics may result in genotoxicity, carcinogenesis, fibrosis, exaggerated inflammatory response, oxidative stress etc. (Deng *et al.* 2017; Schirinzi *et al.* 2017). According to some studies, microplastics have the potential to cross the placental barrier as well (Grafmueller *et al.* 2015; Ragusa *et al.* 2021) which results in various reproductive and teratogenic abnormalities (Swan *et al.* 2005; Swan 2008; Lang *et al.* 2008).

Notably, one of the most important routes of microplastic transfer from environmental matrices to human is through the ingestion of contaminated water. As sustainable development goals target for good health and well-being, there is a requirement to focus on the availability of clean water, free of any contamination to everyone (SDG 2015). Although there are certain laws and regulations targeting curtailment of production and use of primary microplastics (Kentin & Kaarto 2018), there are many more challenges ahead. Effective strategies for microplastic removal from wastewater need to be developed so that surface water sources are not contaminated upon receiving discharge from wastewater treatment plants. In this paper, a short review of various existing removal options for microplastics in wastewater treatment plants has been done.

CHALLENGES IN THE REMOVAL OF MICROPLASTICS

It is important to understand that toxic effects of microplastics are not only due to their inherent nature but also due to various additives and plasticizers that are added during manufacturing. Furthermore, when these microplastics enter the environment, adsorption of various chemicals, pollutants, and microbes takes place on its surface. Thus, it is not only the microplastics which are of concern; rather, the adhered contaminants also pose significant health risks.

Regulation for primary microplastics, which is mostly found in pharmaceuticals and cosmetic products, is scarce, especially in developing countries (Perren *et al.* 2018). Lack of suitable plastic waste management systems further leads to the dumping of plastics mixed with other wastes at landfill sites, resulting into secondary microplastics in the long run. Removal of microplastics becomes challenging owing to their wide occurrence and small size, which usually allows them to escape the sieving/filtration processes. Conventional wastewater treatment plants are not fully efficient for their removal and hence final effluent contains significant amounts of microplastics. Discharge of this effluent into the surface water bodies and/or surface run-off allows microplastics to become mixed with the drinking water supply chain (Murphy *et al.* 2016). In addition, the amount of microplastics that is entrapped in sludge often gets no further treatment and is mostly disposed off onto the land (Yuan *et al.* 2022). This results in the contamination of groundwater and soil as well, through percolation (Tagg *et al.* 2021). Moreover, as the groundwater is generally considered safe for drinking and hence does not pass through any treatment process, microplastics present in the groundwater do not even get the chance of removal and thus makes humans more prone to ingestion of the

Table 1 | Ecological and human health aspects of microplastics

Effect	Description	References
Ecological aspects		
Reduced chlorophyll development	Microplastics concentration ranging from 0.9 to 2.1 mg/L significantly reduces the development of chlorophyll in plants	Prata *et al.* (2018)
Toxicity to earthworms	Low-density polyethylene caused toxicity to earthworms based on observations on mortality, growth, and tunnel formation	Lwanga *et al.* (2016, 2017)
MPs transfer along food chain	MPs can transfer *via* the trophic route and enter the food of humans	Sussarellu *et al.* (2014)
Reduced gamete quality	Lower gamete quality causes less offspring to be produced and less fecundity	Rochman *et al.* (2013)
Marine larvae growth	Sea urchin larvae exposed to polyethylene microspheres showed little effect on larval growth	Kaposi *et al.* (2014)
Hepatic damage	Due to metabolic stress caused by MPs, as well as pollutants accumulating on its surface, liver damage has been found in some organisms	Wright *et al.* (2013)
Reduced feeding or filtering	Animals containing MP in their digestive tracts were found to eat less, resulting in lower energy levels and fat reserves	Lee *et al.* (2013)
Mortality	Due to the combination of physical and physiological effects of MP particles on marine copepod (*Tigriopus japonicas*), fatality is increased	Von Moss *et al.* (2012)
Immune response	MP in animal tissue can induce an immune response leading to inflammation	Kohler (2010)
Increased reactive oxygen species (ROS)	Ingested microplastics have been shown to increase free radicals which leads to cellular and DNA damage	Karami *et al.* (2017)
Disruption in gas exchange on the ocean floor	Accumulation of MPs at the floor of water bodies disrupts the usual gas exchange process, thus affecting the ecosystem and benthic communities	Thompson *et al.* (2004)

Type of microplastic	Particles/Chemical(s) associated	Description	References
Human health aspects			
Polycarbonate plastics, epoxy resins	Bisphenol A (BPA)	Affects brain development leading to loss of sex differentiation in brain structures and behaviour, suspected endocrine disrupting chemical	Rist *et al.* (2018)
Polystyrene used for Styrofoam packaging	Styrene	Endocrine disrupting chemical	Rist *et al.* (2018)
Polyethylene (PE) and polystyrene (PS) particles	–	Exposure to PE and PS causes genotoxicity, apoptosis, and necrosis, leading to tissue damage, fibrosis, and carcinogenesis	Wright & Kelly (2017)
Polyvinyl chloride	Vinyl chloride	Angiosarcoma of liver	Gennaro *et al.* (2008); Talsness *et al.* (2009)
Polyvinyl chloride (in medical tubing)	Di(2-ethylhexyl)phthalate (DEHP)	High levels of BPA in infants	Urban *et al.* (2000)
Polyethylene (PE) particles	–	PE particles up to 50 μm reported in lymph nodes, liver, and spleen, causing immune activation of macrophages and production of cytokines	Doorn *et al.* (1996); USEPA (1989)

Table 2 | Occurrence of microplastics in water reported across the globe

Type of water	Identification method	Size range (μm)	Concentration (particles/L)	Morphology	Composition	References
Marine water	FTIR spectroscopy	–	1.25	Fragments, fibre/line, foam	Polyethylene, polypropylene	Robin *et al.* (2020)
	FTIR spectroscopy	–	–	Fragments, fibre/line, foam	Polyethylene, polypropylene, polystyrene	Karthik *et al.* (2018)
	–	–	0.004–4,137	–	–	Avio *et al.* (2017)
	–	–	–	Pellets, fragments, sponge, fibre	–	Frias *et al.* (2013)
River	FTIR spectroscopy	–	10–520	Film, fragments, fibres	Polyethylene, polyamide, polypropylene, polyethylene terephthalate,	He *et al.* (2020)
	FTIR spectroscopy	–	1.6	–	Polyethylene, polypropylene, polystyrene	Kataoka *et al.* (2019)
	Visual	–	–	Fibres, fragments, flakes	–	Kay *et al.* (2018)
	FTIR spectroscopy	–	50.3–1,600	–	Polypropylene, polyester, rayon, cotton + viscose phenoxy resin, poly(vinyl stearate)	Peng *et al.* (2018)
Ground drinking water sources	FTIR spectroscopy*	50–150	0–7	Fragments	Polyethylene, polyamide, polyester, polyvinylchloride	Mintenig *et al.* (2019)
	Pyrolysis – GC MS*	–	6.4	Fibres	Polyethylene	Panno *et al.* (2019)
Raw wastewater	FTIR spectroscopy	598 ± 89	15.70 ± 5.23	Flake, fibre, film, bead, foam	Polyethylene terephthalate, Polyester, polyethylene, polypropylene, polyurethane, polyvinylchloride, polyamide, polyacrylate	Murphy *et al.* (2016)
	Visual	100–5,000	293,000–320,000	Fibres	–	Dris *et al.* (2015)
Wastewater treatment plant effluent	FTIR spectroscopy	–	0.21	Fragments, films, fibre, foam, pellet	Polyacrylic, polypropylene, polyethylene	Dyachenko *et al.* (2017)
	FTIR spectroscopy	–	0.5–6.9	Fibres, microbeads	Polyester, polyethylene, polypropylene, polystyrene, polyvinylchloride, polyamide, polyacrylate	Talvitie *et al.* (2017)
	Visual	125–355 and above	0.004–0.195	Fragment, pellet, fibre, film, foam	–	Mason *et al.* (2016)
	FTIR spectroscopy	598 ± 89	0.25 ± 0.04	Flake, fibre, film, bead, foam	Polyethylene terephthalate, Polyester, polyethylene, polypropylene, polyurethane, polyvinylchloride, polyamide, polyacrylate	Murphy *et al.* (2016)
Sewage sludge	GC–MS	–	28–12,000	–	Polyethylene terephthalate, polycarbonate, terephthalic acid, polycarbonate	Zhang *et al.* (2019)
	Microscope FTIR spectroscopy	–	1,600–56,400	Fibres	Polyolefin, acrylic fibres, polyethylene, polyamide	Li *et al.* (2018)

(*Continued*)

Table 2 | Continued

Type of water	Identification method	Size range (μm)	Concentration (particles/L)	Morphology	Composition	References
	FTIR spectroscopy	<500			Polyethylene, polypropylene, polyamide, polystyrene	Mintenig *et al.* (2017)
Drinking water treatment plant effluent	FTIR spectroscopy & micro-Raman imaging microscopy	1–10	338 ± 76 to 628 ± 28	Fragments, fibres	Polyethylene terephthalate, polypropylene, polyethylene	Pivokonsky *et al.* (2018)
Tap water	FTIR spectroscopy	100–5,000	0–61	Fibres	–	Kosuth *et al.* (2018)
Plastic bottled water	FTIR spectroscopy	6.5 – >100	0 to >10,000	Fragments, fibres	Polypropylene	Mason *et al.* (2018)
		–	3.57	Fibres	–	Kosuth *et al.* (2018)
	Micro – FTIR spectroscopy	5–20	118 ± 88 (returnable bottles), 14 ± 14 (single-use bottles)	Fragments	Polyethylene terephthalate, polypropylene	Schymanski *et al.* (2018)
	Micro – Raman spectroscopy	<5	4,890 ± 5,430 (returnable bottles) 2,649 ± 2,857 (single-use bottles)	–	Polyethylene	Oßmann *et al.* (2018)
Glass bottled water	FTIR spectroscopy	6.5 – >100	1,410 14.8 (>100 μm) 1,396 (6.5–100 μm)	Fragment, fibre, pellet, film, foam	–	Mason *et al.* (2018)
	Micro – FTIR spectroscopy	>100	50 ± 52	–	Polyamide, polyethylene, polypropylene	Schymanski *et al.* (2018)
	Micro – Raman spectroscopy	<5	6,292 ± 10,521	–	Polyethylene, styrene – butadiene – copolymer	Oßmann *et al.* (2018)
Cardboard bottled water	Micro – FTIR spectroscopy	>100	11 ± 8	Fibres	Cellulose, polyethylene, polypropylene	Schymanski *et al.* (2018)

same. Where low- and middle-income countries are still struggling for full-scale solid waste and wastewater management, addition of new technologies for removal of microplastics is an extra burden in the supply chain. Therefore, the new treatment options need to be cost effective as well as simple to incorporate.

EXISTING REMOVAL TECHNIQUES FOR MICROPLASTICS IN WATER

Removal of microplastics through conventional wastewater treatment systems

Wastewater treatment processes involve treatment mainly at three stages *viz.* primary, secondary, and tertiary. In many instances a preliminary step is also employed to remove the large floating materials and grits, however this step is generally considered under the primary treatment. Overall, primary treatment involves physical settling of insoluble solids from the wastewater stream through various mechanisms such as screening, grit removal, oil and grease removal, and sedimentation (with coagulation). At this stage, the contaminants which can readily float or settle under the action of gravity are removed. Secondary (biological) treatment targets degradation of the biological content of the wastewater using either the attached growth system or the suspended growth system. Tertiary (advance) treatment is the final treatment stage in which chemical disinfection, filtration (ultrafiltration/microfiltration/nanofiltration), reverse osmosis, ozonation etc. are employed. The removal efficiency in wastewater treatment plants is calculated based upon the particle concentration of microplastics (*viz.* the number of microplastic particles/litre) in the influent and effluent.

Considering the stage-wise removal of microplastics, approximately 35–59% of the microplastics are removed during the preliminary treatment and 50–98% are removed during the primary treatment in wastewater treatment plants (Sun *et al.* 2019). The main mechanism involved in removal of microplastics at this stage is the skimming and settling of the entrapped microplastics during gravity separation. The shape of the microplastic particles also matters, as the fibres are removed more easily than the fragments in these treatments (Magnusson & Noren 2014; Talvitie *et al.* 2016; Ziajahromi *et al.* 2017). Plants that discharge the effluent only after the secondary treatment are more prone to release microplastics into the environment (Carr *et al.* 2016), although some advanced techniques such as membrane bioreactors may reduce the microplastic concentration to some extent (Lares *et al.* 2018). It has also been reported that overall microplastic removal can be achieved up to 88% in the absence of tertiary treatment and up to 97% with tertiary treatment (Sun *et al.* 2019). With time, the performance of reactors also goes down, which results in lower removal efficiency (Leslie *et al.* 2017). A summarized evaluation of microplastic removal processes is shown in Table 3.

Removal of microplastics through modifications in conventional wastewater treatment methods

Removal of microplastics using existing water treatment systems is highly desirable compared to any new technology because of the familiarity with the system, ease of operation and low costs involved. Therefore, Ma *et al.* investigated the aluminium (Al)- and iron (Fe)-based salts ($AlCl_3 \cdot 6H_2O$ and $FeCl_3 \cdot 6H_2O$, respectively) for coagulation during the wastewater treatment process and achieved microplastic removal up to 36.89% for the particle sizes <0.5 mM using Al-based salts (Ma *et al.* 2019). A striking feature was that higher efficiency could be obtained for small-sized particles, while it decreased gradually for large-sized microplastics (4.51% for sizes between 2 and 5 mm). Fe-based salts, however, were poor in performance for microplastics removal. The main mechanisms involved in removal were charge neutralization and sweep flocculation. As the size of floc formed using Al-based salts was less compared to that of Fe-based salts, it rendered more surface area and hence higher removal efficiency. Furthermore, Al-based salts provided higher positive zeta potential as compared to Fe-based salts, which rendered easy charge neutralization of the microplastic particles with a negative zeta potential. The microplastic particles entangled over the flocs were then removed using ultrafiltration membranes. The removal efficiency of Al-based flocs is also affected by the pH and polyacrylamide (PAM). It was seen that pH did not play much of a role in the removal of microplastics at the low dose of Al salt (0.5 mM $AlCl_3 \cdot 6H_2O$), however, at high coagulant doses (5 mM $AlCl_3 \cdot 6H_2O$) the removal efficiency decreased upon increasing the pH. Similarly, addition of PAM did not affect the removal of microplastics at low doses of $AlCl_3 \cdot 6H_2O$, but it resulted in enhanced removal at higher doses. Nevertheless, the addition of PAM or other similar additives should be minimal in water treatment processes, considering their non-biodegradable nature and enhanced risk upon increase in acrylamide monomer content (Stephens 1991).

Electrocoagulation is another viable solution for the removal of microplastics. Potentially, electrocoagulation does not require chemicals thus making it environment friendly. Electrocoagulation involves the production of coagulants electrically. Usually, Fe^{+2} and/or Al^{+3} ions of the metal electrodes react with the OH^- ions produced after the electrolysis and generate metal hydroxide coagulants. Microplastics, being solid particles, become destabilized in the presence of these coagulants and subsequently are entrapped in the sludge blanket produced by the coagulants. Flocs containing microplastic particles can then be removed from the water. Perren *et al.* have reported that approximately 90% of the microbeads were removed using the Al-based electrocoagulation technique (Perren *et al.* 2018). The major mechanism involved was charge neutralization and flocculation (Table 3).

Microplastics can also be removed during the filtration stage of the wastewater treatment (Michielssen *et al.* 2016). Microfiltration (0.1–1 μm), ultrafiltration (2–100 μm), and nanofiltration (~2 nm) have been utilized for the removal of microplastics (Poerio *et al.* 2019). Wang *et al.* integrated the low-cost biochar in the sand filter systems to improve the efficiency of the filtration, which can result in removal of microplastics (Wang *et al.* 2020a). Up to 10 μm diameter size of microplastic particles could be removed with efficiency of more than 95%. This immobilization was achieved through three major mechanisms *viz.* 'stuck', 'trapped', and 'entangled' as visualized through environmental scanning electron microscopy. 'Stuck' represented the containment of the microplastic particles in the gaps between the filter particles owing to the small size of the latter. 'Trapping' is achieved when the filter material is porous in structure and flow of the water is slow. Furthermore, 'entanglement' occurs when the deep holes of the filter paper provide restriction for the microplastic particles against the water flow. Similarly, Wang *et al.* employed granular activated carbon filters in the wastewater treatment plants and achieved the reduction in microplastics' abundance in treated effluent of up to 56.8–60.9% (Wang *et al.* 2020b). Moreover, the performance was more suitable for the small-sized microplastic particles, as approximately 73.7–98.5% of the captured particles were in the size

Table 3 | Various removal methods of microplastics

Process description	Major mechanism	Lowest size of microplastic particle removed/finest mesh	Efficiency (%)	Advantages	Challenges	References
Wastewater treatment plant processes	Skimming, settling of the entrapped microplastics	300 μm	99.9	Conventional process, no additional cost	Not possible to remove MPs of size <300 μm	Magnusson & Noren (2014)
	Primary, secondary, and tertiary	100 μm	99.9	Conventional process, no additional cost	Not possible to remove MPs of size <100 μm	Carr *et al.* (2016)
	Secondary treatment Tertiary treatment Membrane bioreactor	20 μm	95.6 97.2 99.4	MBR process exhibited greater overall efficiency	Complete retention is not possible	Michielssen *et al.* (2016)
	Membrane bioreactor (MBR)	250 μm	99.3	MBR process helped to retain more microplastics compared to conventional activated sludge process	Not possible to remove MPs of size <250 μm	Lares *et al.* (2018)
Al and Fe salt	Coagulation	<0.5 mm	45.34±3.93	Simple process, does not require additional set-up	Low efficiency	Ma *et al.* (2019)
Electrocoagulation	Charge neutralization, flocculation	–	90 (pH 3–10) 99.24 (pH 7.5)	Does not rely on chemicals or microorganisms, energy efficient	Operation time needs to lowered down	Perren *et al.* (2018)
Filtration with biochar	Morphologically controlled mechanism (Stuck, trapped, entangled)	10 μm	>95	Low cost and efficient	Process is slow and results in obstruction of the pores with time; costly; regeneration is tough	Wang *et al.* (2020a)
Filtration with granular activated carbon (combined with coagulation and sedimentation)	Physical properties (size and shape)	1–5 μm	56.8–60.9	Efficient to remove plastic particles of nano-size range	Process is slow and results in obstruction of the pores with time; costly; regeneration is tough	Wang *et al.* (2020b)
Pulse clarification with filtration	Entrapment in sludge blanket formed due to coagulation floats	<100 μm	85	Removal efficiency is comparable to the other treatment plants having advanced processes	Complete retention is not possible	Sarkar *et al.* (2021)
Algal masses	Electrostatic charge onto the microplastic particles and algal surfaces	20 μm	94.5	No chemical, electrical, and mechanical operations	Efficiency will vary owing to physiological and topographical differences on the seaweed surface	Sundbæk *et al.* (2018)

(*Continued*)

Table 3 | Continued

Process description	Major mechanism	Lowest size of microplastic particle removed/finest mesh	Efficiency (%)	Advantages	Challenges	References
Bioinspired molecules	Mechanical capture mechanism driven by the hydrophobic and van der Waals interactions	–	–	Flexible; possibility to remove different types of plastic particles in wastewater stream	Method yet to be established for practical purposes	Herbort & Schuhen (2017)
Zr metal organic framework (MOF)-based foams	Entrapment	–	95.5 ± 1.2	High performance, excellent durability	Flexibility and robustness of the MOF-based foams; Removal efficiency affected by the particle size and zeta potential	Chen *et al.* (2020)
Photocatalytic micromotors	Phoretic interaction and shovelling/pushing interactions	–	–	Self-propelled devices, works efficiently independent of the fuel	Selectivity of micromotors for microplastics is crucial	Wang *et al.* (2019)

range of 1–5 μm. The main mechanism involved here was postulated to be synergistic combination of physical adsorption and biodegradation (Zheng *et al.* 2018).

Pulse clarification is another technique which helps to remove microplastics in wastewater treatment plants. Sarkar *et al.* demonstrated that pulse clarification and filtration are cumulatively able to remove up to 85% of microplastics (Sarkar *et al.* 2021). The mechanism involved here is the entrapment of microplastic particles in the sludge blanket formed due to the coagulation (H_2SO_4, $Al_2(SO_4)_3.24H_2O$). Pulsation helps to keep the sludge blanket in expansion which eases the entrapment. Finally, filtration further adds in the removal of microplastics.

However, filtration processes have some inherent shortcomings such as: (1) the process is slow and overtime it results in obstruction of the pores, (2) the process works under pressure and hence costly, (3) membrane regeneration is a cumbersome process. Furthermore, although conventional wastewater treatment processes can help in the removal of microplastics to some extent, none of these processes are specifically targeted to remove/degrade the microplastics (Iyare *et al.* 2020). Significant amounts of the removed microplastics remain in sludge, which further contaminates the land and subsequently groundwater. Therefore, alternative options that are targeted for the microplastics' removal need to be explored.

EMERGING TECHNIQUES FOR THE REMOVAL OF MICROPLASTICS

Algal masses

Microalgae may offer a possibility for the removal of microplastics, as it has been seen that microalgae colonize microplastic particles thus altering the buoyancy of the aggregates. This results in differential sedimentations rates as compared to the un-aggregated particles (Lagarde *et al.* 2016). This property may be utilized for the removal of microplastics by incorporating the microalgae. Sundbæk *et al.* employed a marine seaweed, namely *Fucus vesiculosus*, for the possible removal of microplastics through translocation in the algal tissues (Sundbæk *et al.* 2018). Due to narrow channels in the algal cells, the movement of microplastics was restricted and thus the plastic particles were captured. Efficiency of as high as ~94.5% was observed, especially in the dissected areas of the algae. As the dissected areas ooze out anionic polysaccharide substances, the adherence of plastic particles is enhanced (Martins *et al.* 2013). Sorption of microplastic particles has also been reported by the other researchers (Bhattacharya *et al.* 2010; Nolte *et al.* 2017). The major mechanism for the sorption of microplastic particles onto the algal surface is the electrostatic charge. The positively charged microplastic particles tend to sorb more onto the algae owing to the presence of anionic polysaccharide substances in the algal cell wall (Nolte *et al.* 2017).

Bioinspired molecules

In order to remove microplastic particles, bioinspired molecules have been developed by some researchers. Herbort and Schuhen demonstrated the concept for the removal of hydrophobic microplastic particles through the combination of organic and inorganic molecular building blocks (Herbort & Schuhen 2017). These bioinspired molecules consist of an inclusion unit (IU) and a capture unit (CU), which are combined together to form an inclusion compound (IC) (Figure 2). In this, IU is the alkoxysilyl functionalized bioinspired component of the molecule, and the CU component has the ability to bond with the various materials through functional groups. Upon capture of the guest molecules (microplastics) in the inclusion cavity, the embedded water molecules are displaced. These released water molecules further combine with other surrounding water molecules through van der Waals forces (Tu *et al.* 2016). Thus, the cavity created by the release of water molecules is filled up by the guest molecules, thus allowing the IC to help in the removal of guest molecules.

Metal organic framework (MOF)-based foams

Metal organic frameworks (MOFs) are porous structures that are the combination of metals and organic ligands. Having high surface area, porosity, and versatile functionality, these chemical moieties help in the entrapment of a variety of pollutants (Zhang *et al.* 2016; Kobielska *et al.* 2018; Mon *et al.* 2018). In order to entrap the microplastics, the material should have sufficient porosity, a suitable framework to capture the pollutant, and high durability. Thus, MOFs might work well as these have such properties. Chen *et al.* employed melamine foam to load zirconium (Zr) MOFs over it. Zr-based UiO-66-X (X=H, NH_2, OH, Br, and NO_2) MOFs were prepared using 1,4-dicarboxybenzene ligand with different functional groups (Chen *et al.* 2020). This framework was used to test microplastic removal in a simulated suspension of polyvinylidiene fluoride (PVDF), polymethyl methacrylate (PMMA), and polystyrene (PS) microplastics. The particle size and zeta potential of different microplastics affected the removal performance. As high as $95.5 \pm 1.2\%$ removal could be achieved using the designed MOF along with having high recyclability (Chen *et al.* 2020).

Photocatalytic micromotors

Micromotors are the miniature self-propelled devices that have attracted researchers across the globe owing to their application in removal of oil (Guix *et al.* 2012; Gao *et al.* 2013; Mou *et al.* 2015), metals/metalloids (Uygun *et al.* 2016; Vilela *et al.* 2016; Wang *et al.* 2016; Villa *et al.* 2018), and various organics (Soler *et al.* 2013; Mushtaq *et al.* 2015; Simmchen *et al.* 2017; Zhang *et al.* 2017). Therefore, Wang *et al.* employed photocatalytic TiO_2-based micromotors

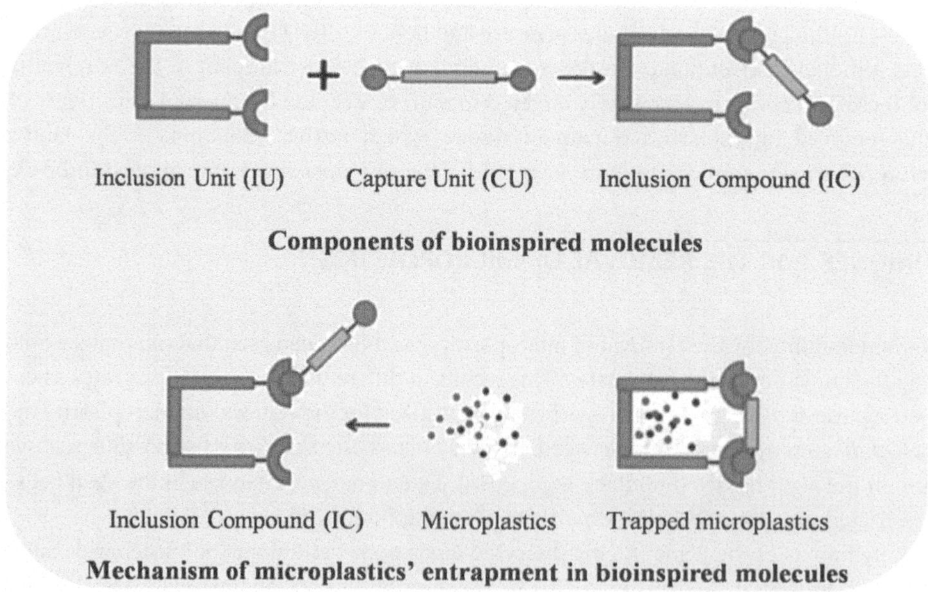

Figure 2 | Entrapment of microplastics using bioinspired molecules. (*Adapted from* Herbort & Schuhen 2017).

($Au@mag@TiO_2$, mag=Ni, Fe) for the removal of microplastics (Wang *et al.* 2019). Being photocatalytic, these micromotors could move themselves in water due to the photocatalytic reactions taking place on the particles. Removal was achieved by phoretic interaction and shovelling/pushing interactions. Phoretic interaction prevailed for individual micromotors, while the shovelling mechanism dominated when a chained assembly of micromotors was used. For the application in environmental systems, the chained assembly of micromotors is the better choice as it works efficiently independent of the fuel (Wang *et al.* 2019).

Advanced oxidation processes

Advanced oxidation processes are one of the most suitable methods for the mineralization of various recalcitrant organic contaminants. Reactive oxygen species, such as hydroxyl radical, sulphate radical are usually employed to mineralize a variety of organic moieties. Therefore, microplastic degradation was also studied using this method. Integrated carbocatalytic oxidation and hydrothermal hydrolysis was used for the degradation of cosmetic microplastics (Kang *et al.* 2019). In this process, manganese carbide nanoparticles were encapsulated in N-doped carbon nanotubes, which activated peroxymonosulphate resulting in degradation of microplastics. Removal efficiency of up to 50 wt% was achieved depending upon the various reaction parameters, such as catalyst dose, microplastic concentration, temperature, and peroxymonosulfate dosage. Ariza-Tarazona and colleagues demonstrated the photocatalytic degradation of polyethylene microplastics in aqueous and solid matrices using the n-TiO_2 semiconductor. The degradation was estimated by the weight loss. It was observed that in the presence of light and n-TiO_2 semiconductor, the total mass of microplastic particles was reduced, indicating the microplastic photocatalytic degradation (Ariza-Tarazona *et al.* 2019). In aqueous medium, the mass loss of polyethylene particles was reported to be 6.40% during the first 18 hours of visible irradiation, while in the solid matrix the mass loss was 1.85% in the first 16 hours of irradiation. The higher mass loss of polyethylene particles in aqueous medium was attributed to the higher concentration of hydroxyl radicals, which promoted the degradation process. Similarly, Tofa and co-workers used zinc oxide nanorods for the photocatalytic degradation of low-density polyethylene (LDPE) residues (Tofa *et al.* 2019). The degradation was confirmed by the increased brittleness, presence of cavities, wrinkles, and cracks on the surface of LDPE after photocatalysis. The degree of degradation was found to be directly proportional to the catalyst surface area.

WAY FORWARD

Microplastic pollution in environmental matrices is a global challenge. Where it is very obvious that reduction in the production and usage of plastics, and safe disposal of the same are the keys to administer this pollution, it is also a reality that it will take time to deal with the immensity and urgency of the problem at present. Therefore, development of efficient techniques for removal of microplastics from water is essential to avoid a number of health problems.

Although there are some prevailing treatment techniques, the efficacy thereof is still a challenge. Moreover, it is also necessary to make these technologies available at minimal cost to make it affordable. Photocatalytic treatment techniques are effective for the removal of microplastics; however, application of this techniques might not be feasible in every case owing to high cost and complexities involved. Mechanical removal of microplastics through self-propelled micromotors is yet to be tested for practical purposes. Similarly, requirement of power for electrocoagulation techniques might be a challenge in low-income settings. In this context, development of small filtration units may prove to be a suitable approach. These units should be able to fit into domestic water treatment equipment so that microplastics can be removed along with all the other impurities of water. These types of filtration units are also recommended for purifying groundwater. Furthermore, realizing that the sources and sinks of microplastics are varied, any single technology will not be able to provide the complete solution. Based on the source from where the microplastic is to be removed, a combination of methods such as biological entrapment (algal masses) and filtration may be used.

COMPETING INTERESTS

The authors declare there are no competing interests.

FUNDING

Authors are thankful to the Indian Council of Medical Research (ICMR), New Delhi for their financial support (project grant number ICMR-NIREH/BPL/IMP-PJ-44/2021-22/469: Principal Investigator – Surya Singh, ICMR – NIREH, Bhopal).

DATA AVAILABILITY STATEMENT

All relevant data are included in the paper or its Supplementary Information.

REFERENCES

Amrutha, K. & Warrier, A. K. 2020 The first report on the source-to-sink characterization of microplastic pollution from a riverine environment in tropical India. *Science of the Total Environment* **739**, 140377. doi:10.1016/j.scitotenv.2020.140377.

Ariza-Tarazona, M. C., Villarreal-Chiu, J. F., Barbieri, V., Siligardi, C. & Cedillo-Gonzalez, E. I. 2019 New strategy for microplastic degradation: green photocatalysis using a protein-based porous N-TiO_2 semiconductor. *Ceramic International* **45**, 9618–9624. doi:10.1016/j.ceramint.2018.10.208.

Arthur, C., Baker, J. & Bamford, H. 2009 Occurrence, effects, and fate of microplastic marine debris. In: *Proceedings of the International Research Workshop*, Sep. 9–11, 2008. NOAA Technical Memorandum NOS-OR&R-30.

Avio, C. G., Gorbi, S. & Regoli, F. 2017 Plastics and microplastics in the oceans: from emerging pollutants to emerged threat. *Marine Environmental Research* **128**, 2–11. doi:10.1016/j.marenvres.2016.05.012.

Barboza, L. G. A. & Gimenez, B. C. G. 2015 Microplastics in the marine environment: current trends and future perspectives. *Marine Pollution Bulletin* **97**, 5–12. doi:10.1016/j.marpolbul.2015.06.008.

Barrett, J., Chase, Z., Zhang, J., Holl, M. M. B., Willis, K., Williams, A., Hardesty, B. D. & Wilcox, C. 2020 Microplastic pollution in deep-sea sediments from the Great Australian Bight. *Frontiers in Marine Science* **7**, 576170. doi:10.3389/fmars.2020.576170.

Bhattacharya, P., Lin, S., Turner, J. P. & Ke, P. C. 2010 Physical adsorption of charged nanoparticles affects algal photosynthesis. *Journal of Physical Chemistry C* **114**, 16556–16561. doi:10.1021/jp1054759.

Blettler, M. C. M. & Wantzen, K. M. 2019 Threats underestimated in freshwater plastic pollution: mini-review. *Water, Air, and Soil Pollution* **230**, 174. doi:10.1007/s11270-019-4220-z.

Carr, S. A., Liu, J. & Tesoro, A. G. 2016 Transport and fate of microplastic particles in wastewater treatment plants. *Water Research* **91**, 174–182. doi:10.1016/j.watres.2016.01.002.

Chen, Y. J., Chen, Y., Miao, C., Wang, Y. R., Gao, G. K., Yang, R. X., Zhu, H. J., Wang, J. H., Li, S. L. & Lan, Y. Q. 2020 Metal-organic framework based foams for efficient microplastic removal. *Journal of Material Chemistry A* **8**, 14644–14652. doi:10.1039/D0TA04891G.

Conti, G. O., Ferrante, M., Banni, M., Favara, C., Nicolosi, I., Cristaldi, A., Fiore, M. & Zuccarello, P. 2020 Micro- and nano-plastics in edible fruits and vegetables. The first diet risks assessment for the general population. *Environmental Research* **187**, 109677. doi:10.1016/j.envres.2020.109677.

Deng, Y., Zhang, Y., Lemos, B. & Ren, H. 2017 Tissue accumulation of microplastics in mice and biomarker responses suggest widespread health risks of exposure. *Scientific Reports* **7**, 46687. doi:10.1038/srep46687.

Doorn, P. F., Campbell, P. A. & Amstutz, H. C. 1996 Metal versus polyethylene wear particles in total hip replacements: a review. *Clinical Orthopaedics and Related Research* **329S**, S206–S216.

Dris, R., Gasperi, J., Rocher, V., Saad, M., Renault, N. & Tassin, B. 2015 Microplastic contamination in an urban area: a case study in Greater Paris. *Environmental Chemistry* **12**, 592–599. doi:10.1071/EN14167.

Dyachenko, A., Mitchell, J. & Arsem, N. 2017 Extraction and identification of microplastic particles from secondary wastewater treatment plant (WWTP) effluent. *Analytical Methods* **9**, 1412–1418. doi:10.1039/C6AY02397E.

Ensign, L. M., Cone, R. & Hanes, J. 2012 Oral drug delivery with polymeric nanoparticles: the gastrointestinal mucus barriers. *Advanced Drug Delivery Reviews* **64**, 557–570. doi:10.1016/j.addr.2011.12.009.

European Commission Report 2017 *Intentionally Added Microplastics in Products, Doc. Ref. 39168 Final Report 17271i3*. Available from: https://ec.europa.eu/environment/chemicals/reach/pdf/39168%20Intentionally%20added%20microplastics%20-%20Final%20report%2020171020.pdf (accessed 21 October 2021).

Frias, J. P. G. L., Antunes, J. C. & Sobral, P. 2013 Local marine litter survey – a case study in Alcobaca municipality, Portugal. *Journal of Integrated Coastal Zone Management* **13**, 169–179. doi:10.5894/rgci395.

Galloway, T. S. & Lewis, C. N. 2016 Marine microplastics spell big problems for future generations. *Proceedings of National Academy of Sciences USA* **113**, 2331–2333. doi:10.1073/pnas.1600715113.

Gao, W., Feng, X., Pei, A., Gu, Y., Li, J. & Wang, J. 2013 Seawater-driven magnesium based Janus micromotors for environmental remediation. *Nanoscale* **5**, 4696–4700. doi:10.1039/C3NR01458D.

Gennaro, V., Ceppi, M., Crosignani, P. & Montanaro, F. 2008 Re-analysis of updated mortality among vinyl and polyvinyl chloride workers: confirmation of historical evidence and new findings. *BMC Public Health* **8**, 21.

Grafmueller, S., Manser, P., Diener, L., Diener, P. A., Maeder-Althaus, X., Maurizi, L., Jochum, W., Krug, H. F., Bureki-Thurnherr, T., Mandach, U. V. & Wick, P. 2015 Bidirectional transfer study of polystyrene nanoparticles across the placental barrier in an *ex vivo* human placental perfusion model. *Environmental Health Perspectives* **123**, 1280–1286. doi:10.1289/ehp.1409271.

Guix, M., Orozco, J., Garcia, M., Gao, W., Sattayasamitsathit, S., Merkoci, A., Escarpa, A. & Wang, J. 2012 Super hydrophobicalkanethiol-coated microsubmarines for effective removal of oil. *ACS Nano* **6**, 4445–4451. doi:10.1021/nn301175b.

He, B., Goonetilleke, A., Ayoko, G. A. & Rintoul, L. 2020 Abundance, distribution patterns, and identification of microplastics in Brisbane River sediments, Australia. *Science of the Total Environment* **700**, 134467. doi:10.1016/j.scitotenv.2019.134467.

Herbort, A. F. & Schuhen, K. 2017 A concept for the removal of microplastics from the marine environment with innovative host-guest relationships. *Environmental Science and Pollution Research* **24**, 11061–11065. doi:10.1007/s11356-016-7216-x.

Iyare, P. U., Ouki, S. K. & Bond, T. 2020 Microplastics removal in wastewater treatment plants: a critical review. *Environmental Science: Water Research and Technology* **6**, 2664–2675. doi:10.1039/D0EW00397B.

Kang, J., Zhou, L., Duan, X., Sun, H., Ao, Z. & Wang, S. 2019 Degradation of cosmetic microplastics *via* functionalized carbon nanosprings. *Matter* **1**, 745–758. doi:10.1016/j.matt.2019.06.004.

Kaposi, K. L., Mos, B., Kelaher, B. P. & Dworjanyn, S. A. 2014 Ingestion of microplastic has limited impact on a marine larva. *Environmental Science & Technology* **48**, 1638–1645.

Karami, A., Golieskardi, A., Choo, C. K., Larat, V., Galloway, T. S. & Salamatinia, B. 2017 The presence of microplastics in commercial salts from different countries. *Scientific Reports* **7**, 46173.

Karthik, R., Robin, R. S., Purvaja, R., Ganguly, D., Anandavelu, I., Raghuraman, R., Hariharan, G., Ramakrishna, A. & Ramesh, R. 2018 Microplastics along the beaches of southeast coast of India. *Science of the Total Environment* **645**, 1388–1399. doi:10.1016/j.scitotenv.2018.07.242.

Kataoka, T., Nihei, Y., Kudou, K. & Hinata, H. 2019 Assessment of the sources and inflow processes of microplastics in the river environments of Japan. *Environmental Pollution* **244**, 958–965. doi:10.1016/j.envpol.2018.10.111.

Kay, P., Hiscoe, R., Moberley, I., Bajic, L. & McKeena, N. 2018 Wastewater treatment plants as a source of microplastics in river catchments. *Environmental Science and Pollution Research* **25**, 20264–20267. doi:10.1007/s11356-018-2070-7.

Kentin, E. & Kaarto, H. 2018 An EU ban on microplastics in cosmetic products and the right to regulate. *Review of European, Comparative & International Environmental Law (RECIEL)* **27**, 254–266. doi:10.1111/reel.12269.

Kobielska, P. A., Howarth, A. J., Farha, O. K. & Nayak, S. 2018 Metal-organic frameworks for heavy metal removal from water. *Coordination Chemistry Reviews* **358**, 92–107. doi:10.1016/j.ccr.2017.12.010.

Kohler, A. 2010 Cellular fate of organic compounds in marine invertebrates. *Comparative Biochemistry and Physiology – Part A* **157** (Supplement), S8–S11.

Kosuth, M., Mason, S. A. & Wattenberg, E. V. 2018 Anthropogenic contamination of tap water, beer, and sea salt. *PLOS One* **13**, 1–18. e0194970. doi:10.1371/journal.pone.0194970.

Kyrikou, I. & Briassoulis, D. 2007 Biodegradation of agricultural plastic films: a critical review. *Journal of Polymers and the Environment* **15**, 125–150. doi:10.1007/s10924-007-0053-8.

Lagarde, F., Olivier, O., Zanella, M., Daniel, P., Hiard, S. & Caruso, A. 2016 Microplastic interactions with freshwater microalgae: hetero-aggregation and changes in plastic density appear strongly dependent on polymer type. *Environmental Pollution* **215**, 331–339. doi:10.1016/j.envpol.2016.05.006.

Lang, I. A., Galloway, T. S., Scarlett, A., Henley, W. E., Depledge, M., Wallace, R. B. & Melzer, D. 2008 Association of urinary Bisphenol A concentration with medical disorders and laboratory abnormalities in adults. *Journal of American Medical Association* **300**, 1303–1310. doi:10.1001/jama.300.11.1303.

Lares, M., Ncibi, M. C., Sillanpaa, M. & Sillanpaa, M. 2018 Occurrence, identification and removal of microplastic particles and fibers in conventional activated sludge process and advanced MBR technology. *Water Research* **133**, 236–246. doi:10.1016/j.watres.2018.01.049.

Lechner, A. & Ramler, D. 2015 The discharge of certain amounts of industrial microplastic from a production plant into the River Danube is permitted by the Austrian legislation. *Environmental Pollution* **200**, 159–160. doi:10.1016/j.envpol.2015.02.019.

Lee, K. W., Shim, W. J., Kwon, O. Y. & Kang, J. H. 2013 Size-dependent effects of micro polystyrene particles in the marine copepod *Tigriopus japonicas*. *Environmental Science & Technology* **47**, 11278–11283.

Leslie, H. A., Brandsma, S. H., van Velzen, M. J. & Vethaak, A. D. 2017 Microplastics en route: field measurements in the Dutch river delta and Amsterdam canals, wastewater treatment plants, North Sea sediments and biota. *Environmental International* **101**, 133–142. doi:10.1016/j.envint.2017.01.018.

Li, X., Chen, L., Mei, Q., Dong, B., Dai, X., Ding, G. & Zeng, E. Y. 2018 Microplastics in sewage sludge from the wastewater treatment plants in China. *Water Research* **142**, 75–85. doi:10.1016/j.watres.2018.05.034.

Liebezeit, G. & Liebezeit, E. 2013 Non-pollen particulates in honey and sugar. *Food Additives and Contaminants: Part A* **30**, 2136–2140. doi:10.1080/19440049.2013.843025.

Lwanga, E. H., Gertsen, H., Gooren, H., Peters, P., Salanki, T., Ploeg, M. V., Besseling, E., Koelmans, A. A. & Geissen, V. 2016 Microplastics in the terrestrial ecosystem: implications for *Lumbricus terrestris* (Oligochaeta, Lumbricidae). *Environmental Science & Technology* **50**, 2685–2691.

Lwanga, E. H., Vega, J. M., Quej, V. K., Chi, J. L. A., Cid, L. S. D., Chi, C., Segura, G. E., Gertsen, H., Salanki, T., Ploeg, M. V., Koelmans, A. A. & Geissen, V. 2017 Field evidence for transfer of plastic debris along a terrestrial food chain. *Scientific Reports* **7**, 14071.

Ma, B., Xue, W., Hu, C., Liu, H., Qu, J. & Li, L. 2019 Characteristics of microplastic removal *via* coagulation and ultrafiltration during drinking water treatment. *Chemical Engineering Journal* **359**, 159–167. doi:10.1016/j.cej.2018.11.155.

Magnusson, K. & Noren, F. 2014 *Screening of Microplastic Particles in and Downstream A Wastewater Treatment Plant*. IVL Swedish Environmental Research Institute. Vol. C55, Stockholm, Sweden; pp. 1–22.

Martins, M. J. F., Mota, C. F. & Pearson, G. A. 2013 Sex-biased gene expression in the brown alga *Fucus vesiculosus*. *BMC Genomics* **14**, 294–326. doi:10.1186/1471-2164-14-294.

Mason, S. A., Garneau, D., Sutton, R., Chu, Y., Ehmann, K., Barnes, J., Fink, P., Papazissimos, D. & Rogers, D. 2016 Microplastic pollution is widely detected in US municipal wastewater treatment plant effluent. *Environmental Pollution* **218**, 1045–1054. doi:10.1016/j.envpol.2016.08.056.

Mason, S. A., Welch, V. G. & Neratko, J. 2018 Synthetic polymer contamination in bottled water. *Frontiers in Chemistry* **6**, 1–11. doi:10.3389/fchem.2018.00407.

McCormick, A., Hoellein, T. J., Mason, S. A., Schluep, J. & Kelly, J. J. 2014 Microplastic is an abundant and distinct microbial habitat in an urban river. *Environmental Science and Technology* **48**, 11863–11871. doi:10.1021/es503610r.

Michielssen, M. R., Michielssen, E. R., Ni, J. & Duhaime, M. B. 2016 Fate of microplastics and other small anthropogenic litter (SAL) in wastewater treatment plants depends on unit processes employed. *Environmental Science: Water Research & Technology* **2**, 1064–1073. doi:10.1039/C6EW00207B.

Mintenig, S. M., Int-Veen, I., Loder, M. G. J., Primpke, S. & Gerdts, G. 2017 Identification of microplastic in effluents of waste water treatment plants using focal plane array-based micro-Fourier transform infrared imaging. *Water Research* **108**, 365–372. doi:10.1016/j.watres.2016.11.015.

Mintenig, S. M., Loder, M. G. J., Primpke, S. & Gerdts, G. 2019 Low numbers of microplastics detected in drinking water from ground water sources. *Science of the Total Environment* **648**, 631–635. doi:10.1016/j.scitotenv.2018.08.178.

Mon, M., Bruno, R., Ferrando-Soria, J., Armentano, D. & Pardo, E. 2018 Metal-organic framework technologies for water remediation: towards a sustainable ecosystem. *Journal of Material Chemistry A* **6**, 4912–4947. doi:10.1039/C8TA00264A.

Mou, F., Pan, D., Chen, C., Gao, Y., Xu, L. & Guan, J. 2015 Magnetically modulated pot-like $MnFe_2O_4$ micromotors: nano-particle assembly fabrication and their capability for direct oil removal. *Advanced Functional Materials* **25**, 6173–6181. doi:10.1002/adfm.201502835.

Murphy, F., Ewins, C., Carbonnier, F. & Quinn, B. 2016 Wastewater treatment works (WwTW) as a source of microplastics in the aquatic environment. *Environmental Science and Technology* **50**, 5800–5808. doi:10.1021/acs.est.5b05416.

Mushtaq, F., Guerrero, M., Sakar, M. S., Hoop, M., Lindo, A. M., Sort, J., Chen, X., Nelson, B. J., Pellicer, E. & Pane, S. 2015 Magnetically driven Bi_2O_3/BiOCl based hybrid microrobots for photocatalytic water remediation. *Journal of Materials Chemistry A* **3**, 23670–23676. doi:10.1039/C5TA05825B.

Nolte, T. M., Hartmann, N. B., Kleijn, J. M., Garnes, J., van de Meent, D., Hendriks, A. J. & Baun, A. 2017 The toxicity of plastic nanoparticles to green algae as influenced by surface modification, medium hardness, and cellular adsorption. *Aquatic Toxicology* **183**, 11–20. doi:10.1016/j.aquatox.2016.12.005.

Novotna, K., Cermakova, L., Pivokonska, L., Cajthaml, T. & Pivokonsky, M. 2019 Microplastics in drinking water treatment – current knowledge and research needs. *Science of the Total Environment* **667**, 730–740. doi:10.1016/j.scitotenv.2019.02.431.

Okoffo, E. D., O'Brien, S., O'Brian, J. W., Tscharke, B. & Thomas, K. V. 2019 Wastewater treatment plants as a source of plastics in the environment: a review of occurrence, methods for identification, quantification and fate. *Environmental Science: Water Research and Technology* **5**, 1908–1931. doi:10.1039/C9EW00428A.

Oβmann, B. E., Sarau, G., Holtmannspotter, H., Pischetsrieder, M., Christiansen, S. H. & Dicke, W. 2018 Small sized microplastics and pigmented particles in bottled mineral water. *Water Research* **141**, 307–316. doi:10.1016/j.watres.2018.05.027.

Panno, S. V., Kelly, W. R., Scott, J., Zheng, W., McNeish, R. E., Holm, N., Hoellein, T. J. & Baranski, E. L. 2019 Microplastic contamination in Karst groundwater systems. *Groundwater* **57**, 189–196. doi:10.1111/gwat.12862.

Peng, G., Xu, P., Zhu, B., Bai, M. & Li, D. 2018 Microplastics in freshwater river sediments in Shanghai, China: a case study of risk assessment in mega-cities. *Environmental Pollution* **234**, 448–456. doi:10.1016/j.envpol.2017.11.034.

Perren, W., Wojtasik, A. & Cai, Q. 2018 Removal of microbeads from wastewater using electrocoagulation. *ACS Omega* **3**, 3357–3364. doi:10.1021/acsomega.7b02037.

Pivokonsky, M., Cermakova, L., Novotna, K., Peer, P., Cajthaml, T. & Janda, V. 2018 Occurrence of microplastics in raw and treated drinking water. *Science of the Total Environment* **643**, 1644–1651. doi:10.1016/j.scitotenv.2018.08.102.

Poerio, T., Piacentini, E. & Mazzei, R. 2019 Membrane processes for microplastic removal. *Molecules* **24**, 4148. doi:10.3390/molecules24224148.

Prata, J. C., Lavorante, B. R. B. O., Montenegro, M. C. B. S. M. & Guihermino, L. 2018 Influence of microplastics on the toxicity of the pharmaceuticals procainamide and doxycycline on the marine microalgae *Tetraselm ischuii. Aquatic Toxicology* **197**, 143–152.

Prata, J. C., Costa, J. P., Lopes, I., Duarte, A. C. & Santos, T. R. 2020 Environmental exposure to microplastics: an overview on possible human health effects. *Science of the Total Environment* **702**, 134455. doi:10.1016/j.scitotenv.2019.134455.

Ragusa, A., Svelato, A., Santacroce, C., Catalano, P., Notarstefano, V., Carnevali, O., Papa, F., Rongioletti, M. C. A., Baiocco, F., Draghi, S., D'Amore, E., Rinaldo, D., Matta, M. & Giorgini, E. 2021 Plasticenta: first evidence of microplastics in human placenta. *Environment International* **146**, 106274. doi:10.1016/j.envint.2020.106274.

Richardson, S. D. & Kimura, S. Y. 2017 Emerging environmental contaminants: challenges facing our next generation and potential engineering solutions. *Environmental Technology & Innovation* **8**, 40–56. doi:10.1016/j.eti.2017.04.002.

Rist, S., Almroth, B. C., Hartmann, N. B. & Karlsson, T. M. 2018 A critical perspective on early communications concerning human health aspects of microplastics. *Science of the Total Environment* **626**, 720–726.

Robin, R. S., Karthik, R., Purvaja, R., Ganguly, D., Anandavelu, I., Mugilarasan, M. & Ramesh, R. 2020 Holistic assessment of microplastics in various coastal environmental matrices, southwest coast of India. *Science of the Total Environment* **703**, 134947. doi:10.1016/j.scitotenv.2019.134947.

Rochman, C. M., Hoh, E., Kurobe, T. & Teh, S. J. 2013 Ingested plastic transfers hazardous chemicals to fish and induces hepatic stress. *Scientific Reports* **3**, 1–7.

Sarkar, D. J., Sarkar, S. D., Das, B. K., Praharaj, J. K., Mahajan, D. K., Purokait, B., Mohanty, T. R., Mohanty, D., Gogoi, P., Kumar, S., Behera, B. K., Manna, R. K. & Samanta, S. 2021 Microplastics removal efficiency of drinking water treatment plant with pulse clarifier. *Journal of Hazardous Materials* **413**, 125347. doi:10.1016/j.jhazmat.2021.125347.

Schirinzi, G. F., Perez-Pomeda, I., Sanchis, J., Rossini, C., Farre, M. & Barcelo, D. 2017 Cytotoxic effects of commonly used nanomaterials and microplastics on cerebral and epithelial human cells. *Environmental Research* **159**, 579–587. doi:10.1016/j.envres.2017.08.043.

Schymanski, D., Goldbeck, C., Humpf, H. U. & Furst, P. 2018 Analysis of microplastics in water by micro-Raman spectroscopy: release of plastic particles from different packaging into mineral water. *Water Research* **129**, 154–162. doi:10.1016/j.watres.2017.11.011.

SDG 2015 *Sustainable Development Goals*. United Nations. Available from: https://sdgs.un.org/goals (accessed 24 June 2021).

Simmchen, J., Baeza, A., Miguel-Lopez, A., Stanton, M. M., Vallet-Regi, M., Ruiz-Molina, D. & Sanchez, S. 2017 Dynamics of novel photoactive AgCl micromotors and their environmental applications. *Chem Nano Mat* **3**, 65. doi:10.1002/cnma.201600300.

Smith, M., Love, D. C., Rochman, C. M. & Neff, R. A. 2018 Microplastics in seafood and the implications for human health. *Current Environmental Health Reports* **5**, 375–386. doi:10.1007/s40572-018-0206-z.

Soler, L., Magdanz, V., Fomin, V. M., Sanchez, S. & Schmidt, O. G. 2013 Self-propelled micromotors for cleaning polluted water. *ACS Nano* **7**, 9611–9620. doi:10.1021/nn405075d.

Stephens, S. H. 1991 Final report on the safety assessment of polyacrylamide. *Journal of the American College of Toxicology* **10**, 193–203. doi:10.3109/10915819109078629.

Sun, J., Dai, X., Wang, Q., Loosdrecht, M. C. M. V. & Ni, B. J. 2019 Microplastics in wastewater treatment plants: detection, occurrence, and removal. *Water Research* **152**, 21–37. doi:10.1016/j.watres.2018.12.050.

Sundbæk, K. B., Koch, I. D. W., Villaro, C. G., Rasmussen, N. S., Holdt, S. L. & Hartmann, N. B. 2018 Sorption of fluorescent polystyrene microplastic particles to edible seaweed *Fucus vesiculosus*. *Journal of Applied Phycology* **30**, 2923–2927. doi:10.1007/s10811-018-1472-8.

Sussarellu, R., Soudant, P., Lambert, C., Fabioux, C., Corporeau, C. & Laot, C. 2014 Microplastics: effect on oyster physiology and reproduction. In: *Platform Presentation, International Workshop on Fate and Impact of Microplastics in Marine Ecosystems (MICRO2014)*, January 13–15, 2014, Plouzane, France.

Swan, S. H., Main, K. M., Liu, F., Stewart, S. L., Kruse, R. L., Calafat, A. M., Mao, C. S., Redmon, J. B., Ternand, C. L., Sullivan, S. & Teague, J. L. & Study for future families research team 2005 Decrease in anogenital distance among male infants with prenatal phthalate exposure. *Environmental Health Perspectives* **113**, 1056–1061. doi:10.1289/ehp.8100.

Swan, S. H. 2008 Environmental phthalate exposure in relation to reproductive outcomes and other health end-points in humans. *Environmental Research* **108**, 177–184. doi:10.1016/j.envres.2008.08.007.

Tagg, A. S., Brandes, E., Fischer, F., Fischer, D., Brandt, J. & Labrenz, M. 2021 Agricultural application of microplastic-rich sewage sludge leads to further uncontrolled contamination. *Science of the Total Environment*. In Press. doi:10.1016/j.scitotenv.2021.150611.

Talsness, C. E., Andrade, A. J. M., Kuriyama, S. N., Taylor, J. A. & vomSaal, F. S. 2009 Components of plastic: experimental studies in animals and relevance for human health. *Philosophical Transactions of Royal Society B* **364**, 2079–2096.

Talvitie, J., Mikola, A., Setala, O., Heinonen, M. & Koistinen, A. 2016 How well is microlitter purified from wastewater? – a detailed study on the stepwise removal of microlitter in a tertiary level wastewater treatment plant. *Water Research* **109**, 164–172. doi:10.1016/j.watres.2016.11.046.

Talvitie, J., Mikola, A., Koistinen, A. & Setala, O. 2017 Solutions to microplastic pollution – removal of microplastics from wastewater effluent with advanced wastewater treatment technologies. *Water Research* **123**, 401–407. doi:10.1016/j.watres.2017.07.005.

Thompson, R. C., Olsen, Y., Mitchell, R. P., Davis, A., Rowland, S. J., John, A. W. G., McGonigle, D. & Russell, A. E. 2004 Lost at sea: where is all the plastic? *Science* **304**, 838. doi:10.1126/science.1094559.

Tofa, T. S., Kunjali, K. L., Paul, S. & Dutta, J. 2019 Visible light photocatalytic degradation of microplastic residues with zinc oxide nanorods. *Environmental Chemistry Letters* **17**, 1341–1346. doi:10.1007/s10311-019-00859-z.

Tu, Y., Peng, F., Adawy, A., Men, Y., Abdelmohsen, L. K. E. A. & Wilson, D. A. 2016 Mimicking the cell: bio-inspired functions of supramolecular assemblies. *Chemical Reviews* **116**, 2023–2078. doi:10.1021/acs.chemrev.5b00344.

Urban, R. M., Jacobs, J. J., Tomlinson, M. J., Gavrilovic, J., Black, J. & Peoch, M. 2000 Dissemination of wear particles to the liver, spleen, and abdominal lymph nodes of patients with hip or knee replacement. *Journal of Bone and Joint Surgery* **82-A**, 457–477.

USEPA 1989 *USEPA Risk Assessment Guidance for Superfund. Human Health Evaluation Manual, (Part A) [R], 1, Office of Emergency and Remedial Response*. EPA/540/189/002, Washington, DC, USA.

Uygun, D. A., Jurado-Sanchez, B., Uygun, M. & Wang, J. 2016 Self-propelled chelation platforms for efficient removal of toxic metals. *Environmental Science: Nano* **3**, 559–566. doi:10.1039/C6EN00043F.

Vilela, D., Parmar, J., Zeng, Y., Zhao, Y. & Sanchez, S. 2016 Graphene based microbots for toxic heavy metal removal and recovery from water. *Nano Letters* **16**, 2860–2866. doi:10.1021/acs.nanolett.6b00768.

Villa, K., Palenzuela, C. L. M., Sofer, Z., Matejkova, S. & Pumera, M. 2018 Metal-free visible light photoactivated C_3N_4 bubble-propelled tubular micromotors with inherent fluorescence and on/off capabilities. *ACS Nano* **12**, 12482–12491. doi:10.1021/acsnano.8b06914.

von Moss, N., Burkhardt-Holm, P. & Kohler, A. 2012 Uptake and effects of microplastics on cells and tissue of the blue mussel *Mytilus edulis* L. after an experimental exposure. *Environmental Science & Technology* **46**, 11327–11335.

Wang, H., Khezri, B. & Pumera, M. 2016 Catalytic DNA functionalized self-propelled micromachines for environmental remediation. *Chem* **1**, 473–481. doi:10.1016/j.chempr.2016.08.009.

Wang, L., Kaeppler, A., Fischer, D. & Simmchen, J. 2019 Photocatalytic TiO_2 micromotors for removal of microplastics and suspended matter. *ACS Applied Materials & Interfaces* **11**, 32937–32944. doi:10.1021/acsami.9b06128.

Wang, Z., Sedighi, M. & Lea-Langton, A. 2020a Filtration of microplastic spheres by biochar: removal efficiency and immobilisation mechanisms. *Water Research* **184**, 116165. doi:10.1016/j.watres.2020.116165.

Wang, Z., Lin, T. & Chen, W. 2020b Occurrence and removal of microplastics in an advanced drinking water treatment plant (ADWTP). *Science of the Total Environment* **700**, 134520. doi:10.1016/j.scitotenv.2019.134520.

Wright, S. L., Thompson, R. C. & Galloway, T. S. 2013 The physical impacts of microplastics on marine organisms: a review. *Environmental Pollution* **178**, 483–492.

Wright, S. L. & Kelly, F. J. 2017 Plastic and human health: a micro issue? *Environmental Science and Technology* **51**, 6634–6647. doi:10.1021/acs.est.7b00423.

Yuan, F., Zhao, H., Sun, H., Sun, Y., Zhao, J. & Xia, T. 2022 Investigation of microplastics in sludge from five wastewater treatment plants in Nanjing, China. *Journal of Environmental Management* **301**, 113793. doi:10.1016/j.jenvman.2021.113793.

Zhang, Y., Yuan, S., Feng, X., Li, H., Zhou, J. & Wang, B. 2016 Preparation of nanofibrous metal–organic framework filters for efficient air pollution control. *Journal of American Chemical Society* **138**, 5785–5788. doi:10.1021/jacs.6b02553.

Zhang, Q., Dong, R., Wu, Y., Gao, W., He, Z. & Ren, B. 2017 Light driven Au-WO_3@C Janus micromotors for rapid photodegradation of dye pollutants. *ACS Applied Materials & Interfaces* **9**, 4674–4683. doi:10.1021/acsami.6b12081.

Zhang, J., Wang, L., Halden, R. U. & Kannan, K. 2019 Polyethylene terephthalate and polycarbonate microplastics in sewage sludge collected from the United States. *Environmental Science and Technology Letters* **6**, 650–655. doi:10.1021/acs.estlett.9b00601.

Zheng, J., Lin, T., Chen, W., Tao, H., Tan, Y. & Ma, B. 2018 Removal of precursors of typical nitrogenous disinfection byproducts in ozonation integrated with biological activated carbon (O_3/BAC). *Chemosphere* **209**, 68–77. doi:10.1016/j.chemosphere.2018.06.018.

Ziajahromi, S., Neale, P. A., Rintoul, L. & Leusch, D. L. 2017 Wastewater treatment plants as a pathway for microplastics: development of a new approach to sample wastewater-based microplastics. *Water Research* **112**, 93–99. doi:10.1016/j.watres.2017.01.042.

First received 9 August 2021; accepted in revised form 21 October 2021. Available online 29 October 2021

IWA Publishing's authorised EU representative for General Product Safety Regulations is Diane D'Arras, 15 rue Duret, 75116 Paris, France, e-mail: safety@iwap.co.uk.

Printed and bound by CPI Group (UK) Ltd, Croydon, CR0 4YY
12/05/2026
02108868-0004